CAMBRIDGE TRACTS IN MATHEMATICS

General Editors

B. BOLLOBÁS, W. FULTON, F. KIRWAN,
P. SARNAK, B. SIMON, B. TOTARO

217 Defocusing Nonlinear Schrödinger Equations

CAMBRIDGE TRACTS IN MATHEMATICS

GENERAL EDITORS

B. BOLLOBÁS, W. FULTON, F. KIRWAN,
P. SARNAK, B. SIMON, B. TOTARO

A complete list of books in the series can be found at www.cambridge.org/mathematics.
Recent titles include the following:

182. Nonlinear Markov Processes and Kinetic Equations. By V. N. KOLOKOLTSOV
183. Period Domains over Finite and p-adic Fields. By J.-F. DAT, S. ORLIK, and M. RAPOPORT
184. Algebraic Theories. By J. ADÁMEK, J. ROSICKÝ, and E.M. VITALE
185. Rigidity in Higher Rank Abelian Group Actions I: Introduction and Cocycle Problem. By A. KATOK and V. NIŢICĂ
186. Dimensions, Embeddings, and Attractors. By J. C. ROBINSON
187. Convexity: An Analytic Viewpoint. By B. SIMON
188. Modern Approaches to the Invariant Subspace Problem. By I. CHALENDAR and J. R. PARTINGTON
189. Nonlinear Perron–Frobenius Theory. By B. LEMMENS and R. NUSSBAUM
190. Jordan Structures in Geometry and Analysis. By C.-H. CHU
191. Malliavin Calculus for Lévy Processes and Infinite-Dimensional Brownian Motion. By H. OSSWALD
192. Normal Approximations with Malliavin Calculus. By I. NOURDIN and G. PECCATI
193. Distribution Modulo One and Diophantine Approximation. By Y. BUGEAUD
194. Mathematics of Two-Dimensional Turbulence. By S. KUKSIN and A. SHIRIKYAN
195. A Universal Construction for Groups Acting Freely on Real Trees. By I. CHISWELL and T. MÜLLER
196. The Theory of Hardy's Z-Function. By A. IVIĆ
197. Induced Representations of Locally Compact Groups. By E. KANIUTH and K. F. TAYLOR
198. Topics in Critical Point Theory. By K. PERERA and M. SCHECHTER
199. Combinatorics of Minuscule Representations. By R. M. GREEN
200. Singularities of the Minimal Model Program. By J. KOLLÁR
201. Coherence in Three-Dimensional Category Theory. By N. GURSKI
202. Canonical Ramsey Theory on Polish Spaces. By V. KANOVEI, M. SABOK, and J. ZAPLETAL
203. A Primer on the Dirichlet Space. By O. EL-FALLAH, K. KELLAY, J. MASHREGHI, and T. RANSFORD
204. Group Cohomology and Algebraic Cycles. By B. TOTARO
205. Ridge Functions. By A. PINKUS
206. Probability on Real Lie Algebras. By U. FRANZ and N. PRIVAULT
207. Auxiliary Polynomials in Number Theory. By D. MASSER
208. Representations of Elementary Abelian p-Groups and Vector Bundles. By D. J. BENSON
209. Non-homogeneous Random Walks. By M. MENSHIKOV, S. POPOV and A. WADE
210. Fourier Integrals in Classical Analysis (Second Edition). By C. D. SOGGE
211. Eigenvalues, Multiplicities and Graphs. By C. R. JOHNSON and C. M. SAIAGO
212. Applications of Diophantine Approximation to Integral Points and Transcendence. By P. CORVAJA and U. ZANNIER
213. Variations on a Theme of Borel. By S. WEINBERGER
214. The Mathieu Groups. By A. A. IVANOV
215. Slenderness I: Foundations. By R. DIMITRIC
216. Justification Logic. By S. ARTEMOV and M. FITTING

Defocusing Nonlinear Schrödinger Equations

BENJAMIN DODSON
The Johns Hopkins University

CAMBRIDGE
UNIVERSITY PRESS

University Printing House, Cambridge CB2 8BS, United Kingdom

One Liberty Plaza, 20th Floor, New York, NY 10006, USA

477 Williamstown Road, Port Melbourne, VIC 3207, Australia

314–321, 3rd Floor, Plot 3, Splendor Forum, Jasola District Centre,
New Delhi – 110025, India

79 Anson Road, #06–04/06, Singapore 079906

Cambridge University Press is part of the University of Cambridge.

It furthers the University's mission by disseminating knowledge in the pursuit of education, learning, and research at the highest international levels of excellence.

www.cambridge.org
Information on this title: www.cambridge.org/9781108472081
DOI: 10.1017/9781108590518

© Benjamin Dodson 2019

This publication is in copyright. Subject to statutory exception
and to the provisions of relevant collective licensing agreements,
no reproduction of any part may take place without the written
permission of Cambridge University Press.

First published 2019

Printed and bound in Great Britain by Clays Ltd, Elcograf S.p.A.

A catalogue record for this publication is available from the British Library.

Library of Congress Cataloging-in-Publication Data
Names: Dodson, Benjamin, 1983– author.
Title: Defocusing nonlinear Schrödinger equations / Benjamin Dodson
(The Johns Hopkins University).
Description: Cambridge ; New York, NY : Cambridge University Press, 2019. |
Series: Cambridge tracts in mathematics | Includes bibliographical references and index.
Identifiers: LCCN 2018047994 | ISBN 9781108472081 (hardback)
Subjects: LCSH: Gross-Pitaevskii equations. | Schrodinger equation. |
Differential equations, Nonlinear.
Classification: LCC QC174.26.W28 D63 2019 | DDC 530.12/4–dc23
LC record available at https://lccn.loc.gov/2018047994

ISBN 978-1-108-47208-1 Hardback

Cambridge University Press has no responsibility for the persistence or accuracy of URLs for external or third-party internet websites referred to in this publication and does not guarantee that any content on such websites is, or will remain, accurate or appropriate.

I dedicate this book to my family, especially my wife Priscilla and my daughter Ella. I also dedicate this book to God, without whom this would not have been possible.

Contents

	Preface	page ix
	Acknowledgments	xii
1	**A First Look at the Mass-Critical Problem**	1
	1.1 Linear Schrödinger Equation and Preliminaries	1
	1.2 Strichartz Estimates	10
	1.3 Small Data Mass-Critical Problem	24
	1.4 A Large Data Global Well-Posedness Result	33
2	**The Cubic NLS in Dimensions Three and Four**	41
	2.1 The Cubic NLS in Three and Four Dimensions with Small Data	41
	2.2 Scattering for the Radial Cubic NLS in Three Dimensions with Large Data	47
	2.3 The Radially Symmetric, Cubic Problem in Four Dimensions	52
3	**The Energy-Critical Problem in Higher Dimensions**	64
	3.1 Small Data Energy-Critical Problem	64
	3.2 Profile Decomposition for the Energy-Critical Problem	72
	3.3 Global Well-Posedness and Scattering When $d \geq 5$	85
	3.4 Interaction Morawetz Estimate	96
4	**Mass-Critical NLS Problem in Higher Dimensions**	102
	4.1 Bilinear Estimates	102
	4.2 Mass-Critical Profile Decomposition	109
	4.3 Radial Mass-Critical Problem in Dimensions $d \geq 2$	128
	4.4 Nonradial Mass-Critical Problem in Dimensions $d \geq 3$	146

5	**Low-Dimensional Well-Posedness Results**	164
	5.1 The Energy-Critical Problem in Dimensions Three and Four	164
	5.2 Three-Dimensional Energy-Critical Problem	178
	5.3 The Mass-Critical Problem When $d = 1$	191
	5.4 The Two-Dimensional Mass-Critical Problem	214
	References	236
	Index	241

Preface

In this book we study the semilinear Schrödinger equation with power-type nonlinearity. This equation has been an active area of research in dispersive partial differential equations. The study of semilinear Schrödinger equations is useful in its own right, since it has many applications to physics. It also provides a great deal of insight into other dispersive partial differential equations and geometric partial differential equations. The study of this equation combines tools from harmonic analysis, microlocal analysis, functional analysis, and topology. It is a truly fascinating topic.

This book is mainly focused on the mass-critical (or L^2-critical) problem with initial data in L^2, and the energy-critical (or $\dot{H}^1$-critical) problem with initial data in $\dot{H}^1$. These problems have been shown to be globally well posed and scattering in the defocusing case for critical initial data, and moreover, these results are sharp. Presentation of the proofs of these results is the goal of this book.

Chapter 1 commences the study of the mass-critical problem. A natural approach to a nonlinear problem is to treat it as a perturbation of a linear problem. Thus, Chapter 1 begins with an examination of the dispersive properties of solutions to the linear equation. In the section that follows, the dispersive estimates are utilized to prove Strichartz estimates for the linear equation. These estimates are very important since they are invariant under time translation of the linear solution operator. In Section 1.3 these estimates are used to prove that the mass-critical problem is scattering when the mass is small. The argument utilizes a standard fixed point argument. Chapter 1 then concludes with a proof of scattering for solutions to the mass-critical problem for data lying in a subspace of $L^2\left(\mathbf{R}^d\right)$, but with no size restriction on the initial data. The proof combines conserved quantities with perturbative arguments.

Chapter 2 addresses the cubic nonlinear Schrödinger equation in dimensions three and four, where it is mass supercritical, and either energy subcritical ($d = 3$) or energy critical ($d = 4$). There the small data problem is complicated by the need to differentiate the nonlinearity, especially in three dimensions, where the fractional product rule is used. However, the fact that the nonlinearity is cubic makes the analysis far more tractable than the analysis of mass-supercritical problems in higher dimensions. In Section 2.2 the three-dimensional cubic problem is discussed. This problem is $\dot{H}^{1/2}$-critical, but study of this problem has provided a great deal of insight into both the mass-critical and energy-critical problems (see for example Kenig and Merle (2010) and Lin and Strauss (1978)). In this section a Morawetz estimate of Lin and Strauss (1978) is proved, directly yielding a scattering result. In the final section of Chapter 2, the scattering result of Bourgain (1999) is presented. This work is widely considered to be the seminal paper for the techniques discussed in this book, although Grillakis (2000) contemporaneously proved global well-posedness in three dimensions for radial initial data. Bourgain (1999) actually proved scattering for the energy-critical, radially symmetric problem in dimensions $d = 3$ and 4. In keeping with the theme of cubic problems for this chapter, only the $d = 4$ result is discussed. The three-dimensional proof may be proved using the same argument as the four-dimensional proof, with only minor modifications.

The energy-critical problem in higher dimensions is addressed in Chapter 3. The first order of business is to prove small data scattering in dimensions $d \geq 5$. Thus, the perturbative energy-critical results of Tao and Visan (2005) are presented in the first section. Section 3.2 discusses Keraani's's (2001) profile decomposition for the energy-critical problem, commencing the discussion of the concentration compactness method with regard to the Schrödinger equation. In Section 3.3 the profile decomposition is used to prove that the defocusing, energy-critical nonlinear Schrödinger equation is scattering in dimensions $d \geq 5$. This result was originally proved in Visan (2007) (see also Visan (2006)). However, in this book the result is proved using a slight modification of the argument in Killip and Visan (2010). The argument utilizes the double Duhamel argument and the interaction Morawetz estimate. The interaction Morawetz estimates of Colliander et al. (2004) and Tao et al. (2007a) are proved in Section 3.4.

Higher-dimensional mass-critical results are presented in Chapter 4. Bilinear estimates are essential to the study of the mass-critical problem. The interaction Morawetz estimates lend themselves quite well to bilinear estimates, a fact that was well exploited by Planchon and Vega (2009). Section 4.1 discusses both the Fourier analytic approach to bilinear estimates found in

Bourgain (1998), as well as the interaction Morawetz approach of Planchon and Vega (2009). Both approaches are used to study the mass-critical problem in Chapters 4 and 5. The subsequent section presents the mass-critical profile decomposition of Tao *et al.* (2008). This profile decomposition crucially relies on the bilinear estimates of Tao (2003), which will not be proved in this book. Then Section 4.3 presents the scattering results of Killip *et al.* (2009), Killip *et al.* (2008), and Tao *et al.* (2007b) for radial data. Section 4.4 then presents the scattering result of Dodson (2012) for nonradial data.

Chapter 5 addresses the defocusing, energy-critical and mass-critical problems in low dimensions. The scattering results of Colliander *et al.* (2008) ($d = 3$) and Ryckman and Visan (2007) ($d = 4$) for the defocusing, energy-critical problem are presented in the first two sections. The $d = 4$ case is completely worked out in the first section, while some of the more technically difficult parts of the $d = 3$ case are presented in Section 5.2. Sections 5.3 and 5.4 follow a similar pattern for the mass-critical problem. There the scattering results of Dodson (2016a) ($d = 1$) and Dodson (2016b) ($d = 2$) are presented for the defocusing, mass-critical problem. The one-dimensional result of Dodson (2016a) is completely worked out in Section 5.3, while some of the technically difficult aspects of the two-dimensional result are resolved in Section 5.4.

In the author's opinion the material would be sufficient for a one-semester course for graduate students who have taken a course in real analysis at the graduate level. The author assumes that the reader is familiar with basic measure theory and integration, such as can be found in Lieb and Loss (2001) or Taylor (2006). The author also assumes that the reader is familiar with basic functional analysis concepts such as Banach spaces, Hölder spaces, and Fréchet spaces. Conway (1990) and Yosida (1980) provide good overall introductions to the field of functional analysis. Finally, the book uses many techniques from the field of harmonic analysis such as interpolations and stationary phase analysis. Grafakos (2004), Muscalu and Schlag (2013), Sogge (1993), and Stein (1993) provide good introductions to harmonic analysis.

The term "nonlinear Schrodinger equation" will often by abbreviated NLS.

Acknowledgments

During the writing of this book, the author was supported by a von Neumann Fellowship, NSF grants 1500424 and 1764358, as well as the NSF postdoctoral fellowship 1103914. Some of this book was written while the author was at the University of California–Berkeley, and some of this book was written while the author was at Johns Hopkins University. The author also wrote parts of this book while a guest at MSRI, Cergy-Pontoise University, the IHES in Paris, the HIM in Bonn, and the IAPCM in Beijing.

The author also wishes to thank the following mathematicians who provided many helpful comments on preliminary versions of this book: Aynur Bulut, Jim Colliander, Magdalena Czubak, Chenjie Fan, Luiz Farah, Justin Holmer, Herbert Koch, Anudeep Kumar, Andrew Lawrie, Jonas Lührmann, Jeremy Marzuola, Dana Mendelson, Changxing Miao, Jason Murphy, Camil Muscalu, Nathan Pennington, Svetlana Roudenko, Paul Smith, Chris Sogge, Daniel Tataru, Michael Taylor, Guixiang Xu, Jianwei Yang, Xueying Yu, and Zehua Zhao.

The author would also like to thank the students in his classes for their helpful comments and suggestions on all or the parts of this book that were presented in lectures, as well as some of Camil Muscalu's students at Cornell University for their helpful comments. The author also sincerely apologizes to anyone who should have been listed here but was omitted, it was certainly unintentional.

1
A First Look at the Mass-Critical Problem

1.1 Linear Schrödinger Equation and Preliminaries

Formally, the solution to the linear Schrödinger equation,

$$iu_t + \Delta u = 0, \quad u(0,x) = u_0, \tag{1.1}$$

may be given by $e^{it\Delta}u_0$. Here Δ is a self-adjoint operator on any L^2-based Sobolev space, so $i\Delta$ is skew-adjoint on any L^2-based Sobolev space, and thus the operator $e^{it\Delta}$ is a perfectly well-defined unitary group. (See Section A.9 of Taylor (2011) for a proper introduction to unitary groups.)

When viewed from a Fourier-analytic perspective, the spectral theory of (1.1), and thus the operator $e^{it\Delta}$, is readily apparent.

Definition 1.1 (Fourier transform) For $f \in L^1(\mathbf{R}^d)$ let

$$(\mathscr{F}f)(\xi) = \hat{f}(\xi) = (2\pi)^{-d/2} \int e^{-ix\cdot\xi} f(x)\, dx.$$

The Fourier transform intertwines translation and multiplication. Let T_{x_0} be the translation operator

$$T_{x_0} f = f(x - x_0).$$

Then by a change of variables, for $f \in L^1(\mathbf{R}^d)$,

$$\mathscr{F}(T_{x_0}f)(\xi) = (2\pi)^{-d/2} \int e^{-ix\cdot\xi} f(x - x_0)\, dx = e^{-ix_0\cdot\xi}\hat{f}(\xi) \tag{1.2}$$

and

$$T_{\xi_0}\mathscr{F}(f)(\xi) = \mathscr{F}(f)(\xi - \xi_0) = (2\pi)^{-d/2} \int e^{-ix\cdot(\xi-\xi_0)} f(x)\, dx$$
$$= \mathscr{F}\left(e^{ix\cdot\xi_0} f\right)(\xi). \tag{1.3}$$

Equations (1.2) and (1.3) are formally equivalent to

$$\mathscr{F}(\partial_x^\alpha f)(\xi) = (2\pi)^{-d/2} \int e^{-ix\cdot\xi}(\partial_x^\alpha f(x))dx$$
$$= (2\pi)^{-d/2} \int e^{-ix\cdot\xi}(i\xi)^\alpha \hat{f}(\xi)d\xi = (i\xi)^\alpha \hat{f}(\xi) \quad (1.4)$$

and

$$\mathscr{F}(x^\alpha f)(\xi) = (2\pi)^{-d/2} \int x^\alpha e^{-ix\cdot\xi} f(x)dx$$
$$= (2\pi)^{-d/2} i^\alpha \int \partial_\xi^\alpha (e^{-ix\cdot\xi}) f(x)dx$$
$$= (i\partial_\xi)^\alpha \hat{f}(\xi), \quad (1.5)$$

respectively, where $\alpha = (\alpha_1, \ldots, \alpha_d)$ is a multi-index, $\alpha_i \geq 0$ for all $1 \leq i \leq d$. Thus if $u(t,x)$ solves (1.1), then the Fourier transform of u (formally) solves

$$i\partial_t(\hat{u}(t,\xi)) - |\xi|^2 \hat{u}(t,\xi) = 0, \quad \hat{u}(0,\xi) = \hat{u}_0(\xi), \quad (1.6)$$

so for each $\xi \in \mathbf{R}^d$, (1.6) gives an ordinary differential equation in time whose solution is

$$\hat{u}(t,\xi) = e^{-it|\xi|^2} \hat{u}_0(\xi). \quad (1.7)$$

It is possible to formally compute $u(t,x)$ from $\hat{u}(t,\xi)$ using the inverse Fourier transform.

Definition 1.2 (Inverse Fourier transform) If $g \in L^1(\mathbf{R}^d)$, let

$$(\mathscr{F}^{-1}g)(x) = \check{g}(x) = (2\pi)^{-d/2} \int g(\xi) e^{ix\cdot\xi} d\xi.$$

The quantity $\mathscr{F}^{-1}\mathscr{F}$ is the identity on a set of functions that is dense in $L^p(\mathbf{R}^d)$ for any $1 \leq p < \infty$.

Definition 1.3 (Schwartz space) Let $\mathscr{S}(\mathbf{R}^d)$ be the set of functions such that

$$\mathscr{S}(\mathbf{R}^d) = \left\{ f \in C^\infty(\mathbf{R}^d) : \sup_{x \in \mathbf{R}^d} |x|^\beta |\partial_x^\alpha f| \leq C(\alpha,\beta) < \infty \, \forall \alpha, \beta \in (\mathbf{Z}_{\geq 0})^d \right\}. \quad (1.8)$$

Remark The Schwartz space of smooth functions is actually a Fréchet space (but not a Banach space).

Clearly $\mathscr{S}(\mathbf{R}^d) \subset L^\infty(\mathbf{R}^d)$, and since $(1+|x|^2)^{-d}$ is integrable on $\mathbf{R}^d$, $\mathscr{S}(\mathbf{R}^d) \subset L^1(\mathbf{R}^d)$, so $\mathscr{F}$ is well defined on $\mathscr{S}(\mathbf{R}^d)$. Furthermore, (1.4) and (1.5) imply

$$\mathscr{F}, \mathscr{F}^{-1} : \mathscr{S}(\mathbf{R}^d) \to \mathscr{S}(\mathbf{R}^d),$$

so $\mathscr{F}^{-1}\mathscr{F}$ is well defined on $\mathscr{S}(\mathbf{R}^d)$.

1.1 Linear Schrödinger Equation and Preliminaries

Lemma 1.4 (Fourier inversion for Schwartz functions) *If $f \in \mathscr{S}(\mathbf{R}^d)$, then*

$$f(x) = \left(\mathscr{F}^{-1}\hat{f}\right)(x) = (2\pi)^{-d/2} \int e^{ix\cdot\xi}\hat{f}(\xi)\,d\xi.$$

Proof For any $\hat{f}(\xi) \in \mathscr{S}(\mathbf{R}^d)$, a direct calculation shows that

$$(2\pi)^{-d/2}\int e^{ix\cdot\xi}\hat{f}(\xi)\,d\xi = (2\pi)^{-d/2}\lim_{\varepsilon\searrow 0}\int e^{-\varepsilon|\xi|^2}e^{ix\cdot\xi}\hat{f}(\xi)\,d\xi, \qquad (1.9)$$

uniformly in x. Since f is integrable, Fubini's theorem implies that for any $\varepsilon > 0$,

$$\begin{aligned}(1.9) &= (2\pi)^{-d}\int\int e^{i(x-y)\cdot\xi}e^{-\varepsilon|\xi|^2}f(y)\,dy\,d\xi \\ &= (2\pi)^{-d}\int f(y)\int e^{i(x-y)\cdot\xi}e^{-\varepsilon|\xi|^2}\,d\xi\,dy \\ &= (4\pi\varepsilon)^{-d/2}\int f(y)e^{-\frac{|x-y|^2}{4\varepsilon}}\,dy. \end{aligned} \qquad (1.10)$$

The last equality follows from the computation

$$\left(\int e^{-x^2}dx\right)^2 = \int e^{-x^2-y^2}dx\,dy = 2\pi\int_0^\infty e^{-r^2}r\,dr = \pi,$$

and completing the square in the exponent. A change of variables implies that for any $\varepsilon > 0$, $x \in \mathbf{R}^d$,

$$(4\pi\varepsilon)^{-d/2}\int e^{-\frac{|x-y|^2}{4\varepsilon}}\,dy = 1,$$

so

$$\lim_{\varepsilon\searrow 0}(4\pi\varepsilon)^{-d/2}e^{-\frac{|x-y|^2}{4\varepsilon}} = \delta(y-x).$$

For any $x \in \mathbf{R}^d$, $\delta(y-x)$ is a tempered distribution, and therefore if $f \in \mathscr{S}(\mathbf{R}^d)$, for any $x \in \mathbf{R}^d$,

$$\lim_{\varepsilon\searrow 0}(4\pi\varepsilon)^{-d/2}\int f(y)e^{-\frac{|x-y|^2}{4\varepsilon}}\,dy = f(x).$$

Further computations also show that this convergence is uniform in $\mathbf{R}^d$, and is uniform in all the seminorms in (1.8). Thus, $\mathscr{F}^{-1}\mathscr{F}$ is the identity on $\mathscr{S}(\mathbf{R}^d)$. A similar argument shows that $\mathscr{F}\mathscr{F}^{-1}$ is the identity on $\mathscr{S}(\mathbf{R}^d)$. □

Therefore, the solution to (1.1), formally given by

$$\mathscr{F}^{-1}\left(e^{-it|\xi|^2}\hat{u}_0(\xi)\right), \qquad (1.11)$$

is well defined for any $t \in \mathbf{R}$ when u_0 is a Schwartz function.

Let $\langle \cdot, \cdot \rangle$ denote the L^2 inner product of functions,

$$\langle f, g \rangle = \mathrm{Re} \int f(x)\overline{g(x)}dx.$$

For any $f, g \in \mathscr{S}(\mathbf{R}^d)$,

$$\begin{aligned}
\langle \mathscr{F}f, g \rangle &= (2\pi)^{-d/2} \mathrm{Re} \int \overline{g(\xi)} \int e^{-ix\cdot\xi} f(x)\,dx\,d\xi \\
&= (2\pi)^{-d/2} \mathrm{Re} \int f(x) \overline{\left(\int e^{ix\cdot\xi} g(\xi)\,d\xi \right)} dx \\
&= \langle f, \mathscr{F}^{-1}g \rangle.
\end{aligned} \quad (1.12)$$

Taking $g = \mathscr{F}f$, $f \in \mathscr{S}(\mathbf{R}^d)$ implies

$$\|\mathscr{F}f\|_{L^2(\mathbf{R}^d)} = \|f\|_{L^2(\mathbf{R}^d)}. \quad (1.13)$$

Since $\mathscr{S}(\mathbf{R}^d)$ is dense in $L^2(\mathbf{R}^d)$, (1.12) may be extended to the Parseval identity,

$$\langle \mathscr{F}f, g \rangle = \langle f, \mathscr{F}^{-1}g \rangle \quad \text{for any } f, g \in L^2(\mathbf{R}^d),$$

and (1.13) may be extended to the Plancherel identity,

$$\|f\|_{L^2(\mathbf{R}^d)} = \|\mathscr{F}f\|_{L^2(\mathbf{R}^d)} \quad \text{for any } f \in L^2(\mathbf{R}^d). \quad (1.14)$$

Since $|e^{it|\xi|^2}| = 1$, (1.7) and (1.14) imply that if u solves (1.1), then

$$\|u(t,x)\|_{L^2(\mathbf{R}^d)} = \left\| e^{it|\xi|^2} \hat{u}_0(\xi) \right\|_{L^2(\mathbf{R}^d)} = \|\hat{u}_0(\xi)\|_{L^2(\mathbf{R}^d)} = \|u_0\|_{L^2(\mathbf{R}^d)}. \quad (1.15)$$

Moreover, since $\mathscr{S}(\mathbf{R}^d)$ is dense in $L^2(\mathbf{R}^d)$, (1.15) implies that (1.11) is well defined for any $u_0 \in L^2(\mathbf{R}^d)$.

Next, using the Fourier and inverse Fourier transforms, if $u_0 \in \mathscr{S}(\mathbf{R}^d)$,

$$\begin{aligned}
u(t,x) &= (2\pi)^{-d} \lim_{\varepsilon \searrow 0} \int e^{-\varepsilon|\xi|^2} e^{-it|\xi|^2} e^{ix\cdot\xi} \int e^{-iy\cdot\xi} u_0(y)\,dy\,d\xi \\
&= (2\pi)^{-d} \lim_{\varepsilon \searrow 0} \int \left(\int e^{-\varepsilon|\xi|^2} e^{-it|\xi|^2} e^{i(x-y)\cdot\xi}\,d\xi \right) u_0(y)\,dy \\
&= \lim_{\varepsilon \searrow 0} \int K(t, x-y, \varepsilon) u_0(y)\,dy.
\end{aligned} \quad (1.16)$$

1.1 Linear Schrödinger Equation and Preliminaries

Completing the square in the exponent,

$$K(t, x-y, \varepsilon) = (2\pi)^{-d} \int e^{-\varepsilon|\xi|^2} e^{-it|\xi|^2} e^{i(x-y)\cdot\xi} d\xi$$

$$= (2\pi)^{-d} \int e^{-\varepsilon|\xi|^2} e^{-it|\xi - \frac{x-y}{2t}|^2} e^{i\frac{|x-y|^2}{4t}} d\xi, \quad (1.17)$$

so by stationary phase analysis (see Chapter 8 of Stein (1993)), letting $\varepsilon \searrow 0$,

$$u(t, x) = (4\pi t)^{-d/2} e^{-i\frac{d\pi}{4}} \int e^{i\frac{|x-y|^2}{4t}} u_0(y) dy. \quad (1.18)$$

Again, since $\mathscr{S}(\mathbf{R}^d)$ is dense in $L^p(\mathbf{R}^d)$, for any $1 \leq p < \infty$, (1.18) gives the dispersive estimate

$$\left\| e^{it\Delta} u_0 \right\|_{L^\infty(\mathbf{R}^d)} \leq \frac{1}{|4\pi t|^{d/2}} \|u_0\|_{L^1(\mathbf{R}^d)}. \quad (1.19)$$

Then by (1.15), (1.19), and the Riesz–Thorin interpolation theorem (see for example Bergh and Löfstrom (1976)), for $2 \leq p \leq \infty$, if p' is the Lebesgue dual exponent satisfying $\frac{1}{p} + \frac{1}{p'} = 1$,

$$\left\| e^{it\Delta} u_0 \right\|_{L^p(\mathbf{R}^d)} \leq \frac{1}{|4\pi t|^{d(1/2 - 1/p)}} \|u_0\|_{L^{p'}(\mathbf{R}^d)}. \quad (1.20)$$

Therefore, for any $u_0 \in \mathscr{S}(\mathbf{R}^d)$,

$$\int_{\mathbf{R}^d} |u(t, x)|^2 dx$$

remains constant, while at the same time,

$$\sup_{x \in \mathbf{R}^d} |u(t, x)|^2 \to 0$$

as $t \to \pm\infty$. Thus $|u(t, x)|^2$ spreads out as $t \to \pm\infty$. For this reason (1.19) is called a dispersive estimate and (1.1) is called a dispersive equation.

The reader who is familiar with the wave equation should observe that the dispersion in (1.19) is faster by a factor of $t^{-1/2}$ than for the wave equation in the same dimension. For example, consider the wave equation in one dimension, and choose some $f \in \mathscr{S}(\mathbf{R})$. The solution to

$$u_{tt} - \Delta u = 0, \quad u(0, x) = f(x), \quad u_t(0, x) = 0 \quad (1.21)$$

is given by

$$u(t, x) = \frac{1}{2} f(x+t) + \frac{1}{2} f(x-t).$$

Such a solution consists of two traveling waves, and thus does not disperse at all.

On the other hand, the solution to the linear Schrödinger equation with $u_0 = f$ will have L^∞-norm that decays at the rate of $t^{-1/2}$. The reason for this difference is that the wave equation obeys finite propagation speed and the Huygens principle. Thus, a solution to the one-dimensional problem (1.21) cannot disperse at all, since it travels at either speed 1 to the right or speed 1 to the left. For the linear Schrödinger equation, following (1.16), and formally taking $\varepsilon \searrow 0$,

$$(2\pi)^{-d} \int e^{-it|\xi|^2} e^{ix\cdot\xi} \hat{f}(\xi) d\xi = (2\pi)^{-d} \int e^{-it|\xi - \frac{x}{2t}|^2} e^{i\frac{|x|^2}{4t}} \hat{f}(\xi) d\xi,$$

and thus formally $\hat{f}(\xi)$ travels with velocity 2ξ. This computation can be made rigorous using the Littlewood–Paley decomposition.

Definition 1.5 (Littlewood–Paley decomposition) Let $\psi(\xi)$ be a smooth, decreasing, radial function supported on $|\xi| \le 2$, $\psi(\xi) = 1$ on $|\xi| \le 1$, and let

$$\phi_k(\xi) = \psi(2^{-k-1}\xi) - \psi(2^{-k}\xi). \tag{1.22}$$

Then $\phi_k(\xi)$ is a radial, smooth function supported on the annulus $2^k \le |\xi| \le 2^{k+2}$. Also, for $\xi \ne 0$,

$$\sum_{k=-\infty}^{\infty} \phi_k(\xi) = 1. \tag{1.23}$$

Let P_k be the Fourier multiplier given by $\phi_k(\xi)$; that is,

$$P_k f = \mathscr{F}^{-1}\left(\phi_k(\xi)\hat{f}(\xi)\right). \tag{1.24}$$

Also let

$$\widetilde{P}_k = P_{k-2} + P_{k-1} + P_k + P_{k+1} + P_{k+2} \text{ and } \tilde{\phi}_k = \phi_{k-2} + \phi_{k-1} + \phi_k + \phi_{k+1} + \phi_{k+2}.$$

Since $\phi_k(\xi)$ is supported on $2^k \le |\xi| \le 2^{k+2}$, (1.23) implies

$$\widetilde{P}_k P_k = P_k.$$

Define

$$P_{\le k} = \sum_{j \le k} P_j, \quad P_{<k} = \sum_{j<k} P_j, \quad P_{\ge k} = \sum_{j \ge k} P_j, \quad P_{>k} = \sum_{j>k} P_j.$$

Also, for any $N > 0$, define

$$(P_N f)(x) = \mathscr{F}^{-1}\left(\phi_0\left(\frac{\xi}{N}\right)\hat{f}(\xi)\right),$$

$$(P_{<N} f)(x) = \mathscr{F}^{-1}\left(\psi\left(\frac{\xi}{N}\right)\hat{f}(\xi)\right), \tag{1.25}$$

$$(P_{\ge N} f)(x) = \mathscr{F}^{-1}\left(\left(1 - \psi\left(\frac{\xi}{N}\right)\right)\hat{f}(\xi)\right).$$

1.1 Linear Schrödinger Equation and Preliminaries

When there is no confusion it is convenient to write $f_k = P_k f$, $f_N = P_N f$, $f_{\leq N} = P_{\leq N} f$, $f_{\geq N} = P_{\geq N} f$. In this book, a lowercase Latin letter always refers to the projection (1.24) and an uppercase Latin letter always refers to the projection (1.25).

Theorem 1.6 (Littlewood–Paley theorem) *For any $1 < p < \infty$,*

$$\|f\|_{L^p(\mathbf{R}^d)} \sim_{p,d} \left\| \left(\sum_{j \in \mathbf{Z}} |P_j f|^2 \right)^{1/2} \right\|_{L^p(\mathbf{R}^d)}. \tag{1.26}$$

Proof See Stein (1970). $\square$

A solution to (1.1) that is localized to frequencies $|\xi| \sim 2^k$ travels at speed $\sim 2^k$.

Theorem 1.7 *Let K_k be the kernel of $P_k e^{it\Delta}$, where P_k is the Littlewood–Paley projection defined in Definition 1.5. That is, as in (1.11) and (1.16), let*

$$u(t,x) = e^{it\Delta} P_k u_0(x) = \int K_k(t, x-y) u_0(y) \, dy,$$

where

$$K_k(t,x) = (2\pi)^{-d} \int e^{ix\cdot\xi} e^{-it|\xi|^2} \phi_k(\xi) \, d\xi. \tag{1.27}$$

If $|x| \leq 2^{k+4}|t|$,

$$|K_k(t,x)| \lesssim_d \frac{2^{dk}}{(1+2^{2k}|t|)^{d/2}}. \tag{1.28}$$

If $|x| > 2^{k+4}|t|$, then for any M,

$$|K_k(t,x)| \lesssim_{d,M} \frac{2^{dk}}{(1+2^k|x|)^M}, \tag{1.29}$$

and if $|x| < 2^{k-4}|t|$, then for any M,

$$|K_k(t,x)| \lesssim_{d,M} \frac{2^{dk}}{(1+2^{2k}|t|)^M}. \tag{1.30}$$

Remark The kernel for the Littlewood–Paley projection operator is given by $K_j(0,x)$, where

$$P_j f(x) = \int K_j(0, x-y) f(y) \, dy. \tag{1.31}$$

To simplify notation, let

$$P_0 f(x) = \int K(x-y) f(y) \, dy, \tag{1.32}$$

and call $K(x)$ the Littlewood–Paley kernel. By (1.29), for any M,

$$|K(x)| \lesssim_M \frac{1}{(1+|x|)^M}, \qquad (1.33)$$

and for any $j \in \mathbf{Z}$, (1.22) implies

$$|K_j(x)| = 2^{jd}|K(2^j x)| \lesssim_M \frac{2^{jd}}{(1+2^j|x|)^M}. \qquad (1.34)$$

Proof For $|t| \leq 2^{-2k}$, direct integration and the support of $\phi_k(\xi)$ implies

$$(2\pi)^{-d} \int e^{ix\cdot\xi} e^{-it|\xi|^2} \phi_k(\xi) d\xi \lesssim_d 2^{dk}.$$

For $|t| > 2^{-2k}$ and $|x| \leq 2^{k+4}|t|$, the same stationary phase argument that gives (1.19) implies

$$|K_k(t,x)| \lesssim \frac{1}{|t|^{d/2}}.$$

Next suppose $|x| > 2^{k+4}|t|$. Then

$$\int e^{ix\cdot\xi} e^{-it|\xi|^2} \phi_k(\xi) d\xi = \int \phi_k(\xi) \frac{(-ix+2it\xi)\cdot\nabla_\xi}{|x-2t\xi|^2} \left(e^{ix\cdot\xi - it|\xi|^2} \right),$$

so for any M,

$$(1.27) = \int \phi_k(\xi) \left(\frac{(-ix+2it\xi)\cdot\nabla_\xi}{|x-2t\xi|^2} \right)^M \left(e^{ix\cdot\xi - it|\xi|^2} \right) d\xi. \qquad (1.35)$$

Integrating by parts, when $|x| > 2^{k+4}|t|$, $|x - 2t\xi| \sim |x|$, so

$$(1.35) \lesssim_M \frac{2^{kd}|t|^M}{|x|^{2M}} + \frac{2^{kd}}{|x|^M 2^{kM}} \lesssim \frac{2^{kd}}{|x|^M 2^{kM}}.$$

This proves (1.29). Equation (1.30) also follows from (1.35) since $|x - 2t\xi| \sim |t||\xi|$ when $|x| < 2^{k-4}|t|$. $\square$

Now choose $\chi(x) \in C_0^\infty(\mathbf{R})$ such that χ is supported on $|x| \leq 2$ and

$$\sum_{m \in \mathbf{Z}} \chi(x-m) = 1 \qquad (1.36)$$

for all $x \in \mathbf{R}$. If $u_0 \in L^1(\mathbf{R})$, then for any N,

$$|x|^N \chi(x) u_0(x) \in L^1(\mathbf{R}).$$

Therefore, by (1.4) and (1.5),

$$\hat{f}(\xi) = \chi(\xi) \mathscr{F}(\chi(x) u_0(x)) \qquad (1.37)$$

1.1 Linear Schrödinger Equation and Preliminaries

is a Schwartz function. By the Sobolev embedding theorem, if f is given by (1.37),

$$\left\|e^{it\Delta}f\right\|_{L^\infty(\mathbf{R})} \lesssim \left\|\chi(x)u_0\right\|_{L^1(\mathbf{R})}.$$

Following the computations in the proof of Theorem 1.7,

$$e^{it\Delta}f = (2\pi)^{-1}\int e^{ix\xi}e^{-it\xi^2}\chi(\xi)\left[\int e^{-iy\xi}\chi(y)u_0(y)dy\right]d\xi$$

$$= (2\pi)^{-1}\int\left[\int \chi(\xi)e^{i(x-y)\xi}e^{-it\xi^2}d\xi\right]\chi(y)u_0(y)dy$$

$$= (2\pi)^{-1}\int\left[\int e^{-it\left(\xi-\frac{x-y}{2t}\right)^2}\chi(\xi)d\xi\right]e^{i\frac{(x-y)^2}{4t}}\chi(y)u_0(y)dy.$$

Plugging in $t=1$ and making stationary phase arguments,

$$e^{it\Delta}f\big|_{t=1} = \int \widetilde{K}(x-y)\chi(y)u_0(y)dy,$$

where for any M,

$$|\widetilde{K}(x-y)| \lesssim_M \frac{1}{(1+|x-y|)^M}. \tag{1.38}$$

Now utilize the Galilean transformation.

Lemma 1.8 (Galilean transformation) *If u solves the linear Schrödinger equation (1.1), then for any $\xi_0 \in \mathbf{R}^d$,*

$$e^{-it|\xi_0|^2}e^{ix\cdot\xi_0}u(t,x-2t\xi_0) \tag{1.39}$$

solves (1.1) with initial data $e^{ix\cdot\xi_0}u_0(x)$.

For any $m \in \mathbf{Z}$, set

$$\hat{f}_m(\xi) = \chi(\xi-m)\mathscr{F}(\chi(x)u_0(x)),$$

and then by (1.38) and (1.39),

$$e^{it\Delta}f_m\big|_{t=1} = \int \widetilde{K}_m(x-y)\chi(y)u_0(y)dy,$$

where for any M,

$$|\widetilde{K}_m(x-y)| \lesssim_M \frac{1}{(1+|x-2tm-y|)^M}.$$

Therefore,

$$\left\|e^{it\Delta}(\chi u_0)\big|_{t=1}\right\|_{L^\infty(\mathbf{R})} \lesssim \left\|\chi(x)u_0(x)\right\|_{L^1(\mathbf{R})}. \tag{1.40}$$

Then by (1.36) and (1.40),
$$\left\|e^{it\Delta}u_0|_{t=1}\right\|_{L^\infty(\mathbf{R})} \lesssim \|u_0\|_{L^1(\mathbf{R})}. \tag{1.41}$$

Taking a d-dimensional partition of unity and decomposing
$$\hat{u}_0(\xi) = \sum_{m\in\mathbf{Z}^d} \chi(\xi - m)\mathscr{F}(\chi(x)u_0(x))$$
proves the same in any dimension when $t = 1$.

Time reversal symmetry and the scaling symmetry generalize (1.41) to any t.

Lemma 1.9 (Scaling symmetry) *If u solves (1.1), then for any $\lambda > 0$,*
$$\lambda^{d/2}u(\lambda^2 t, \lambda x) \tag{1.42}$$
solves (1.1) with initial data $\lambda^{d/2}u_0(\lambda x)$.

Remark The invariance in (1.42) is called the mass-critical scaling.

Taking $\lambda = t^{1/2}$, (1.41) implies
$$\lambda^{d/2}\|u(\lambda^2,\lambda x)\|_{L^\infty(\mathbf{R}^d)} \lesssim \lambda^{d/2}\|u_0(\lambda x)\|_{L^1(\mathbf{R}^d)} = \lambda^{-d/2}\|u_0\|_{L^1(\mathbf{R}^d)},$$
and therefore,
$$\|u(t,x)\|_{L^\infty(\mathbf{R}^d)} \lesssim t^{-\frac{d}{2}}\|u_0\|_{L^1(\mathbf{R}^d)}. \tag{1.43}$$

1.2 Strichartz Estimates

The inhomogeneous version of (1.1) is given by
$$iu_t + \Delta u = F, \quad u(0,x) = 0. \tag{1.44}$$

By unitary group theory, the solution to (1.44) is formally given by
$$u(t) = -i\int_0^t e^{i(t-\tau)\Delta}F(\tau)d\tau.$$

The dispersive estimate (1.20) implies a space-time integrability estimate for a solution to (1.44).

Lemma 1.10 (Inhomogeneous estimate) *Suppose $d(\frac{1}{2} - \frac{1}{q}) < 1$; $\tilde{p}, p < \infty$, and*
$$d\left(\frac{1}{2} - \frac{1}{q}\right) = \frac{1}{\tilde{p}} + \frac{1}{p}.$$

Then
$$\left\|\int_{-\infty}^t e^{i(t-\tau)\Delta}F(\tau)d\tau\right\|_{L_t^p L_x^q(\mathbf{R}\times\mathbf{R}^d)} \lesssim_{d,p,\tilde{p},q} \|F\|_{L_t^{\tilde{p}'} L_x^{q'}(\mathbf{R}\times\mathbf{R}^d)}. \tag{1.45}$$

1.2 Strichartz Estimates

Proof The inequality (1.45) is a two-dimensional integral in time, and (1.20) implies that the operator $e^{i(t-\tau)\Delta}F(\tau)$ is singular when $t = \tau$. Restricting τ to $\tau < 0$ and t to $t > 0$ reduces the singularity from a one-dimensional singularity to a singularity only at the origin. By (1.20) and Hölder's inequality, for any j, $k \in \mathbf{Z}$,

$$\left\| \int_{-2^{k+1}}^{-2^k} e^{i(t-\tau)\Delta} F(\tau)\, d\tau \right\|_{L_t^p L_x^q([2^j, 2^{j+1}] \times \mathbf{R}^d)} \lesssim \frac{2^{j/p} 2^{k/\tilde{p}}}{(2^j + 2^k)^{d\left(\frac{1}{2} - \frac{1}{q}\right)}} \|F\|_{L_t^{\tilde{p}'} L_x^{q'}([-2^{k+1}, -2^k] \times \mathbf{R}^d)},$$

so for any $j \in \mathbf{Z}$,

$$\left\| \int_{-\infty}^{0} e^{i(t-\tau)\Delta} F(\tau)\, d\tau \right\|_{L_t^p L_x^q([2^j, 2^{j+1}] \times \mathbf{R}^d)} \lesssim \sum_{k \in \mathbf{Z}} \frac{2^{j/p} 2^{k/\tilde{p}}}{(2^j + 2^k)^{d\left(\frac{1}{2} - \frac{1}{q}\right)}} \|F\|_{L_t^{\tilde{p}'} L_x^{q'}([-2^{k+1}, -2^k] \times \mathbf{R}^d)}. \quad (1.46)$$

Since $\frac{1}{p} + \frac{1}{\tilde{p}} < 1$, $\frac{1}{p} < 1 - \frac{1}{\tilde{p}} = \frac{1}{\tilde{p}'}$, so by Young's inequality and (1.46),

$$\left\| \int_{-\infty}^{0} e^{i(t-\tau)\Delta} F(\tau)\, d\tau \right\|_{L_t^p L_x^q([0,\infty) \times \mathbf{R}^d)} \lesssim_{p,q,d} \|F\|_{L_t^{\tilde{p}'} L_x^{q'}((-\infty, 0] \times \mathbf{R}^d)}. \quad (1.47)$$

Now for almost every $(t, \tau) \in \{[0,1] \times [0,1] : \tau < t\}$, there exists a unique $j, k \in \mathbf{Z}$, $j \leq 0$, $0 \leq k \leq 2^{-j} - 2$ such that $\tau \in I_k^j = [k2^j, (k+1)2^j]$ and $t \in I_{k+1}^j = [(k+1)2^j, (k+2)2^j]$. This decomposition trades off closeness to the singularity,

$$\text{dist}\left(\bigcup_{k=0}^{2^{-j}-2} I_k^j \times I_{k+1}^j, \{(t, \tau) : t = \tau\} \right) \lesssim 2^j, \quad (1.48)$$

with decreasing measure,

$$\mu\left(\bigcup_{k=0}^{2^{-j}-1} I_k^j \times I_{k+1}^j \right) \lesssim 2^j. \quad (1.49)$$

Remark This observation was pointed out to the author by Manoussos Grillakis and Matei Machedon.

Suppose without loss of generality that $\|F\|_{L_t^{\tilde{p}'} L_x^{q'}(\mathbf{R} \times \mathbf{R}^d)} = 1$. Following the argument proving the Christ–Kiselev lemma in Christ and Kiselev (2001) and

Smith and Sogge (2000), define the map

$$G : \mathbf{R} \to [0,1],$$

$$G(t) = \int_{-\infty}^{t} \|F(\tau)\|_{L_x^{q'}(\mathbf{R}^d)}^{\tilde{p}'} d\tau,$$

and for any $j \le 0$, $0 \le k \le 2^{-j} - 1$, let

$$I_k^j = G^{-1}\left([k2^j, (k+1)2^j)\right).$$

Inequality (1.47) implies

$$\left\| \int_{I_k^j} e^{i(t-\tau)\Delta} F(\tau) d\tau \right\|_{L_t^p L_x^q\left(I_{k+1}^j \times \mathbf{R}^d\right)} \lesssim \|F\|_{L_t^{\tilde{p}'} L_x^{q'}\left(I_k^j \times \mathbf{R}^d\right)} = 2^{j/\tilde{p}'}.$$

Since the I_k^j intervals are disjoint, (1.47) and $d\left(\frac{1}{2} - \frac{1}{q}\right) < 1$ imply

$$\left\| \int_{-\infty}^{t} e^{i(t-\tau)\Delta} F(\tau) d\tau \right\|_{L_t^p L_x^q(\mathbf{R}\times\mathbf{R}^d)} \lesssim_{q,d} \sum_{j\le 0} 2^{j/\tilde{p}'} 2^{-j/p}$$

$$= \sum_{j\le 0} 2^{j} 2^{-j/\tilde{p}} 2^{-j/p} \lesssim_{\tilde{p},p} 1. \quad (1.50)$$

This proves the inhomogeneous estimate. □

Taking $p = \tilde{p}$,

$$\left\| \int_{-\infty}^{t} e^{i(t-\tau)\Delta} F(\tau) d\tau \right\|_{L_t^p L_x^q(\mathbf{R}\times\mathbf{R}^d)} \lesssim_{p,q,d} \|F\|_{L_t^{p'} L_x^{q'}(\mathbf{R}\times\mathbf{R}^d)} \quad (1.51)$$

when $p > 2$ and $\frac{2}{p} = d(\frac{1}{2} - \frac{1}{q})$.

Definition 1.11 (Admissible pair) A pair (p,q) is called admissible if

$$\frac{2}{p} = d\left(\frac{1}{2} - \frac{1}{q}\right), \quad (1.52)$$

and $4 \le p \le \infty$ when $d = 1$; $2 < p \le \infty$ when $d = 2$; or $2 \le p \le \infty$ when $d \ge 3$. Let

$$\mathscr{A}_d = \{(p,q) : (p,q) \text{ is admissible }\}.$$

If $p > 2$, (p,q) is called a non-endpoint admissible pair, and when $p = 2$, (p,q) is called an endpoint admissible pair. Inequality (1.51) directly yields a non-endpoint estimate for the space-time norm of a solution to (1.1).

Theorem 1.12 (Non-endpoint Strichartz estimate) *If (p,q) is an admissible pair with $p > 2$,*

$$\|e^{it\Delta} u_0\|_{L_t^p L_x^q(\mathbf{R}\times\mathbf{R}^d)} \lesssim_{p,q,d} \|u_0\|_{L^2(\mathbf{R}^d)} \quad (1.53)$$

1.2 Strichartz Estimates

and

$$\left\| \int_{\mathbf{R}} e^{-i\tau\Delta} F(\tau) d\tau \right\|_{L^2(\mathbf{R}^d)} \lesssim_{p,q,d} \|F\|_{L_t^{p'} L_x^{q'}(\mathbf{R}\times\mathbf{R}^d)}. \tag{1.54}$$

The seminal result in this direction was proved in Strichartz (1977) for $p = q$. The non-endpoint estimate $(p > 2)$ was subsequently proved by Ginibre and Velo (1992) and Yajima (1987). See also Tao (2006).

Proof The proof uses a $T-T^*$ argument on the L^2 inner product and (1.51). Since $L^2(\mathbf{R}^d)$ is a Hilbert space and Δ is self-adjoint, by (1.51),

$$\begin{aligned}
\left\| \int_{\mathbf{R}} e^{-i\tau\Delta} F(\tau) d\tau \right\|_{L^2(\mathbf{R}^d)}^2 &= \int_{\mathbf{R}} \int_{\mathbf{R}} \langle e^{-i\tau\Delta} F(\tau), e^{-it\Delta} F(t) \rangle_{L^2} dt\, d\tau \\
&= \int_{\mathbf{R}} \int_{\mathbf{R}} \langle e^{i(t-\tau)\Delta} F(\tau), F(t) \rangle_{L^2} dt\, d\tau \\
&\lesssim_{p,q,d} \|F\|_{L_t^{p'} L_x^{q'}(\mathbf{R}\times\mathbf{R}^d)}^2. \tag{1.55}
\end{aligned}$$

This proves (1.54). Relation (1.53) follows by duality. $\square$

This result is actually pretty close to sharp. It is natural to place an L^2-norm in the right-hand side of (1.53). This is because a space-time norm over $\mathbf{R} \times \mathbf{R}^d$ is invariant under the action of the operator $e^{it_0\Delta}$ for a fixed $t_0 \in \mathbf{R}$, so it is appropriate to seek a norm for the initial data that is also invariant under the action of $e^{it_0\Delta}$, such as L^2. Moreover, any p, q, and d such that

$$\|e^{it\Delta} f\|_{L_t^p L_x^q(\mathbf{R}\times\mathbf{R}^d)} \leq C(p,q,d) \|f\|_{L^2(\mathbf{R}^d)}, \tag{1.56}$$

must satisfy certain algebraic relations. For example, (1.15) shows that

$$\|e^{it\Delta} f\|_{L_t^\infty L_x^2(\mathbf{R}\times\mathbf{R}^d)} = \|f\|_{L^2(\mathbf{R}^d)}, \tag{1.57}$$

but that (1.56) cannot hold for any $1 \leq p < \infty$ and $q = 2$. Next, under (1.42),

$$\|u_0(x)\|_{L^2(\mathbf{R}^d)} = \|\lambda^{d/2} u_0(\lambda x)\|_{L^2(\mathbf{R}^d)},$$

while

$$\|\lambda^{d/2} u(\lambda^2 t, \lambda x)\|_{L_t^p L_x^q(\mathbf{R}\times\mathbf{R}^d)} = \lambda^{d/2 - 2/p - d/q} \|u(t,x)\|_{L_t^p L_x^q(\mathbf{R}\times\mathbf{R}^d)}.$$

Taking $\lambda \to 0$ or $\lambda \to \infty$ shows that (1.56) fails unless (p,q) satisfies (1.52).

Remark A counterexample to (1.56) for $p < 2$ could be constructed via N evenly spaced, nonintersecting, identical bump functions that are Galilean translated, so that the $L_t^p L_x^q$-norm of the solution is of the order $N^{1/p}$, while the L^2-norm of the initial data is of the order $N^{1/2}$.

When $d = 1$, $2 \leq q \leq \infty$ restricts p to $p \geq 4$. Therefore, it only remains to consider the case when $p = 2$, (p,q) satisfies (1.52), and $d \geq 2$. When $d = 2$, there are known counterexamples to the endpoint result of Theorem 1.13. These counterexamples are all nonradial and involve Brownian motion. In fact, these counterexamples also exclude the weaker result of replacing $L_t^2 L_x^\infty$ with $L_t^2(I, BMO)$. See Montgomery-Smith (1998) for more information and Tao (2000) for a positive result for radially symmetric functions.

When $d \geq 3$, Keel and Tao (1998) proved the endpoint case of Theorem 1.12, yielding the sharp result for Strichartz estimates.

Theorem 1.13 (Strichartz estimates) *Suppose (p,q) and $(\tilde{p},\tilde{q})$ are admissible pairs, and $I \subset \mathbf{R}$ is a possibly infinite time interval. Then*

$$\left\|e^{it\Delta}u_0\right\|_{L_t^{\tilde{p}} L_x^{\tilde{q}}(I \times \mathbf{R}^d)} \lesssim_{\tilde{p},\tilde{q},d} \|u_0\|_{L^2(\mathbf{R}^d)}, \tag{1.58}$$

$$\left\|\int_{\mathbf{R}} e^{-it\Delta} F(t)\, dt\right\|_{L^2(\mathbf{R}^d)} \lesssim_{p,q,d} \|F\|_{L_t^{p'} L_x^{q'}(I \times \mathbf{R}^d)}, \tag{1.59}$$

and

$$\left\|\int_{\tau<t,\tau\in I} e^{i(t-\tau)\Delta} F(\tau)\, d\tau\right\|_{L_t^{\tilde{p}} L_x^{\tilde{q}}(I \times \mathbf{R}^d)} \lesssim_{p,q,\tilde{p},\tilde{q}} \|F\|_{L_t^{p'} L_x^{q'}(I \times \mathbf{R}^d)}. \tag{1.60}$$

The constant in (1.58)–(1.60) is independent of (p,q) and $(\tilde{p},\tilde{q})$ when $d \neq 2$.

Proof The argument presented here is not precisely the same as in Keel and Tao (1998), but relies on the same components as their argument, the first of which was an atomic decomposition of L^p spaces. The decomposition used here is similar.

Suppose $f \in L^{q'}(\mathbf{R}^d)$ for some $1 \leq q' \leq 2$ is a continuous function. Decompose

$$f = \sum_{k\in\mathbf{Z}} f_k(x) = \sum_{k\in\mathbf{Z}} \mu_k(x) f(x)$$

$$= \sum_{k\in\mathbf{Z}} \mu_{\{x : 2^k \|f\|_{L^{q'}} < |f(x)| \leq 2^{k+1} \|f\|_{L^{q'}}\}}(x) f(x), \tag{1.61}$$

where $\mu_A(x)$ is the characteristic function of the set A,

$$\mu_A(x) = \begin{cases} 1 & \text{if } x \in A, \\ 0 & \text{if } x \notin A. \end{cases}$$

Since the characteristic functions $\mu_k(x)$ are supported on disjoint sets, for any $1 \leq p < \infty$,

$$\|f\|_{L^p(\mathbf{R}^d)}^p = \sum_k \|f_k\|_{L^p(\mathbf{R}^d)}^p. \tag{1.62}$$

1.2 Strichartz Estimates

Because all the $\mu_k(x)$ have support with finite measure, by (1.62),

$$\int |f_k(x)| dx \lesssim 2^{-k(q'-1)} \|f\|_{L^{q'}(\mathbf{R}^d)}^{1-q'} \left(\int |f_k(x)|^{q'} dx \right)$$

$$\lesssim 2^{-k(q'-1)} \frac{\|f_k\|_{L^{q'}(\mathbf{R}^d)}^{q'}}{\|f\|_{L^{q'}(\mathbf{R}^d)}^{q'-1}} \lesssim 2^{-k(q'-1)} \|f_k\|_{L^{q'}(\mathbf{R}^d)}. \qquad (1.63)$$

Because $\mu_k(x) f(x)$ lies in L^∞ for any k,

$$\left(\int |f_k(x)|^2 dx \right)^{1/2} \lesssim \left(\int |f_k(x)|^{q'} \|f\|_{L^{q'}(\mathbf{R}^d)}^{2-q'} 2^{k(2-q')} dx \right)^{1/2}$$

$$\lesssim 2^{k\left(1-\frac{q'}{2}\right)} \|f_k\|_{L^{q'}(\mathbf{R}^d)}^{q'/2} \|f\|_{L^{q'}(\mathbf{R}^d)}^{1-q'/2}. \qquad (1.64)$$

This decomposition is already useful for proving a result analogous to Christ and Kiselev (2001) for non-endpoint Strichartz estimates.

Take $d = 1$. For any $(p,q) \in \mathscr{A}_1$, there exists $0 \le \theta \le 1$ such that

$$\frac{1}{p'} = \frac{\theta}{4/3} + \frac{1-\theta}{1}, \quad \frac{1}{q'} = \frac{\theta}{1} + \frac{1-\theta}{2}.$$

Making a simple calculation, $\frac{1}{p'} = \frac{4-\theta}{4}, \frac{1}{q'} = \frac{1+\theta}{2}, \frac{1}{q} = \frac{1-\theta}{2}$, and $2\left(1 - \frac{1}{p'}\right) = \frac{1}{q'} - \frac{1}{2}$. Let

$$F^{(1)}(t) = \sum_{k: 2^k \le \|F(t)\|_{L^{q'}}^{\frac{1+\theta}{4-\theta}}} F_k(t,x) \quad \text{and} \quad F^{(2)}(t) = \sum_{k: 2^k > \|F(t)\|_{L^{q'}}^{\frac{1+\theta}{4-\theta}}} F_k(t,x).$$

By (1.62) and (1.64),

$$\|F^{(1)}(t)\|_{L^2_x(\mathbf{R})} \lesssim \left(\sum_{2^k \le \|F(t)\|_{L^{q'}}^{\frac{1+\theta}{4-\theta}}} 2^{k(2-q')} \cdot \|F_k(t)\|_{L^{q'}_x(\mathbf{R})}^{q'} \right)^{1/2} \|F(t)\|_{L^{q'}_x(\mathbf{R})}^{1-\frac{q'}{2}}$$

$$\lesssim \|F(t)\|_{L^{q'}_x(\mathbf{R})}^{q'/2} \|F(t)\|_{L^{q'}_x(\mathbf{R})}^{(2-q') \cdot \frac{1+\theta}{2(4-\theta)}} \|F(t)\|_{L^{q'}_x(\mathbf{R})}^{1-\frac{q'}{2}}$$

$$= \|F(t)\|_{L^{q'}_x(\mathbf{R})} \|F(t)\|_{L^{q'}_x(\mathbf{R})}^{\frac{p'}{2} \cdot \frac{2-q'}{2q'}}$$

$$= \|F(t)\|_{L^{q'}_x(\mathbf{R})}^{p'}. \qquad (1.65)$$

Also by (1.63), (1.64), and Hölder's inequality,

$$\|F^{(2)}(t)\|_{L_x^1(\mathbf{R})} \lesssim \sum_{k:2^k > \|F(t)\|_{L_x^{q'}}^{\frac{1+\theta}{4-\theta}}} 2^{-k(q'-1)} \|F_k(t)\|_{L_x^{q'}(\mathbf{R})}$$

$$\lesssim \left(\sum_k \|F_k(t)\|_{L_x^{q'}(\mathbf{R})}^{q'}\right)^{1/q'} \left(\sum_{k:2^k > \|F(t)\|_{L_x^{q'}}^{\frac{1+\theta}{4-\theta}}} 2^{-kq'}\right)^{1/q}$$

$$\lesssim \|F(t)\|_{L_x^{q'}(\mathbf{R})} \|F(t)\|_{L_x^{q'}(\mathbf{R})}^{-\frac{p'}{2q}}. \quad (1.66)$$

Then by direct calculation, $\frac{p'}{2q} = \frac{1-\theta}{4-\theta}$, so $1 - \frac{p'}{2q} = \frac{3}{4-\theta} = \frac{3}{4}p'$. Therefore, by (1.65) and (1.66), if $(p,q) \in \mathscr{A}_1$,

$$L_t^{p'} L_x^{q'} \subset L_t^{4/3} L_x^1 + L_t^1 L_x^2, \quad (1.67)$$

and furthermore,

$$\|F^{(1)}\|_{L_t^1 L_x^2(\mathbf{R}\times\mathbf{R})} + \|F^{(2)}\|_{L_t^{4/3} L_x^1(\mathbf{R}\times\mathbf{R})} \lesssim \|F\|_{L_t^{p'} L_x^{q'}(\mathbf{R}\times\mathbf{R})}, \quad (1.68)$$

with constant independent of (p',q'). Therefore, (1.51), (1.53), and (1.54) imply

$$\left\|\int_0^t e^{i(t-\tau)\Delta} F(\tau) d\tau\right\|_{L_t^\infty L_x^2 \cap L_t^4 L_x^\infty(\mathbf{R}\times\mathbf{R})}$$

$$\lesssim \|F^{(1)}\|_{L_t^1 L_x^2(\mathbf{R}\times\mathbf{R})} + \|F^{(2)}\|_{L_t^{4/3} L_x^1(\mathbf{R}\times\mathbf{R})} \lesssim \|F\|_{L_t^{p'} L_x^{q'}(\mathbf{R}\times\mathbf{R})}.$$

This directly implies (1.59), and then by duality (1.58). By interpolation, (1.60) also holds for $d = 1$.

If $d = 2$ set $\tilde{p} = \min(p, \tilde{p})$ and choose $\tilde{q}$ satisfying $(\tilde{p}, \tilde{q}) \in \mathscr{A}_2$. Making an argument almost identical to (1.61)–(1.68),

$$L_t^{\tilde{p}'} L_x^{\tilde{q}'} \subset L_t^{\tilde{p}'} L_x^{\tilde{q}'} + L_t^1 L_x^2, \quad (1.69)$$

and again by (1.51), (1.53), and (1.54),

$$\left\|\int_0^t e^{i(t-\tau)\Delta} F(\tau) d\tau\right\|_{L_t^\infty L_x^2 \cap L_t^{\tilde{p}} L_x^{\tilde{q}}(\mathbf{R}\times\mathbf{R}^d)}$$

$$\lesssim_{\tilde{p},\tilde{q}} \|F_1\|_{L_t^1 L_x^2(\mathbf{R}\times\mathbf{R}^d)} + \|F_2\|_{L_t^{\tilde{p}'} L_x^{\tilde{q}'}(\mathbf{R}\times\mathbf{R}^d)} \lesssim \|F\|_{L_t^{p'} L_x^{q'}(\mathbf{R}\times\mathbf{R}^d)}.$$

Therefore, to complete the proof of Theorem 1.13, it only remains to prove that when $d \geq 3$,

$$\left\|\int_0^t e^{i(t-\tau)\Delta} F(\tau) d\tau\right\|_{L_t^2 L_x^{\frac{2d}{d-2}}(\mathbf{R}\times\mathbf{R}^d)} \lesssim \|F\|_{L_t^2 L_x^{\frac{2d}{d+2}}(\mathbf{R}\times\mathbf{R}^d)}. \quad (1.70)$$

1.2 Strichartz Estimates

Indeed, assuming (1.70) is true, if $F \in L_t^2 L_x^{\frac{2d}{d+2}}(\mathbf{R} \times \mathbf{R}^d)$, then as in (1.55),

$$\left\langle \int_{\mathbf{R}} e^{-it\Delta} F(t)\, dt, \int_{\mathbf{R}} e^{-i\tau\Delta} F(\tau)\, d\tau \right\rangle_{L^2} = \int_{\mathbf{R}} \left\langle \int_{\mathbf{R}} e^{i(t-\tau)\Delta} F(\tau)\, d\tau, F(t) \right\rangle_{L^2} dt$$
$$\lesssim \|F\|^2_{L_t^2 L_x^{\frac{2d}{d+2}}(\mathbf{R}\times\mathbf{R}^d)},$$

which proves

$$\left\| \int_0^t e^{i(t-\tau)\Delta} F(\tau) \right\|_{L_t^\infty L_x^2(\mathbf{R}\times\mathbf{R}^d)} \lesssim \|F\|_{L_t^2 L_x^{\frac{2d}{d+2}}(\mathbf{R}\times\mathbf{R}^d)}. \tag{1.71}$$

The estimate

$$\left\| \int_0^t e^{i(t-\tau)\Delta} F(\tau) \right\|_{L_t^2 L_x^{\frac{2d}{d-2}}(\mathbf{R}\times\mathbf{R}^d)} \lesssim \|F\|_{L_t^1 L_x^2(\mathbf{R}\times\mathbf{R}^d)}$$

is the dual to (1.71), and

$$\left\| \int_0^t e^{i(t-\tau)\Delta} F(\tau) \right\|_{L_t^\infty L_x^2(\mathbf{R}\times\mathbf{R}^d)} \lesssim \|F\|_{L_t^1 L_x^2(\mathbf{R}\times\mathbf{R}^d)} \tag{1.72}$$

follows from (1.15).

As in (1.67) and (1.69), for any $(p,q) \in \mathscr{A}_d$, $d \geq 3$, $L_t^{p'} L_x^{q'} \subset L_t^1 L_x^2 + L_t^2 L_x^{\frac{2d}{d+2}}$, with

$$\left\| F^{(1)}(t) \right\|_{L_t^1 L_x^2(\mathbf{R}\times\mathbf{R}^d)} + \left\| F^{(2)}(t) \right\|_{L_t^2 L_x^{\frac{2d}{d+2}}(\mathbf{R}\times\mathbf{R}^d)} \lesssim_d \|F\|_{L_t^{p'} L_x^{q'}(\mathbf{R}\times\mathbf{R}^d)},$$

and combining (1.70)–(1.72),

$$\left\| \int_0^t e^{i(t-\tau)\Delta} F(\tau) \right\|_{L_t^2 L_x^{\frac{2d}{d-2}} \cap L_t^\infty L_x^2(\mathbf{R}\times\mathbf{R}^d)} \lesssim \|F\|_{L_t^1 L_x^2(\mathbf{R}\times\mathbf{R}^d) + L_t^2 L_x^{\frac{2d}{d+2}}(\mathbf{R}\times\mathbf{R}^d)}$$
$$\lesssim \|F\|_{L_t^{p'} L_x^{q'}(\mathbf{R}\times\mathbf{R}^d)}.$$

Following Keel and Tao (1998), observe that to prove (1.70), it suffices to prove that for any $F, G \in L_t^2 L_x^{\frac{2d}{d+2}}$,

$$\int_{\mathbf{R}} \int_0^t \langle e^{i(t-\tau)\Delta} F(\tau), G(t) \rangle\, dt\, d\tau \lesssim \|F\|_{L_t^2 L_x^{\frac{2d}{d+2}}(\mathbf{R}\times\mathbf{R}^d)} \|G\|_{L_t^2 L_x^{\frac{2d}{d+2}}(\mathbf{R}\times\mathbf{R}^d)},$$

and without loss of generality it suffices to prove that for any $F \in L_t^2 L_x^{\frac{2d}{d+2}}$,

$$\int_{\mathbf{R}} \int_0^t \langle e^{i(t-\tau)\Delta} F(\tau), F(t) \rangle\, dt\, d\tau \lesssim \|F\|^2_{L_t^2 L_x^{\frac{2d}{d+2}}(\mathbf{R}\times\mathbf{R}^d)}. \tag{1.73}$$

To prove (1.73), Keel and Tao (1998) combined the atomic decomposition of L^p spaces with the Whitney decomposition in time in a very powerful way. Recall that in the proof of Lemma 1.10, it was enough to use (1.47) combined with the decomposition of $\{[0,1] \times [0,1] : \tau < t\}$ into unions of rectangles $I_k^j \times I_{k+1}^j$. Examining such rectangles more closely, however, it is apparent that most of the volume of $I_k^j \times I_{k+1}^j$ is occupied by (τ,t) pairs for which $|\tau - t| \sim 2^j$. This is potentially quite useful, via the dispersive estimate (1.20). The Whitney decomposition enables this fact to be exploited quite effectively.

Definition 1.14 (Dyadic interval) A dyadic interval is an interval of length λ, where λ is some dyadic number, and the coordinates of the endpoints are integer multiples of λ.

A general Whitney covering result may be found in Grafakos (2004). In this book it will only be necessary to use a dyadic Whitney decomposition in the special case $\Omega = \{(t, \tau) \in \mathbf{R}^2 : t < \tau\}$.

Lemma 1.15 (Dyadic Whitney decomposition) *There exists a partition of Ω into a family $\mathscr{Q}$ of essentially disjoint squares $Q = I \times J$, where I and J are dyadic intervals, with the property that* $\mathrm{dist}\,(Q, \partial \Omega) \sim \mathrm{diam}\,(Q)$ *for any* $Q \in \mathscr{Q}$.

Proof Take the set of all dyadic subintervals of $\mathbf{R}$. If I_1, I_2, and I_0 are dyadic intervals, $|I_1| = |I_2| = \frac{1}{2}|I_0|$, and $I_1 \cup I_2 = I_0$, then I_1 and I_2 are said to be descendants of I_0 and I_0 is said to be the parent of I_1 and I_2. Say $I \sim J$ if

1. I and J have the same length,
2. I is to the left of J,
3. I and J are nonadjacent but have adjacent parents, that is, $I \subset I_0, J \subset J_0, I_0$ and J_0 are adjacent, dyadic subintervals of $[0,1]$, and $|I_0| = |J_0| = 2|I|$.

Then let $\mathscr{Q}$ be the set of all squares $I \times J$, where $I, J \subset \mathbf{R}$ are dyadic intervals and $I \sim J$. Also let $\mathscr{Q}_\lambda$ be the set of all $I \times J \in \mathscr{Q}$ satisfying $|I| = |J| = \lambda$. For almost every $\tau, t \in \mathbf{R}^2$, $\tau < t$, there exists $I \sim J$ such that $\tau \in I$ and $t \in J$. □

For each t, τ, decompose F according to the atomic decomposition (1.61) and the time intervals according to the Whitney decomposition,

$$\int_{\mathbf{R}} \int_{\tau < t} \langle e^{-i\tau\Delta} F(\tau), e^{-it\Delta} F(t) \rangle \, dt \, d\tau$$
$$= \sum_j \sum_{\substack{I \sim J, \\ |I| = |J| = 2^j}} \int_I \int_J \sum_{k,l} \langle e^{i(t-\tau)\Delta} F_k(\tau), F_l(t) \rangle \, dt \, d\tau. \quad (1.74)$$

1.2 Strichartz Estimates

Plugging $p = 2$ into (1.50) shows that for any fixed j, (1.74) is bounded by $\|F\|^2_{L^2_t L^{\frac{2d}{d+2}}_x}$. However, more can be said. For any k, $\|F_k\|_{L^1 \cap L^\infty}$ is finite, and so for (k,l) far away from $\left(-j\left(\frac{d+2}{4}\right), -j\left(\frac{d+2}{4}\right)\right)$, there is a gain, which may be combined with Young's inequality to prove (1.73). To see this consider two cases separately.

Case 1, $k + l \leq -j\left(\frac{d+2}{2}\right)$. Using (1.64) to compute $\|F_k\|_{L^2}$ when $k \leq -j\left(\frac{d+2}{4}\right)$, (1.15) implies

$$\int_I \int_J \langle e^{i(t-\tau)\Delta} F_k(\tau), F_l(t) \rangle \, dt \, d\tau$$

$$\lesssim 2^{\frac{2k}{d+2}} 2^{\frac{2l}{d+2}} \left(\int_I \|F_k(\tau)\|^{\frac{d}{d+2}}_{L^{\frac{2d}{d+2}}_x(\mathbf{R}^d)} \|F(\tau)\|^{\frac{2}{d+2}}_{L^{\frac{2d}{d+2}}_x(\mathbf{R}^d)} \, d\tau \right)$$

$$\times \left(\int_J \|F_l(t)\|^{\frac{d}{d+2}}_{L^{\frac{2d}{d+2}}_x(\mathbf{R}^d)} \|F(t)\|^{\frac{2}{d+2}}_{L^{\frac{2d}{d+2}}_x(\mathbf{R}^d)} \, dt \right),$$

so by Hölder's inequality, if $|I| = |J| = 2^j$,

$$\int_I \int_J \langle e^{i(t-\tau)\Delta} F_k(\tau), F_l(t) \rangle \, dt \, d\tau$$

$$\lesssim 2^{\frac{2k}{d+2}} 2^{\frac{2l}{d+2}} 2^j \left(\int_I \|F_k(\tau)\|^{\frac{2d}{d+2}}_{L^{\frac{2d}{d+2}}_x(\mathbf{R}^d)} \|F(\tau)\|^{\frac{4}{d+2}}_{L^{\frac{2d}{d+2}}_x(\mathbf{R}^d)} \, d\tau \right)^{1/2}$$

$$\times \left(\int_J \|F_l(t)\|^{\frac{2d}{d+2}}_{L^{\frac{2d}{d+2}}_x(\mathbf{R}^d)} \|F(t)\|^{\frac{4}{d+2}}_{L^{\frac{2d}{d+2}}_x(\mathbf{R}^d)} \, dt \right)^{1/2}.$$

By Young's inequality, the Cauchy–Schwarz inequality, and (1.62), for $k, l \leq -j\left(\frac{d+2}{4}\right)$,

$$\sum_j \sum_{\substack{I \sim J, \\ |I|=|J|=2^j}} \sum_{\substack{k,l \leq \\ -j\left(\frac{d+2}{4}\right)}} 2^{\frac{2k}{d+2}} 2^{\frac{2l}{d+2}} 2^j \left(\int_I \|F_k(\tau)\|^{\frac{2d}{d+2}}_{L^{\frac{2d}{d+2}}_x(\mathbf{R}^d)} \|F(\tau)\|^{\frac{4}{d+2}}_{L^{\frac{2d}{d+2}}_x(\mathbf{R}^d)} \, d\tau \right)^{1/2}$$

$$\times \left(\int_J \|F_l(t)\|^{\frac{2d}{d+2}}_{L^{\frac{2d}{d+2}}_x(\mathbf{R}^d)} \|F(t)\|^{\frac{4}{d+2}}_{L^{\frac{2d}{d+2}}_x(\mathbf{R}^d)} \, dt \right)^{1/2}$$

$$\lesssim \sum_j \sum_{\substack{k,l \leq \\ -j\left(\frac{d+2}{4}\right)}} 2^{\frac{2k}{d+2}} 2^{\frac{2l}{d+2}} 2^j \left(\int_{\mathbf{R}} \|F_k(\tau)\|^{\frac{2d}{d+2}}_{L^{\frac{2d}{d+2}}_x(\mathbf{R}^d)} \|F(\tau)\|^{\frac{4}{d+2}}_{L^{\frac{2d}{d+2}}_x(\mathbf{R}^d)} \, d\tau \right)^{1/2}$$

$$\times \left(\int_{\mathbf{R}} \|F_l(t)\|_{L_x^{\frac{2d}{d+2}}(\mathbf{R}^d)}^{\frac{2d}{d+2}} \|F(t)\|_{L_x^{\frac{2d}{d+2}}(\mathbf{R}^d)}^{\frac{4}{d+2}} dt \right)^{1/2}$$

$$\lesssim \sum_j \sum_{\substack{k \leq \\ -j(\frac{d+2}{4})}} 2^{\frac{4k}{d+2}} 2^j \left(\int_{\mathbf{R}} \|F_k(t)\|_{L_x^{\frac{2d}{d+2}}(\mathbf{R}^d)}^{\frac{2d}{d+2}} \|F(t)\|_{L_x^{\frac{2d}{d+2}}(\mathbf{R}^d)}^{\frac{4}{d+2}} d\tau \right)$$

$$\lesssim \|F\|_{L_t^2 L_x^{\frac{2d}{d+2}}(\mathbf{R} \times \mathbf{R}^d)}^2.$$

For $k+l \leq -j\left(\frac{d+2}{2}\right)$ and $k > -j\left(\frac{d+2}{4}\right)$ or $l > -j\left(\frac{d+2}{4}\right)$, use the non-endpoint Strichartz estimates of Theorem 1.12. Suppose without loss of generality that $k > -j\left(\frac{d+2}{4}\right)$. For any $\varepsilon > 0$, by (1.54) and interpolation,

$$\int_I \int_J \langle e^{i(t-\tau)\Delta} F_k(\tau), F_l(t) \rangle \, dt \, d\tau$$

$$\lesssim_\varepsilon 2^{\frac{2l}{d+2}} 2^{\frac{2k\varepsilon}{d+2}} \left(\int_I \|F_k(\tau)\|_{L_x^{\frac{2d}{d+2}}(\mathbf{R}^d)}^{\frac{d}{d+2}} \|F(\tau)\|_{L_x^{\frac{2d}{d+2}}(\mathbf{R}^d)}^{\frac{2}{d+2}} d\tau \right)^\varepsilon$$

$$\times \left(\int_I \|F_k(\tau)\|_{L_x^{\frac{2d}{d+2}}(\mathbf{R}^d)}^2 d\tau \right)^{\frac{1-\varepsilon}{2}} \left(\int_J \|F_l(t)\|_{L_x^{\frac{2d}{d+2}}(\mathbf{R}^d)}^{\frac{d}{d+2}} \|F(t)\|_{L_x^{\frac{2d}{d+2}}(\mathbf{R}^d)}^{\frac{2}{d+2}} dt \right).$$
(1.75)

By Young's inequality, the Cauchy–Schwarz inequality, Hölder's inequality in time, and (1.62),

$$\sum_j \sum_{\substack{I \sim J, \\ |I|=|J|=2^j}} \sum_{\substack{k+l \leq -j(\frac{d+2}{2}), \\ k > -j(\frac{d+2}{4})}} 2^{\frac{2l}{d+2}} 2^{\frac{2k\varepsilon}{d+2}} \left(\int_I \|F_k(\tau)\|_{L_x^{\frac{2d}{d+2}}(\mathbf{R}^d)}^{\frac{d}{d+2}} \|F(\tau)\|_{L_x^{\frac{2d}{d+2}}(\mathbf{R}^d)}^{\frac{2}{d+2}} d\tau \right)^\varepsilon$$

$$\times \left(\int_I \|F_k(\tau)\|_{L_x^{\frac{2d}{d+2}}(\mathbf{R}^d)}^2 d\tau \right)^{\frac{1-\varepsilon}{2}} \left(\int_J \|F_l(t)\|_{L_x^{\frac{2d}{d+2}}(\mathbf{R}^d)}^{\frac{d}{d+2}} \|F(t)\|_{L_x^{\frac{2d}{d+2}}(\mathbf{R}^d)}^{\frac{2}{d+2}} dt \right)$$

$$\lesssim \sum_j \sum_{\substack{I \sim J, \\ |I|=|J|=2^j}} \sum_{\substack{k+l \leq -j(\frac{d+2}{2}), \\ k > -j(\frac{d+2}{4})}} 2^{\frac{2l}{d+2}} 2^{\frac{2k\varepsilon}{d+2}} 2^{j(\frac{1+\varepsilon}{2})}$$

$$\times \left(\int_I \|F_k(\tau)\|_{L_x^{\frac{2d}{d+2}}(\mathbf{R}^d)}^{\frac{2d}{d+2}} \|F(\tau)\|_{L_x^{\frac{2d}{d+2}}(\mathbf{R}^d)}^{\frac{4}{d+2}} d\tau \right)^{\varepsilon/2}$$

1.2 Strichartz Estimates

$$\times \left(\int_I \|F_k(\tau)\|^2_{L_x^{\frac{2d}{d+2}}(\mathbf{R}^d)} d\tau \right)^{\frac{1-\varepsilon}{2}} \left(\int_J \|F_l(t)\|^{\frac{2d}{d+2}}_{L_x^{\frac{2d}{d+2}}(\mathbf{R}^d)} \|F(t)\|^{\frac{4}{d+2}}_{L_x^{\frac{2d}{d+2}}(\mathbf{R}^d)} dt \right)$$

$$\lesssim \sum_j \sum_{\substack{k+l \leq -j(\frac{d+2}{2}), \\ k > -j(\frac{d+2}{4})}} 2^{\frac{2l}{d+2}} 2^{\frac{2k\varepsilon}{d+2}} 2^{j(\frac{1+\varepsilon}{2})}$$

$$\times \left(\int_{\mathbf{R}} \|F_k(\tau)\|^{\frac{2d}{d+2}}_{L_x^{\frac{2d}{d+2}}(\mathbf{R}^d)} \|F(\tau)\|^{\frac{4}{d+2}}_{L_x^{\frac{2d}{d+2}}(\mathbf{R}^d)} d\tau \right)^{\varepsilon/2}$$

$$\times \left(\int_{\mathbf{R}} \|F_k(\tau)\|^2_{L_x^{\frac{2d}{d+2}}(\mathbf{R}^d)} d\tau \right)^{\frac{1-\varepsilon}{2}} \left(\int_{\mathbf{R}} \|F_l(t)\|^{\frac{2d}{d+2}}_{L_x^{\frac{2d}{d+2}}(\mathbf{R}^d)} \|F(t)\|^{\frac{4}{d+2}}_{L_x^{\frac{2d}{d+2}}(\mathbf{R}^d)} dt \right)$$

$$\lesssim \|F\|^2_{L_t^2 L_x^{\frac{2d}{d+2}}}.$$

The case when $k+l \leq -j\left(\frac{d+2}{2}\right)$ and $l > -j\left(\frac{d+2}{4}\right)$ can be treated in identical fashion.

Case 2, $k+l \geq -j\left(\frac{d+2}{2}\right)$. Combining (1.20) with (1.63),

$$\int_I \int_J \langle e^{i(t-\tau)\Delta} F_k(\tau), F_l(t) \rangle \, dt \, d\tau \lesssim 2^{\frac{k(2-d)}{d+2}} 2^{\frac{l(2-d)}{d+2}} 2^{-dj/2}$$

$$\times \left(\int_I \|F_k(\tau)\|_{L_x^{\frac{2d}{d+2}}(\mathbf{R}^d)} d\tau \right) \left(\int_J \|F_l(t)\|_{L_x^{\frac{2d}{d+2}}(\mathbf{R}^d)} dt \right),$$

so by Hölder's inequality, if $|I| = |J| = 2^j$,

$$\int_I \int_J \langle e^{i(t-\tau)\Delta} F_k(\tau), F_l(t) \rangle \, dt \, d\tau \lesssim 2^{\frac{k(2-d)}{d+2}} 2^{\frac{l(2-d)}{d+2}} 2^{j(\frac{2-d}{2})}$$

$$\times \left(\int_I \|F_k(\tau)\|^2_{L_x^{\frac{2d}{d+2}}(\mathbf{R}^d)} d\tau \right)^{1/2} \left(\int_J \|F_l(t)\|^2_{L_x^{\frac{2d}{d+2}}(\mathbf{R}^d)} dt \right)^{1/2}. \quad (1.76)$$

By Young's inequality, the Cauchy–Schwarz inequality, and (1.62), when $k, l \geq -j\left(\frac{d+2}{4}\right)$,

$$\sum_j \sum_{\substack{|I|=|J| \\ =2^j}} \sum_{\substack{k,l \geq \\ -j(\frac{d+2}{4})}} 2^{\frac{k(2-d)}{d+2}} 2^{\frac{l(2-d)}{d+2}} 2^{j(\frac{2-d}{2})} \left(\int_I \|F_k(\tau)\|^2_{L_x^{\frac{2d}{d+2}}(\mathbf{R}^d)} d\tau \right)^{1/2}$$

$$\times \left(\int_J \|F_l(t)\|^2_{L_x^{\frac{2d}{d+2}}(\mathbf{R}^d)} dt \right)^{1/2}$$

$$\lesssim \sum_j \sum_{\substack{k,l \geq \\ -j\left(\frac{d+2}{4}\right)}} 2^{\frac{k(2-d)}{d+2}} 2^{\frac{l(2-d)}{d+2}} 2^{j\left(\frac{2-d}{2}\right)} \left(\int_{\mathbf{R}} \|F_k(\tau)\|^2_{L_x^{\frac{2d}{d+2}}(\mathbf{R}^d)} d\tau \right)^{1/2}$$

$$\times \left(\int_{\mathbf{R}} \|F_l(t)\|^2_{L_x^{\frac{2d}{d+2}}(\mathbf{R}^d)} dt \right)^{1/2}$$

$$\lesssim \sum_j \sum_{\substack{k \geq \\ -j\left(\frac{d+2}{4}\right)}} 2^{\frac{2k(2-d)}{d+2}} 2^{j\left(\frac{2-d}{2}\right)} \left(\int_{\mathbf{R}} \|F_k(\tau)\|^2_{L_x^{\frac{2d}{d+2}}(\mathbf{R}^d)} d\tau \right)$$

$$\lesssim \|F\|^2_{L_t^2 L_x^{\frac{2d}{d+2}}(\mathbf{R}\times\mathbf{R}^d)}.$$

When $l < -j\left(\frac{d+2}{4}\right)$ and $k+l \geq -j\left(\frac{d+2}{2}\right)$, combining (1.75) and (1.76),

$$\int_I \int_J \langle e^{i(t-\tau)\Delta} F_k(\tau), F_l(t) \rangle \, dt \, d\tau$$

$$\lesssim_\varepsilon A \left(\int_I \|F_k(\tau)\|^{\frac{2d}{d+2}}_{L_x^{\frac{2d}{d+2}}(\mathbf{R}^d)} \|F(\tau)\|^{\frac{4}{d+2}}_{L_x^{\frac{2d}{d+2}}(\mathbf{R}^d)} d\tau \right)^{\varepsilon/4}$$

$$\times \left(\int_I \|F_k(\tau)\|^2_{L_x^{\frac{2d}{d+2}}(\mathbf{R}^d)} d\tau \right)^{\frac{2-\varepsilon}{4}}$$

$$\times \left(\int_J \|F_l(t)\|^{\frac{2d}{d+2}}_{L_x^{\frac{2d}{d+2}}(\mathbf{R}^d)} \|F(t)\|^{\frac{4}{d+2}}_{L_x^{\frac{2d}{d+2}}(\mathbf{R}^d)} dt \right)^{1/2}, \quad (1.77)$$

where for clarity we have put $A = 2^{l\frac{(1-\varepsilon)}{d+2}} 2^{j\frac{1-\varepsilon}{4}} 2^{l\frac{(2+2\varepsilon-d)}{2(d+2)}} 2^{k\frac{2+2\varepsilon-d}{2(d+2)}} 2^{j\frac{2+2\varepsilon-d}{4}}$. Then by Young's inequality, the Cauchy–Schwarz inequality, (1.77), and (1.62),

$$\sum_j \sum_{\substack{|I|=|J| \\ =2^j}} \sum_{\substack{k+l \geq -j\left(\frac{d+2}{2}\right), \\ l < -j\left(\frac{d+2}{4}\right)}} \int_I \int_J \langle e^{i(t-\tau)\Delta} F_k(\tau), F_l(t) \rangle \, dt \, d\tau \lesssim \|F\|^2_{L_t^2 L_x^{\frac{2d}{d+2}}(\mathbf{R}\times\mathbf{R}^d)}.$$

The case when $k < -j\left(\frac{d+2}{4}\right)$ and $k+l \geq -j\left(\frac{d+2}{2}\right)$ is identical. Therefore, the proof of (1.73) is complete, which completes the proof of Theorem 1.13. $\square$

Definition 1.16 (Strichartz space) Let $S^0(I \times \mathbf{R}^d)$ be the Strichartz space

$$S^0(I \times \mathbf{R}^d) = \left\{ u : \|u\|_{L_t^p L_x^q(I\times\mathbf{R}^d)} < \infty \text{ for all } (p,q) \in \mathscr{A}_d \right\}.$$

1.2 Strichartz Estimates

When $d = 1$ or $d \geq 3$, define the norm

$$\|u\|_{S^0(I \times \mathbf{R}^d)} = \sup_{(p,q) \in \mathscr{A}_d} \|u\|_{L_t^p L_x^q(I \times \mathbf{R}^d)}.$$

When $d = 2$, let $0 < w(p) \leq 1$ be a weight satisfying $w(p) \to 0$ as $p \to 2$, and

$$w(p) \|e^{it\Delta} u_0\|_{L_t^p L_x^q(I \times \mathbf{R}^2)} \lesssim \|u_0\|_{L^2(\mathbf{R}^2)},$$

with bound independent of p. Then let

$$\|u\|_{S^0(I \times \mathbf{R}^2)} = \sup_{(p,q) \in \mathscr{A}_2} w(p) \|u\|_{L_t^p L_x^q(I \times \mathbf{R}^2)}.$$

As in the Sobolev spaces, for $s \in \mathbf{R}$, let

$$\|u\|_{\dot{S}^s(I \times \mathbf{R}^d)} = \||\nabla|^s u\|_{S^0(I \times \mathbf{R}^d)} \tag{1.78}$$

and

$$\|u\|_{S^s(I \times \mathbf{R}^d)} = \|\langle \nabla \rangle^s u\|_{S^0(I \times \mathbf{R}^d)}. \tag{1.79}$$

Let $N^0(I \times \mathbf{R}^d)$ be the space dual to $S^0(I \times \mathbf{R}^d)$. Let $\dot{N}^s(I \times \mathbf{R}^d)$ and $N^s(I \times \mathbf{R}^d)$ be defined in a manner identical to (1.78) and (1.79), respectively

$$\|F\|_{\dot{N}^s(I \times \mathbf{R}^d)} = \||\nabla|^s F\|_{N^0(I \times \mathbf{R}^d)}$$

and

$$\|F\|_{N^s(I \times \mathbf{R}^d)} = \|\langle \nabla \rangle^s F\|_{N^0(I \times \mathbf{R}^d)}.$$

See Tao (2006) for more information on these spaces.

When $d \geq 3$,

$$S^0(I \times \mathbf{R}^d) = L_t^\infty L_x^2(I \times \mathbf{R}^d) \cap L_t^2 L_x^{\frac{2d}{d-2}}(I \times \mathbf{R}^d),$$

$$N^0(I \times \mathbf{R}^d) = L_t^1 L_x^2 + L_t^2 L_x^{\frac{2d}{d+2}},$$

so by Theorem 1.13, when $d \geq 3$,

$$\left\| \int_0^t e^{i(t-\tau)\Delta} F(\tau) d\tau \right\|_{S^0(I \times \mathbf{R}^d)} \lesssim \|F\|_{N^0(I \times \mathbf{R}^d)},$$

$$\left\| \int_0^t e^{i(t-\tau)\Delta} F(\tau) d\tau \right\|_{\dot{S}^s(I \times \mathbf{R}^d)} \lesssim \|F\|_{\dot{N}^s(I \times \mathbf{R}^d)},$$

and
$$\left\| \int_0^t e^{i(t-\tau)\Delta} F(\tau) d\tau \right\|_{S^s(I \times \mathbf{R}^d)} \lesssim \|F\|_{N^s(I \times \mathbf{R}^d)}.$$

The same also holds in dimension $d = 1$.

1.3 Small Data Mass-Critical Problem

Using Strichartz estimates, the mass-critical nonlinear Schrödinger equation

$$iu_t + \Delta u = \mu |u|^{\frac{4}{d}} u = F(u), \quad u(0,x) = u_0(x), \quad \mu = \pm 1, \tag{1.80}$$

with $\|u_0\|_{L^2}$ small, may be treated as a perturbation of the linear Schrödinger equation. It is called mass critical because it is invariant under the scaling symmetry of (1.42). That is, if u solves (1.80) on the interval $[-T,T]$, then for any $\lambda > 0$,

$$\lambda^{d/2} u\left(\lambda^2 t, \lambda x\right) \tag{1.81}$$

solves (1.80) on $\left[-\frac{T}{\lambda^2}, \frac{T}{\lambda^2}\right]$ with initial data

$$\lambda^{d/2} u_0(\lambda x),$$

which preserves the L^2-norm, or mass,

$$\|u_0(x)\|_{L_x^2(\mathbf{R}^d)} = \left\|\lambda^{d/2} u_0(\lambda x)\right\|_{L_x^2(\mathbf{R}^d)}.$$

Theorem 1.17 *For any $d \geq 1$, there exists $\varepsilon_0(d) > 0$, such that if*

$$\|u_0\|_{L^2(\mathbf{R}^d)} \leq \varepsilon_0(d), \tag{1.82}$$

then (1.80) is globally well posed and scattering.

Roughly speaking, global well-posedness and scattering means that (1.80) has a solution, and the solution is close to a solution to the linear Schrödinger equation as $t \to \pm \infty$.

Definition 1.18 (Well-posedness) An initial value problem is said to be well posed on an interval I, $0 \in I \subset \mathbf{R}$, if

1. there exists a unique solution to the initial value problem,
2. the solution is continuous in time,
3. the solution depends continuously on the initial data.

1.3 Small Data Mass-Critical Problem

Definition 1.19 (Scattering) A solution to an initial value problem is said to scatter forward in time if the solution exists on $[0,\infty)$ and there exists u_+ such that

$$u(t) - e^{it\Delta}u_+ \to 0, \tag{1.83}$$

as $t \to +\infty$. A solution to an initial value problem is said to scatter backward in time if the solution exists on $(-\infty, 0]$ and there exists u_- such that

$$u(t) - e^{it\Delta}u_- \to 0, \tag{1.84}$$

as $t \to -\infty$. An initial value problem with initial data in some set is said to be scattering if the problem is globally well posed, scatters both forward and backward in time, and u_+ and u_- depend continuously on the initial data.

In this book, the Banach space in which (1.83) and (1.84) are proved to hold will be the critical L^2-based Sobolev space. This same space will also be used in Definition 1.18 to define continuity in time. As was already mentioned, the critical Sobolev space for (1.80) is $L^2\left(\mathbf{R}^d\right)$.

Proof of Theorem 1.17 For any $d \geq 1$, $p = q = \frac{2(d+2)}{d}$ lies in $\mathscr{A}_d$, so let X be the set of functions

$$X = \left\{ u : \mathbf{R} \times \mathbf{R}^d \to \mathbf{C} : \|u\|_{L^{\frac{2(d+2)}{d}}_{t,x}(\mathbf{R}\times\mathbf{R}^d)} \leq C\varepsilon_0 \right\},$$

for some constant C. By Theorem 1.13 and (1.82), there exists some constant $C(d)$ such that

$$\left\|e^{it\Delta}u_0\right\|_{L^{\frac{2(d+2)}{d}}_{t,x}(\mathbf{R}\times\mathbf{R}^d)} \leq \frac{C\varepsilon_0}{2}. \tag{1.85}$$

Now define the map

$$\Phi(u)(t) = e^{it\Delta}u_0 - i\int_0^t e^{i(t-\tau)\Delta}F(u(\tau))\,d\tau. \tag{1.86}$$

By semigroup theory, if $u \in X$ satisfies

$$\Phi(u)(t) = u(t), \tag{1.87}$$

then u solves (1.80). By the contraction mapping principle, to prove the existence of a unique solution to (1.80) in X, it suffices to prove

1. $\Phi : X \to X$,
2. the map Φ is a contraction.

By Theorem 1.13 and Hölder's inequality, if $u \in X$,

$$\left\| \int_0^t e^{i(t-\tau)\Delta} F(u(\tau)) d\tau \right\|_{L_{t,x}^{\frac{2(d+2)}{d}}(\mathbf{R}\times\mathbf{R}^d)} \lesssim_d \|F(u)\|_{L_{t,x}^{\frac{2(d+2)}{d+4}}(\mathbf{R}\times\mathbf{R}^d)}$$

$$\lesssim \|u\|_{L_{t,x}^{\frac{2(d+2)}{d}}(\mathbf{R}\times\mathbf{R}^d)}^{1+\frac{4}{d}}$$

$$\lesssim (C\varepsilon_0)^{1+\frac{4}{d}}. \tag{1.88}$$

Hence, choosing $\varepsilon_0(d)$ sufficiently small, (1.85) and (1.88) imply $\Phi: X \to X$.

To prove Φ is a contraction, for $u, v \in X$,

$$\|\Phi(u) - \Phi(v)\|_{L_{t,x}^{\frac{2(d+2)}{d}}(\mathbf{R}\times\mathbf{R}^d)} \lesssim \|F(u) - F(v)\|_{L_{t,x}^{\frac{2(d+2)}{d+4}}(\mathbf{R}\times\mathbf{R}^d)}$$

$$\lesssim \||u|^{\frac{4}{d}} + |v|^{\frac{4}{d}}\|_{L_{t,x}^{\frac{d+2}{2}}(\mathbf{R}\times\mathbf{R}^d)} \|u-v\|_{L_{t,x}^{\frac{2(d+2)}{d}}(\mathbf{R}\times\mathbf{R}^d)}$$

$$\lesssim (C\varepsilon_0)^{\frac{4}{d}} \|u-v\|_{L_{t,x}^{\frac{2(d+2)}{d}}(\mathbf{R}\times\mathbf{R}^d)}.$$

For $\varepsilon_0 > 0$ sufficiently small,

$$\|\Phi(u) - \Phi(v)\|_{L_{t,x}^{\frac{2(d+2)}{d}}(\mathbf{R}\times\mathbf{R}^d)} \leq \frac{1}{2} \|u-v\|_{L_{t,x}^{\frac{2(d+2)}{d}}(\mathbf{R}\times\mathbf{R}^d)},$$

proving Φ is a contraction. Therefore, there exists a unique $u \in X$ satisfying (1.87). That u is continuous in time follows directly from Theorem 1.13, the dominated convergence theorem, and the fact that the Schwartz functions are dense in $L_x^2(\mathbf{R}^d)$.

Continuous dependence on initial data follows from a perturbation lemma.

Lemma 1.20 (Perturbation lemma) *Let I be a compact time interval and let w be an approximate solution to (1.80) for some $d \geq 1$. That is, for some e small,*

$$(i\partial_t + \Delta)w = F(w) + e. \tag{1.89}$$

Suppose, for some $\varepsilon_0 > 0$ small, with $0 \leq \varepsilon \leq \varepsilon_0$, that u solves (1.80),

$$\|u\|_{L_{t,x}^{\frac{2(d+2)}{d}}(\mathbf{R}\times\mathbf{R}^d)} \leq \varepsilon_0, \qquad \|e\|_{L_{t,x}^{\frac{2(d+2)}{d+4}}(\mathbf{R}\times\mathbf{R}^d)} \leq \varepsilon, \tag{1.90}$$

and for some $t_0 \in \mathbf{R}$,

$$\left\| e^{i(t-t_0)\Delta}(u(t_0) - w(t_0)) \right\|_{L_{t,x}^{\frac{2(d+2)}{d}}(\mathbf{R}\times\mathbf{R}^d)} \leq \varepsilon. \tag{1.91}$$

1.3 Small Data Mass-Critical Problem

Then

$$\|u-w\|_{L_{t,x}^{\frac{2(d+2)}{d}}(\mathbf{R}\times\mathbf{R}^d)} \lesssim \varepsilon \tag{1.92}$$

and

$$\|(i\partial_t+\Delta)(u-w)+e\|_{L_{t,x}^{\frac{2(d+2)}{d+4}}(\mathbf{R}\times\mathbf{R}^d)} \lesssim \varepsilon. \tag{1.93}$$

Proof If u solves (1.80) and w solves (1.89), then $v = w - u$ solves

$$(i\partial_t+\Delta)v = F(w) - F(u) + e.$$

By Theorem 1.13,

$$\|v\|_{L_{t,x}^{\frac{2(d+2)}{d}}(\mathbf{R}\times\mathbf{R}^d)} \lesssim \|e^{i(t-t_0)\Delta}(w(t_0)-u(t_0))\|_{L_{t,x}^{\frac{2(d+2)}{d}}(\mathbf{R}\times\mathbf{R}^d)}$$
$$+ \|F(w)-F(u)+e\|_{L_{t,x}^{\frac{2(d+2)}{d+4}}(I\times\mathbf{R}^d)}.$$

Calculating,

$$\|F(w)-F(u)\|_{L_{t,x}^{\frac{2(d+2)}{d+4}}(\mathbf{R}\times\mathbf{R}^d)}$$

$$\lesssim \|v\|_{L_{t,x}^{\frac{2(d+2)}{d}}(\mathbf{R}\times\mathbf{R}^d)} \left(\|u\|_{L_{t,x}^{\frac{2(d+2)}{d}}(\mathbf{R}\times\mathbf{R}^d)} + \|v\|_{L_{t,x}^{\frac{2(d+2)}{d}}(\mathbf{R}\times\mathbf{R}^d)} \right)^{\frac{4}{d}}. \tag{1.94}$$

Therefore, by Theorem 1.13, (1.90), (1.91), and (1.94),

$$\|v\|_{L_{t,x}^{\frac{2(d+2)}{d}}(\mathbf{R}\times\mathbf{R}^d)} \lesssim \varepsilon + \|v\|_{L_{t,x}^{\frac{2(d+2)}{d}}(\mathbf{R}\times\mathbf{R}^d)} \left(\varepsilon_0 + \|v\|_{L_{t,x}^{\frac{2(d+2)}{d}}(\mathbf{R}\times\mathbf{R}^d)} \right)^{\frac{4}{d}}, \tag{1.95}$$

which proves (1.92) and (1.93). $\square$

Setting $e = 0$, (1.93) and Theorem 1.13 imply that if u and w solve (1.80) with initial data

$$\|w(t_0)\|_{L_x^2(\mathbf{R}^d)}, \quad \|u(t_0)\|_{L_x^2(\mathbf{R}^d)} \leq \varepsilon_0,$$

then taking $\varepsilon = \|w(t_0) - u(t_0)\|_{L^2}$ implies

$$\|u-w\|_{L_{t,x}^{\frac{2(d+2)}{d}}(\mathbf{R}\times\mathbf{R}^d)} + \|u-w\|_{L_t^\infty L_x^2(\mathbf{R}\times\mathbf{R}^d)} \lesssim \|u(t_0)-w(t_0)\|_{L_x^2(\mathbf{R}^d)}.$$

This proves continuous dependence on initial data, which completes the proof of global well-posedness.

To prove scattering, set

$$u_+ = u_0 - i \int_0^\infty e^{-it\Delta} F(u(t)) \, dt \tag{1.96}$$

and

$$u_- = u_0 + i \int_{-\infty}^0 e^{-it\Delta} F(u(t)) \, dt. \tag{1.97}$$

By (1.54) and (1.88), (1.96) and (1.97) are well defined, and $u_+, u_- \in L_x^2(\mathbf{R}^d)$. Additionally, by the dominated convergence theorem,

$$\lim_{T \to \infty} \|F(u)\|_{L_{t,x}^{\frac{2(d+2)}{d+4}}([T,\infty) \times \mathbf{R}^d)} = 0.$$

Therefore,

$$\left\|e^{iT\Delta} u_+ - u(T)\right\|_{L_x^2(\mathbf{R}^d)} = \left\|\int_T^\infty e^{i(t-\tau)\Delta} F(u(\tau)) \, d\tau\right\|_{L_x^2(\mathbf{R}^d)} \to 0.$$

This proves that each initial data function $\|u_0\|_{L^2} \leq \varepsilon_0$ has a solution that is global in time and scatters forward in time. The proof that the solution also scatters backward in time is identical. Expressions (1.94), (1.95), (1.96), and (1.97) prove that u_+ and u_- depend continuously on the initial data. □

Theorem 1.17 also holds if (1.82) is replaced by the condition

$$\left\|e^{it\Delta} u_0\right\|_{L_{t,x}^{\frac{2(d+2)}{d}}(\mathbf{R} \times \mathbf{R}^d)} \leq \varepsilon_0(d), \tag{1.98}$$

for some $\varepsilon_0(d) > 0$ sufficiently small. By Strichartz estimates, (1.98) is weaker than (1.82), although it does give a Banach norm on u_0. By a similar argument, to show that (1.80) is globally well posed on $[0, \infty)$, and that the solution scatters forward in time, it is enough to show that

$$\left\|e^{it\Delta} u_0\right\|_{L_{t,x}^{\frac{2(d+2)}{d}}([0,\infty) \times \mathbf{R}^d)} \leq \varepsilon_0(d), \tag{1.99}$$

and similarly for $(-\infty, 0]$. To prove local well-posedness of (1.80) on the interval $[-T, T]$ for initial data near u_0, it is enough to show

$$\left\|e^{it\Delta} u_0\right\|_{L_{t,x}^{\frac{2(d+2)}{d}}([-T,T] \times \mathbf{R}^d)} \leq \varepsilon_0(d). \tag{1.100}$$

1.3 Small Data Mass-Critical Problem

Theorem 1.21 (Local well-posedness)

1. *For any $u_0 \in L_x^2(\mathbf{R}^d)$, there exists $T(u_0) > 0$ such that (1.80) is locally well posed on $[-T, T]$. The term $T(u_0)$ depends on the profile of the initial data as well as its size. Moreover, (1.80) is well posed on an open interval $I \subset \mathbf{R}$, $0 \in I$.*
2. *If $\sup(I) < +\infty$, where I is the maximal interval of existence of (1.80),*

$$\lim_{T \nearrow \sup(I)} \|u\|_{L_{t,x}^{\frac{2(d+2)}{d}}([0,T] \times \mathbf{R}^d)} = \infty. \tag{1.101}$$

The corresponding result also holds for $\inf(T) > -\infty$.

Proof Inequality (1.53) and the dominated convergence theorem imply that for any $u_0 \in L_x^2(\mathbf{R}^d)$,

$$\lim_{T \searrow 0} \|e^{it\Delta} u_0\|_{L_{t,x}^{\frac{2(d+2)}{d}}([-T,T] \times \mathbf{R}^d)} = 0.$$

Therefore, there exists $T(u_0) > 0$ sufficiently small such that (1.100) holds, which gives the first part of the theorem.

To prove the second part notice that if $J \subset I$ is a compact subinterval then

$$\|u\|_{L_{t,x}^{\frac{2(d+2)}{d}}(J \times \mathbf{R}^d)} < \infty. \tag{1.102}$$

Indeed, by Definition 1.18, the first part of Theorem 1.21, (1.53), and the dominated convergence theorem, for any $t_0 \in I$, there exists some $T(t_0)$ such that

$$\|u\|_{L_{t,x}^{\frac{2(d+2)}{d}}([t_0 - T(t_0), t_0 + T(t_0)] \times \mathbf{R}^d)} \leq \varepsilon_0.$$

Since $\bigcup_{t_0 \in J}(t_0 - T(t_0), t_0 + T(t_0))$ is an open cover of J, J compact implies the existence of a finite subcover, which in turn implies (1.102). If

$$\lim_{T \nearrow \sup(I)} \|u\|_{L_{t,x}^{\frac{2(d+2)}{d}}([0,T] \times \mathbf{R}^d)} < \infty, \tag{1.103}$$

and $\sup(I) < \infty$, then (1.54), (1.86), (1.87), and the dominated convergence theorem imply that $u(t)$ converges in L^2 to some $f \in L_x^2(\mathbf{R}^d)$ as $t \nearrow \sup(I)$. Then by the first result of this theorem and Lemma 1.20, (1.103) implies that (1.80) is locally well posed on $[0, \sup(I) + \delta)$ for some $\delta > 0$, which contradicts the maximality of I. Therefore, (1.101) also holds. $\square$

If u is a global solution to (1.80), a uniform bound on the $L_{t,x}^{\frac{2(d+2)}{d}}$-norm implies scattering.

Theorem 1.22 *A solution to (1.80) scatters forward in time to some $e^{it\Delta}u_+$, $u_+ \in L_x^2(\mathbf{R}^d)$, if and only if for some $t_0 \in \mathbf{R}$,*

$$\|u\|_{L_{t,x}^{\frac{2(d+2)}{d}}([t_0,\infty)\times\mathbf{R}^d)} < \infty. \tag{1.104}$$

Proof If (1.104) holds, set

$$u_+ = u(t_0) - i\int_{t_0}^{\infty} e^{-i\tau\Delta} F(u(\tau))\,d\tau.$$

Inequality (1.54) implies $u_+ \in L_x^2(\mathbf{R}^d)$, and by the dominated convergence theorem,

$$\lim_{t\nearrow\infty} \|e^{it\Delta}u_+ - u(t)\|_{L_x^2(\mathbf{R}^d)} = 0.$$

Conversely, if u scatters, then (1.53) and the dominated convergence theorem imply that for some $T(\varepsilon_0, u_+)$ sufficiently large,

$$\|e^{it\Delta}u_+\|_{L_{t,x}^{\frac{2(d+2)}{d}}([T,\infty)\times\mathbf{R}^d)} \le \frac{\varepsilon_0}{2} \quad \text{and} \quad \|u(T) - e^{iT\Delta}u_+\|_{L_x^2(\mathbf{R}^d)} \le \frac{\varepsilon_0}{2}. \tag{1.105}$$

Then, by Theorem 1.13, (1.105) implies

$$\|e^{i(t-T)\Delta}u(T)\|_{L_{t,x}^{\frac{2(d+2)}{d}}([T,\infty)\times\mathbf{R}^d)} \lesssim \varepsilon_0,$$

and Theorem 1.21 with (1.99) implies

$$\|u\|_{L_{t,x}^{\frac{2(d+2)}{d}}([T,\infty)\times\mathbf{R}^d)} \lesssim \varepsilon_0.$$

Since $[t_0, T]$ is compact, (1.102) implies that $\|u\|_{L_{t,x}^{\frac{2(d+2)}{d}}([t_0,T]\times\mathbf{R}^d)} < \infty$. $\square$

The Sobolev norm of u_0 in any space more regular than L^2 gives sufficient information on the profile of u_0.

Theorem 1.23 *For any $\varepsilon > 0$, there exists $T(\|u_0\|_{H^\varepsilon(\mathbf{R}^d)}) > 0$ such that (1.80) is locally well posed on $[-T, T]$. Also, if (1.80) has a solution on some maximal interval $I \subset \mathbf{R}$, with $\sup(I) < \infty$, then*

$$\lim_{T\nearrow\sup(I)} \|u\|_{L_t^\infty H^\varepsilon([0,T]\times\mathbf{R}^d)} = +\infty.$$

An identical result holds when $\inf(I) > -\infty$.

Proof By Bernstein's inequality,

$$\|P_{>N}u_0\|_{L_x^2(\mathbf{R}^d)} \lesssim \frac{1}{N^\varepsilon}\|u_0\|_{H^\varepsilon(\mathbf{R}^d)}.$$

1.3 Small Data Mass-Critical Problem

Also, by (1.7), the Sobolev embedding theorem, and Hölder's inequality,

$$\left\|e^{it\Delta}P_{\leq N}u_0\right\|_{L_{t,x}^{\frac{2(d+2)}{d}}([-T,T]\times\mathbf{R}^d)} \lesssim T^{\frac{d}{2(d+2)}}N^{\frac{d}{d+2}-\varepsilon}\|u_0\|_{H^\varepsilon(\mathbf{R}^d)}.$$

Choosing $N\big(\varepsilon_0, \|u_0\|_{H^\varepsilon(\mathbf{R}^d)}, \varepsilon\big)$ sufficiently large and $T = \frac{1}{N^2}$ yields (1.100), which proves the theorem. $\square$

Upgrading Theorem 1.23 to the existence of $T\big(\|u_0\|_{L^2_x(\mathbf{R}^d)}\big) > 0$ for which (1.80) is locally well posed is the goal of the rest of this work, since such a result implies global well-posedness and scattering. To see this, suppose without loss of generality that $T\big(\|u_0\|_{L^2_x(\mathbf{R}^d)}\big) = 1$, that is, suppose that (1.80) is locally well posed on $[-1,1]$ for all initial data with L^2-norm given by $\|u_0\|_{L^2}$. Then fix some $1 \leq T_0 < \infty$, and set

$$v_0 = \lambda^{d/2}u_0(\lambda x),$$

with $\lambda = T_0^{1/2}$. Since $\|v_0\|_{L^2} = \|u_0\|_{L^2}$, $T(\|v_0\|_{L^2}) = 1$, so (1.80) is locally well posed on $[-1,1]$ with initial data v_0. Rescaling,

$$\frac{1}{\lambda^{d/2}}v\left(\frac{t}{\lambda^2}, \frac{x}{\lambda}\right) = u(t,x)$$

gives local well-posedness of (1.80) with initial data u_0 on $[-T_0, T_0]$. Since T_0 is arbitrary, global well-posedness follows. Similar computations also give $\|u\|_{L_{t,x}^{\frac{2(d+2)}{d}}([-T,T]\times\mathbf{R}^d)} < \infty$, with bound independent of T, which by Theorem 1.22 proves scattering.

The same picture appears when viewing (1.80) through the lens of scaling symmetry. For general $s \in \mathbf{R}$,

$$\left\|\lambda^{d/2}u_0(\lambda x)\right\|_{\dot{H}^s(\mathbf{R}^d)} = \lambda^s\|u_0\|_{\dot{H}^s(\mathbf{R}^d)}.$$

Thus, for $s > 0$, proving local well-posedness on a short time interval when $\|u_0\|_{H^s}$ is large is equivalent to proving local well-posedness on a unit interval when $\|u_0\|_{\dot{H}^s}$ is small. This is why in Theorem 1.23, the time interval T has a lower bound that depends on $\|u_0\|_{H^s}$.

When $s < 0$, proving local well-posedness on a unit interval for $\|u_0\|_{\dot{H}^s}$ of size 1 is equivalent to proving local well-posedness on a short time interval when $\|u_0\|_{\dot{H}^s}$ is small. This fact was exploited by Christ et al. (2003, 2008) to prove ill-posedness of (1.80) when $u_0 \in \dot{H}^s$, $s < 0$. In particular, they constructed a solution that is discontinuous in time at $t = 0$.

Thus, it is natural to study the scattering behavior of (1.80) with initial data $u_0 \in L^2_x(\mathbf{R}^d)$; scattering results in L^2 are sharp.

The set of initial data for which scattering occurs is an open set in $L^2(\mathbf{R}^d)$.

Theorem 1.24 (Long-time perturbations) *Let I be a compact time interval and let w be an approximate solution to (1.80) on $I \times \mathbf{R}^d$ in the sense that*

$$(i\partial_t + \Delta)w = F(w) + e.$$

Assume u solves (1.80) and

$$\|u\|_{L_{t,x}^{\frac{2(d+2)}{d}}(I \times \mathbf{R}^d)} \leq M. \tag{1.106}$$

Let $t_0 \in I$, and suppose that $u(t_0)$ is close to $w(t_0)$ in the sense that

$$\|e^{i(t-t_0)\Delta}(u(t_0) - w(t_0))\|_{L_{t,x}^{\frac{2(d+2)}{d}}(I \times \mathbf{R}^d)} \leq \varepsilon.$$

Also assume

$$\|e\|_{L_{t,x}^{\frac{2(d+2)}{d+4}}(I \times \mathbf{R}^d)} \leq \varepsilon.$$

Then there exists $\varepsilon_1(M,d) > 0$ such that if $0 < \varepsilon \leq \varepsilon_1(M,d)$,

$$\|u - w\|_{L_{t,x}^{\frac{2(d+2)}{d}}(I \times \mathbf{R}^d)} \leq C(M,d)\varepsilon \tag{1.107}$$

and

$$\|w\|_{L_{t,x}^{\frac{2(d+2)}{d}}(I \times \mathbf{R}^d)} \leq C(M,d).$$

Proof This argument first appeared in Bourgain (1999) for the energy-critical problem.

Divide I into $\sim \left(1 + \frac{M}{\varepsilon_0}\right)^{\frac{2(d+2)}{d}}$ subintervals J_k such that on each J_k,

$$\|u\|_{L_{t,x}^{\frac{2(d+2)}{d}}(J_k \times \mathbf{R}^d)} \leq \varepsilon_0,$$

where ε_0 is small. For each $J_k = [a_k, b_k]$, by Theorem 1.13 and (1.94),

$$\|v\|_{L_{t,x}^{\frac{2(d+2)}{d}}(J_k \times \mathbf{R}^d)} \lesssim \|e^{i(t-a_k)\Delta}v(a_k)\|_{L_{t,x}^{\frac{2(d+2)}{d}}(J_k \times \mathbf{R}^d)}$$
$$+ \|v\|_{L_{t,x}^{\frac{2(d+2)}{d}}(J_k \times \mathbf{R}^d)} \left(\|v\|_{L_{t,x}^{\frac{2(d+2)}{d}}(J_k \times \mathbf{R}^d)} + \|u\|_{L_{t,x}^{\frac{2(d+2)}{d}}(J_k \times \mathbf{R}^d)} \right)^{\frac{4}{d}} + \varepsilon; \tag{1.108}$$

so, if $\left\|e^{i(t-a_k)\Delta}v(a_k)\right\|_{L_{t,x}^{\frac{2(d+2)}{d}}(\mathbf{R}\times\mathbf{R}^d)} \leq \varepsilon_0$, we have

$$\|v\|_{L_{t,x}^{\frac{2(d+2)}{d}}(J_k\times\mathbf{R}^d)} \lesssim \left\|e^{i(t-a_k)\Delta}v(a_k)\right\|_{L_{t,x}^{\frac{2(d+2)}{d}}(\mathbf{R}\times\mathbf{R}^d)} + \varepsilon.$$

Also, from Theorem 1.13 and (1.108), it follows that

$$\left\|e^{i(t-b_k)\Delta}v(b_k)\right\|_{L_{t,x}^{\frac{2(d+2)}{d}}(\mathbf{R}\times\mathbf{R}^d)} \lesssim \left\|e^{i(t-a_k)\Delta}v(a_k)\right\|_{L_{t,x}^{\frac{2(d+2)}{d}}(\mathbf{R}\times\mathbf{R}^d)} + \varepsilon. \quad (1.109)$$

Iterating (1.109) over each J_k, there exists $\varepsilon_1(M,d)$ sufficiently small such that if $0 < \varepsilon \leq \varepsilon_1$,

$$\|u-w\|_{L_{t,x}^{\frac{2(d+2)}{d}}(I\times\mathbf{R}^d)} \leq C(M,d)\varepsilon.$$

This gives (1.107). Then by (1.106),

$$\|w\|_{L_{t,x}^{\frac{2(d+2)}{d}}(I\times\mathbf{R}^d)} \leq C(M,d). \qquad \square$$

1.4 A Large Data Global Well-Posedness Result

The proof of global well-posedness and scattering for generic large data cannot rely solely on perturbative arguments. Instead, some knowledge of the solution to the nonlinear problem is required, beyond what can be gleaned from the behavior of a solution to the linear equation and perturbation theory.

Suppose u solves the nonlinear Schrödinger equation

$$iu_t + \Delta u = \mathcal{N} = \mu|u|^p u, \quad (1.110)$$

on $I \times \mathbf{R}^d$, where I is a compact interval, $\mu = \pm 1$, and $p > 0$. If Y is a scalar, vector, tensor, or linear operator, then

$$Y(iu_t + \Delta u - \mathcal{N}) = 0, \quad (1.111)$$

and a well-chosen Y can yield important local conservation results.

Conservation laws typically arise via integrating by parts, so it is convenient to assume that $u(t) \in \mathscr{S}(\mathbf{R}^d)$. This is a safe assumption to make since $\mathscr{S}(\mathbf{R}^d)$ is dense in $H^s(\mathbf{R}^d)$ for any $s \in \mathbf{R}$. The local well-posedness result in Theorem 1.21 then implies that the solution $u(t)$ is sufficiently regular, at least for short times.

Remark See Sulem and Sulem (1999) for derivation of conservation laws using Noether's theorem.

Lemma 1.25 (First local conservation law) Let $T_{00}(t,x)$ be the mass density,

$$T_{00}(t,x) = |u(t,x)|^2,$$

and for $j = 1,\ldots,d$ let $T_{0j}(t,x)$ be the momentum density

$$T_{0j}(t,x) = 2\operatorname{Im}(\bar{u}\partial_j u).$$

If u solves (1.110) on $I \times \mathbf{R}^d$, then

$$\partial_t T_{00} + \partial_j T_{0j} = 2\{\mathcal{N},u\}_m = 0, \tag{1.112}$$

where $\{\cdot,\cdot\}_m$ is the mass bracket

$$\{f,g\}_m = \operatorname{Im}(f\bar{g}). \tag{1.113}$$

Remark This work follows the usual convention of summing repeated indices over $j = 1,\ldots,d$. Since we will always be working on $\mathbf{R}^d$ with the Euclidean metric, we will not concern ourselves with whether indices are raised or lowered. See Tao (2006) for more on tensor notation.

Proof Take $Y = -i\bar{u}$ in (1.111).

$$\operatorname{Re}((-i\bar{u})(iu_t + \Delta u - \mathcal{N})) = \operatorname{Re}(\bar{u}(u_t - i\Delta u + i\mathcal{N}))$$
$$= \frac{1}{2}\frac{\partial}{\partial t}|u(t,x)|^2 + \nabla \cdot \operatorname{Im}[\bar{u}\nabla u] - \{\mathcal{N},u\}_m.$$

Since $\{\mathcal{N},u\}_m = \operatorname{Im}(|u|^{p+2}) = 0$, the proof is complete. $\square$

Corollary 1.26 (Conservation of mass) If u solves (1.110) on the interval $I \times \mathbf{R}^d$, then the mass

$$\int_{\mathbf{R}^d} T_{00}(t,x)\,dx = \int |u(t,x)|^2 dx$$

is conserved for all $t \in I$.

Remark The identity

$$\partial_t T_{00}(t,x) + \partial_j T_{0j}(t,x) = 0 \tag{1.114}$$

also implies that the quantity $T_{00}(t,x)$ changes in time at the rate of $T_{0j}(t,x)$. This is why $T_{0j}(t,x)$ is called the momentum. Integrating by parts,

$$\partial_t \int x_j |u(t,x)|^2 dx = \int T_{0j}(t,x)\,dx.$$

Lemma 1.27 (Second local conservation law) Let $\{f,g\}_p$ be the momentum bracket

$$\{f,g\}_p = \operatorname{Re}(f\nabla\bar{g} - g\nabla\bar{f}),$$

1.4 A Large Data Global Well-Posedness Result

and for $j = 1, \ldots, d$ let

$$\{f, g\}_p^j = \mathrm{Re}\left(f \partial_j \bar{g} - g \partial_j \bar{f}\right).$$

Then

$$\partial_t T_{0j} + \partial_k L_{jk} = 2\{\mathcal{N}, u\}_p^j, \qquad (1.115)$$

where $L_{jk}(t,x)$ is the linear part of the energy current

$$L_{jk}(t,x) = L_{kj}(t,x) = -\partial_j \partial_k |u(t,x)|^2 + 4\mathrm{Re}\left(\overline{(\partial_j u)}(\partial_k u)\right). \qquad (1.116)$$

Proof By (1.111),

$$\{(iu_t + \Delta u - \mathcal{N}), u\}_p^j = 0. \qquad (1.117)$$

Expanding (1.117),

$$\frac{\partial}{\partial t} \mathrm{Im}[\bar{u}\nabla u] + 2\partial_k \mathrm{Re}(\nabla u \partial_k \bar{u}) - \frac{1}{2}\nabla \Delta \left(|u|^2\right) - \{\mathcal{N}, u\}_p = 0,$$

which directly gives (1.115). □

Corollary 1.28 (Conservation of momentum) *If u solves (1.110) on $I \times \mathbf{R}^d$ then the total momentum*

$$\int_{\mathbf{R}^d} T_{0j}(t,x)\, dx$$

is a conserved quantity for any $t \in I$.

Proof If $\mathcal{N} = \mu |u|^p u$ then

$$\{\mathcal{N}, u\}_p = -\frac{\mu p}{p+2} \nabla\left(|u(t,x)|^{p+2}\right). \qquad (1.118)$$

Therefore,

$$\partial_t T_{0j} + \partial_k T_{jk} = 0, \qquad (1.119)$$

where

$$T_{jk} = L_{jk} + \delta_{jk} \frac{2\mu p}{p+2} |u(t,x)|^{p+2} \qquad (1.120)$$

and

$$\int \partial_t T_{0j}(t,x)\, dx = \int -\partial_k T_{jk}(t,x)\, dx = 0,$$

which yields Corollary 1.28. □

Plugging $Y = \bar{u}_t$ into (1.111) yields conservation of energy.

Lemma 1.29 (Conservation of energy) *The total energy*

$$E(u(t)) = \frac{1}{2}\int |\nabla u(t,x)|^2 dx + \frac{\mu}{p+2}\int |u(t,x)|^{p+2} dx \qquad (1.121)$$

is conserved.

Proof If $Y = \bar{u}_t$, then integrating by parts,

$$\int \operatorname{Re}(\bar{u}_t(iu_t + \Delta u - \mathcal{N}))dx = \int \operatorname{Re}(\bar{u}_t(\Delta u - \mathcal{N}))$$

$$= -\frac{d}{dt}\left[\int \frac{1}{2}|\nabla u(t,x)|^2 + \frac{\mu}{p+2}|u(t,x)|^{p+2} dx\right]$$

$$= 0.$$

When $\mu = +1$, the energy and the mass are positive definite quantities that together control the H_x^1-norm. This directly yields a global result for large data.

Theorem 1.30 *The defocusing, mass-critical initial value problem*

$$iu_t + \Delta u = |u|^{\frac{4}{d}}u, \quad u(0) \in H^1(\mathbf{R}^d), \qquad (1.122)$$

is globally well posed.

Proof By Theorem 1.21, (1.122) is locally well posed on an open interval I. By the Sobolev embedding theorem, $H_x^1(\mathbf{R}^d) \hookrightarrow L^{\frac{2(d+2)}{d}}(\mathbf{R}^d)$, so $E(u(0)) < \infty$, where $E(u(t))$ is the energy given by (1.121). Then by conservation of energy,

$$\sup_{t \in I}\|u(t)\|_{L_x^{\frac{2(d+2)}{d}}(\mathbf{R}^d)} < \infty. \qquad (1.123)$$

Therefore, for any $T \in I$,

$$\|u\|_{L_{t,x}^{\frac{2(d+2)}{d}}([0,T]\times\mathbf{R}^d)} \lesssim_{E(u(0))} T^{\frac{d}{2(d+2)}},$$

so by (1.101), $\sup(I) = \infty$. A similar computation also shows $\inf(I) = -\infty$. □

Remark Conservation of energy does not imply (1.123) in the case when $\mu = -1$. This is not merely a technical difficulty, but is indicative of a significant separation between the behavior of (1.110) for large data when $\mu = +1$ and when $\mu = -1$. In fact, there are known counterexamples to global well-posedness and scattering of (1.122) with $\mu = -1$. This distinction does not arise in the case of small data, where the nonlinearity is treated perturbatively, ignoring its sign altogether.

1.4 A Large Data Global Well-Posedness Result

The case $\mu = +1$ is called defocusing because in that case a solution to (1.110) will disperse faster than a linear solution disperses. In the case $\mu = -1$, a solution to (1.110) will disperse at a slower rate than a solution to the linear equation, and could, at least in principle, not disperse at all, or even concentrate: see Weinstein (1982, 1989).

The pseudoconformal conservation law quantifies this fact. Observe that (1.18) implies that if u solves (1.110) for $\mathcal{N} = 0$, the quantity

$$\| (x + 2it\nabla) u(t) \|_{L^2(\mathbf{R}^d)} \tag{1.124}$$

is a conserved quantity. Indeed,

$$(x + 2it\nabla) \frac{1}{t^{d/2}} \int e^{i\frac{|x-y|^2}{4t}} f(y) \, dy = \frac{1}{t^{d/2}} \int e^{i\frac{|x-y|^2}{4t}} y f(y) \, dy,$$

and therefore (1.15) implies the conservation of (1.124).

Theorem 1.31 (Pseudoconformal conservation law) *If u solves* (1.110) *on* $I \times \mathbf{R}^d$,

$$\| (x + 2it\nabla) u(t) \|_{L^2(\mathbf{R}^d)}^2 + \frac{8\mu t^2}{p+2} \int |u(t,x)|^{p+2} dx$$
$$= \| xu(0) \|_{L^2(\mathbf{R}^d)}^2 + \int_0^t 4\mu s \left(\int \frac{4-dp}{p+2} |u(s,x)|^{p+2} dx \right) ds. \tag{1.125}$$

Proof By (1.114),

$$\frac{\partial}{\partial t} |x|^2 |u(t,x)|^2 = -2|x|^2 \nabla \cdot \text{Im}[\bar{u}\nabla u](t,x)$$
$$= -2\nabla \cdot \left(|x|^2 \text{Im}[\bar{u}\nabla u](t,x) \right) + 4x \cdot \text{Im}[\bar{u}\nabla u](t,x). \tag{1.126}$$

By (1.119) and (1.120),

$$-4 \frac{\partial}{\partial t} (tx \cdot \text{Im}[\bar{u}\nabla u]) = -4x \cdot \text{Im}[\bar{u}\nabla u] - 4t \frac{\partial}{\partial t} (x \cdot \text{Im}[\bar{u}\nabla u])$$
$$= -4x \cdot \text{Im}[\bar{u}\nabla u] - 2tx_j \cdot \partial_j \Delta \left(|u(t,x)|^2 \right)$$
$$+ 8tx_j \cdot \partial_k \text{Re}((\partial_j \bar{u})(\partial_k u)) + \frac{4\mu pt}{p+2} x_j \cdot \partial_j \left(|u(t,x)|^{p+2} \right).$$

Therefore, integrating by parts,

$$-4 \frac{\partial}{\partial t} t \int x \cdot \text{Im}[\bar{u}\nabla u] \, dx$$
$$= -4 \int x \cdot \text{Im}[\bar{u}\nabla u] \, dx - 8t \int |\nabla u|^2 dx - \frac{4\mu dpt}{p+2} \int |u|^{p+2} dx. \tag{1.127}$$

Finally, by conservation of energy,

$$\frac{d}{dt}\left[4t^2\|\nabla u(t)\|^2_{L^2(\mathbf{R}^d)} + \frac{8\mu t^2}{p+2}\int |u(t,x)|^{p+2}dx\right] = 16tE(u(t)). \quad (1.128)$$

Summing (1.126), (1.127), and (1.128) proves (1.125). □

In the mass-critical case, $p = \frac{4}{d}$, (1.125) implies

$$\|(x+2it\nabla)u(t)\|^2_{L^2(\mathbf{R}^d)} + \frac{8\mu t^2}{p+2}\int |u(t,x)|^{p+2}dx = \|xu(0)\|^2_{L^2(\mathbf{R}^d)}.$$

This implies a scattering result for the defocusing, mass-critical problem.

Theorem 1.32 *The initial value problem*

$$iu_t + \Delta u = |u|^{\frac{4}{d}}u, \qquad u(0,x) = u_0 \in L^2(\mathbf{R}^d) \quad (1.129)$$

has a global solution for all $u_0 \in H^1(\mathbf{R}^d)$. Moreover,

$$\|u\|^{\frac{2(d+2)}{d}}_{L^{\frac{2(d+2)}{d}}_{t,x}(\mathbf{R}\times\mathbf{R}^d)} \lesssim \|u_0\|_{L^2}\|u_0\|_{\dot{H}^1(\mathbf{R}^d)}\|xu_0\|_{L^2(\mathbf{R}^d)}. \quad (1.130)$$

Proof By the Sobolev embedding theorem and interpolation,

$$\|u(0)\|^{\frac{2(d+2)}{d}}_{L^{\frac{2(d+2)}{d}}_x(\mathbf{R}^d)} \lesssim \|u(0)\|^{\frac{2(d+2)}{d}}_{\dot{H}^{\frac{d}{d+2}}(\mathbf{R}^d)} \lesssim \|u(0)\|^2_{\dot{H}^1(\mathbf{R}^d)}\|u(0)\|^{\frac{4}{d}}_{L^2_x(\mathbf{R}^d)}.$$

By conservation of energy and Hölder's inequality, for $T = \|xu_0\|_{L^2(\mathbf{R}^d)}/\|u_0\|_{\dot{H}^1(\mathbf{R}^d)}$,

$$\|u\|^{\frac{2(d+2)}{d}}_{L^{\frac{2(d+2)}{d}}_{t,x}([-T,T]\times\mathbf{R}^d)} \lesssim \|u_0\|^{4/d}_{L^2(\mathbf{R}^d)}\|u_0\|_{\dot{H}^1(\mathbf{R}^d)}\|xu_0\|_{L^2(\mathbf{R}^d)}.$$

For $|t| > T$, the pseudoconformal conservation law implies

$$\frac{4dt^2}{d+2}\int_{\mathbf{R}^d}|u(t,x)|^{\frac{2(d+2)}{d}}dx \leq \|xu_0\|_{L^2(\mathbf{R}^d)}.$$

Integrating over $|t| > T$ completes the proof of (1.130). □

Remark Both sides of (1.130) are invariant under the L^2-critical scaling (1.81).

For the focusing case ($\mu = -1$), let $\tau = -\frac{1}{t}$, $y = \frac{x}{t}$, and

$$v(\tau,y) = \frac{1}{t^{d/2}}e^{i\frac{|x|^2}{4t}}u\left(-\frac{1}{t},\frac{x}{t}\right). \quad (1.131)$$

1.4 A Large Data Global Well-Posedness Result

By direct calculation,

$$\left\|(x+2i\tau\nabla)u(\tau)\right\|^2_{L^2_x(\mathbf{R}^d)} + \frac{8\mu\tau^2}{p+2}\left\|u(\tau)\right\|^{p+2}_{L^{p+2}_x} = 8E(v(\tau)).$$

Also by direct calculation, if u solves (1.129) on the interval $I = (T_1, T_2)$ then $v(\tau, y)$ also solves (1.129) on $\left(-\frac{1}{T_1}, -\frac{1}{T_2}\right)$.

Definition 1.33 (Pseudoconformal transformation) Equation (1.131) is called the pseudoconformal transformation of $u(t,x)$.

The sharp interpolation estimate of Weinstein (1982),

$$\|f\|^{\frac{2(d+2)}{d}}_{L^{\frac{2(d+2)}{d}}_x(\mathbf{R}^d)} \leq \left(\frac{\|f\|_{L^2}}{\|Q\|_{L^2}}\right)^{\frac{2}{d}} \|\nabla f\|^2_{L^2},$$

implies that whenever $\|u_0\|_{L^2} < \|Q\|_{L^2}$,

$$E(u(t)) \gtrsim_{\|u_0\|_{L^2}} \|u(t)\|^{\frac{2(d+2)}{d}}_{L^{\frac{2(d+2)}{d}}}. \tag{1.132}$$

Since (1.131) preserves the L^2-norm, as well as the $L^{\frac{2(d+2)}{d}}_{t,x}$-norm, (1.131)–(1.132) imply that for any $0 < T < \infty$,

$$\|u\|^{\frac{2(d+2)}{d}}_{L^{\frac{2(d+2)}{d}}_{t,x}(\mathbf{R}\times\mathbf{R}^d)} \lesssim_{\|u_0\|_{L^2}} T E(u_0) + \frac{1}{T}\|xu_0\|^2_{L^2}.$$

Taking $T \sim E(u_0)^{1/2}/\|xu_0\|_{L^2}$ concludes the proof of the corresponding focusing result.

Theorem 1.34 *The initial value problem*

$$iu_t + \Delta u = -|u|^{\frac{4}{d}}u, \qquad u(0,x) = u_0 \in L^2(\mathbf{R}^d) \tag{1.133}$$

has a global solution for all $u_0 \in H^1(\mathbf{R}^d)$, $xu_0 \in L^2$, *whenever* $\|u_0\|_{L^2} < \|Q\|_{L^2}$. *Moreover,*

$$\|u\|^{\frac{2(d+2)}{d}}_{L^{\frac{2(d+2)}{d}}_{t,x}(\mathbf{R}\times\mathbf{R}^d)} \lesssim_{\|u_0\|_{L^2(\mathbf{R}^d)}} \|u_0\|_{H^1(\mathbf{R}^d)}\|xu_0\|_{L^2(\mathbf{R}^d)}.$$

Remark The interpolation result of Weinstein (1982) is extremely useful for the focusing nonlinear Schrödinger problem because Q is the positive H^1 solution to the elliptic partial differential equation

$$\Delta Q + |Q|^{\frac{4}{d}}Q = Q, \tag{1.134}$$

and $e^{it}Q$ is certainly not a scattering solution to (1.133). Berestycki and Lions

(1979) proved the existence of a solution to (1.134). Applying the pseudoconformal transformation to $e^{it}Q$ gives a finite time blowup solution to (1.133),

$$u(t,x) = \frac{1}{t^{d/2}} e^{-\frac{i}{t}} e^{i\frac{|x|^2}{4t}} Q\left(\frac{x}{t}\right). \qquad (1.135)$$

Theorem 1.34 suggests that scattering could hold for initial data u_0 satisfying $\|u_0\|_{L^2} < \|Q\|_{L^2}$. See Dodson (2015) for a proof of this fact.

Remark See Holmer and Roudenko (2008) and Holmer et al. (2010) for results for the focusing, cubic nonlinear Schrödinger equation when $d = 3$.

2

The Cubic NLS in Dimensions Three and Four

2.1 The Cubic NLS in Three and Four Dimensions with Small Data

In general, the power-type nonlinear Schrödinger equation

$$iu_t + \Delta u = F(u) = \mu |u|^p u, \quad u(0,x) = u_0, \quad \mu = \pm 1 \tag{2.1}$$

is invariant under the scaling symmetry

$$u(t,x) \mapsto \lambda^{2/p} u(\lambda^2 t, \lambda x). \tag{2.2}$$

The critical Sobolev norm for (2.2) is $\dot{H}^{s_c}$, where

$$s_c = \frac{d}{2} - \frac{2}{p}. \tag{2.3}$$

Thus, the cubic nonlinear Schrödinger equation

$$iu_t + \Delta u = F(u) = \mu |u|^2 u, \quad u(0,x) = u_0$$

is mass critical when $d=2$, $\dot{H}^{1/2}$-critical when $d=3$, and energy critical when $d=4$. Indeed,

$$\|u(0,x)\|_{\dot{H}^{1/2}(\mathbf{R}^3)} = \|\lambda u(0,\lambda x)\|_{\dot{H}^{1/2}(\mathbf{R}^3)},$$

$$\|u(0,x)\|_{\dot{H}^1(\mathbf{R}^4)} = \|\lambda u(0,\lambda x)\|_{\dot{H}^1(\mathbf{R}^4)},$$

and

$$\|u(0,x)\|_{L^4_x(\mathbf{R}^4)} = \|\lambda u(0,\lambda x)\|_{L^4_x(\mathbf{R}^4)}.$$

Global well-posedness and scattering hold for the cubic problem with small initial data in the critical Sobolev spaces in dimensions $d = 3, 4$.

Theorem 2.1 There exists $\varepsilon_0 > 0$ such that

$$iu_t + \Delta u = F(u) = \mu |u|^2 u, \quad u(0,x) = u_0 \in \dot{H}^{s_c}(\mathbf{R}^d) \qquad (2.4)$$

is globally well posed and scattering for $\|u_0\|_{\dot{H}^{s_c}(\mathbf{R}^d)} \leq \varepsilon_0$ when $d = 3, 4$.

Proof Using Strichartz spaces, let

$$X = \{u : \|u\|_{\dot{S}^{s_c}(\mathbf{R}\times\mathbf{R}^d)} \leq C\varepsilon_0\},$$

and define the operator

$$\Phi(u)(t) = e^{it\Delta} u_0 - i \int_0^t e^{i(t-\tau)\Delta} F(u(\tau)) d\tau.$$

By Strichartz estimates, when $d = 4$ and $s_c = 1$,

$$\|\Phi(u)\|_{\dot{S}^1(\mathbf{R}\times\mathbf{R}^4)} \lesssim \|u_0\|_{\dot{H}^1(\mathbf{R}^4)} + \|\nabla F(u)\|_{L^{3/2}_{t,x}(\mathbf{R}\times\mathbf{R}^4)}. \qquad (2.5)$$

By the product rule and the Sobolev embedding theorem, which implies $\|u\|_{L^6_{t,x}} \lesssim \|u\|_{\dot{S}^1}$, for ε_0 sufficiently small,

$$(2.5) \lesssim \|u_0\|_{\dot{H}^1(\mathbf{R}^4)} + \|\nabla u\|_{L^3_{t,x}(\mathbf{R}\times\mathbf{R}^4)} \|u\|^2_{L^6_{t,x}(\mathbf{R}\times\mathbf{R}^4)} \lesssim \varepsilon_0 + (C\varepsilon_0)^3 \lesssim \varepsilon_0. \qquad (2.6)$$

Thus, $\Phi: X \to X$ when $d = 4$.

When $d = 3$, Strichartz estimates imply

$$\|\Phi(u)\|_{\dot{S}^{1/2}(\mathbf{R}\times\mathbf{R}^3)} \lesssim \|u_0\|_{\dot{H}^{1/2}(\mathbf{R}^3)} + \||\nabla|^{1/2} F(u)\|_{L^{10/7}_{t,x}(\mathbf{R}\times\mathbf{R}^3)}.$$

Use the fractional product rule to estimate $|\nabla|^{1/2} F(u)$.

Theorem 2.2 (Fractional product rule) *For $0 < s < 1$, $1 < p < \infty$,*

$$\|fg\|_{\dot{W}^{s,p}(\mathbf{R}^d)} \lesssim \|f\|_{L^{q_1}(\mathbf{R}^d)} \|g\|_{\dot{W}^{s,p_2}(\mathbf{R}^d)} + \|g\|_{L^{r_1}(\mathbf{R}^d)} \|f\|_{\dot{W}^{s,r_2}(\mathbf{R}^d)},$$

where

$$\frac{1}{p} = \frac{1}{q_1} + \frac{1}{q_2} = \frac{1}{r_1} + \frac{1}{r_2},$$

$q_2, r_2 \in (1, \infty)$, and $(q_1, r_1) \in (1, \infty]$. Here $\dot{W}^{s,p}(\mathbf{R}^d)$ is the L^p-based Sobolev space of order s.

Proof See Taylor (2000). □

By the fractional product rule, for any interval $I \subset \mathbf{R}$,

$$\||\nabla|^{1/2} F(u)\|_{L^{10/7}_{t,x}(I\times\mathbf{R}^3)} \lesssim \||\nabla|^{1/2} u\|_{L^{10/3}_{t,x}(I\times\mathbf{R}^3)} \|u\|^2_{L^5_{t,x}(I\times\mathbf{R}^3)}. \qquad (2.7)$$

Therefore, as in (2.6), for $\varepsilon_0 > 0$ sufficiently small, $\Phi: X \to X$ when $d = 3$.

2.1 3D and 4D Cubic NLS with Small Data

To prove that Φ is a contraction, by the fundamental theorem of calculus,

$$F(u) - F(v) = \int_0^1 \frac{d}{dt} F(v + t(u-v)) \, dt = (u-v) \cdot \int_0^1 F'(v + t(u-v)) \, dt. \tag{2.8}$$

By the product rule, when $d = 4$,

$$\|\Phi(u) - \Phi(v)\|_{\dot{S}^1(\mathbf{R} \times \mathbf{R}^4)}$$

$$\lesssim \|u - v\|_{\dot{S}^1(\mathbf{R} \times \mathbf{R}^4)} \left(\|u\|_{L^6_{t,x}(\mathbf{R} \times \mathbf{R}^4)}^2 + \|v\|_{L^6_{t,x}(\mathbf{R} \times \mathbf{R}^4)}^2 \right) + \|u - v\|_{L^6_{t,x}(\mathbf{R} \times \mathbf{R}^4)}$$

$$\times \left(\|u\|_{L^6_{t,x}(\mathbf{R} \times \mathbf{R}^4)} + \|v\|_{L^6_{t,x}(\mathbf{R} \times \mathbf{R}^4)} \right) \left(\|u\|_{\dot{S}^1(\mathbf{R} \times \mathbf{R}^4)} + \|v\|_{\dot{S}^1(\mathbf{R} \times \mathbf{R}^4)} \right)$$

$$\lesssim (C\varepsilon_0^2) \|u - v\|_{\dot{S}^1(\mathbf{R} \times \mathbf{R}^4)}.$$

When $d = 3$, the fractional product rule implies

$$\||\nabla|^{1/2}(F(u) - F(v))\|_{L^{10/7}_{t,x}(\mathbf{R} \times \mathbf{R}^3)}$$

$$\lesssim \||\nabla|^{1/2}(u-v)\|_{L^{10/3}_{t,x}(\mathbf{R} \times \mathbf{R}^3)} \left(\|u\|_{L^5_{t,x}(\mathbf{R} \times \mathbf{R}^3)}^2 + \|v\|_{L^5_{t,x}(\mathbf{R} \times \mathbf{R}^3)}^2 \right)$$

$$+ \|u - v\|_{L^5_{t,x}(\mathbf{R} \times \mathbf{R}^3)} \left(\|u\|_{L^5_{t,x}(\mathbf{R} \times \mathbf{R}^3)} + \|v\|_{L^5_{t,x}(\mathbf{R} \times \mathbf{R}^3)} \right)$$

$$\times \left(\||\nabla|^{1/2} u\|_{L^{10/3}_{t,x}(\mathbf{R} \times \mathbf{R}^3)} + \||\nabla|^{1/2} v\|_{L^{10/3}_{t,x}(\mathbf{R} \times \mathbf{R}^3)} \right)$$

$$\lesssim (C\varepsilon_0)^2 \left(\|u-v\|_{L^5_{t,x}(\mathbf{R} \times \mathbf{R}^3)} + \||\nabla|^{1/2}(u-v)\|_{L^{10/3}_{t,x}(\mathbf{R} \times \mathbf{R}^3)} \right). \tag{2.9}$$

Therefore,

$$\|\Phi(u) - \Phi(v)\|_{\dot{S}^{sc}(\mathbf{R} \times \mathbf{R}^d)} \lesssim (C\varepsilon_0)^2 \|u - v\|_{\dot{S}^{sc}(\mathbf{R} \times \mathbf{R}^d)},$$

proving that Φ is a contraction. Then by the contraction mapping principle, Strichartz estimates, and the product (or fractional product) rule, the solution $u \in C_t^0(\mathbf{R}; \dot{H}_x^{sc}(\mathbf{R}^d)) \cap \dot{S}^{sc}(\mathbf{R} \times \mathbf{R}^d)$ exists and is unique.

The proof of continuous dependence on initial data again relies on a perturbation lemma.

Lemma 2.3 (Perturbation lemma) *Let I be a compact time interval and let w be an approximate solution to* (2.4), *that is, for some e small,*

$$(i\partial_t + \Delta) w = F(w) + e. \tag{2.10}$$

Also suppose

$$\|w\|_{L_t^\infty \dot{H}_x^{sc}(I \times \mathbf{R}^d)} \leq E. \tag{2.11}$$

Let u solve (2.4), and suppose that for some $t_0 \in \mathbf{R}$,

$$\|u(t_0) - w(t_0)\|_{\dot{H}^{s_c}(\mathbf{R}^d)} \leq E', \tag{2.12}$$

for some $E' > 0$. Moreover, assume the smallness condition, $0 \leq \varepsilon \leq \varepsilon_0(E, E')$, along with

$$\|u\|_{L^{d+2}_{t,x}(\mathbf{R} \times \mathbf{R}^d)} \leq \varepsilon_0, \tag{2.13}$$

$$\|\nabla e\|_{\dot{N}^{s_c}(\mathbf{R} \times \mathbf{R}^d)} \leq \varepsilon, \tag{2.14}$$

and

$$\left\|e^{i(t-t_0)\Delta}(u(t_0) - w(t_0))\right\|_{L^{d+2}_{t,x}(\mathbf{R} \times \mathbf{R}^d)} \leq \varepsilon.$$

Then

$$\|u - w\|_{L^{d+2}_{t,x}(\mathbf{R} \times \mathbf{R}^d)} \lesssim \varepsilon \tag{2.15}$$

and

$$\|(i\partial_t + \Delta)(u - w) + e\|_{\dot{N}^{s_c}(\mathbf{R} \times \mathbf{R}^d)} \lesssim \varepsilon. \tag{2.16}$$

Remark The proofs of Theorem 2.1 and Lemma 2.3 are presented in Holmer and Roudenko (2008). See also Guevara (2014) and Kenig and Merle (2010).

Proof If u solves (2.4) and w solves (2.10), then $v = w - u$ solves

$$(i\partial_t + \Delta)v = F(w) - F(u) + e.$$

By Strichartz estimates, the Sobolev embedding theorem, and Duhamel's principle,

$$\|v\|_{L^{d+2}_{t,x}(\mathbf{R} \times \mathbf{R}^d)} \lesssim \left\|e^{i(t-t_0)\Delta}(w(t_0) - u(t_0))\right\|_{L^{d+2}_{t,x}(\mathbf{R} \times \mathbf{R}^d)}$$
$$+ \|F(w) - F(u) + e\|_{\dot{N}^{s_c}(\mathbf{R} \times \mathbf{R}^d)}$$

and

$$\|v\|_{\dot{S}^{s_c}(\mathbf{R} \times \mathbf{R}^d)} \lesssim \|w(t_0) - u(t_0)\|_{\dot{H}^{s_c}(\mathbf{R}^d)} + \|F(w) - F(u) + e\|_{\dot{N}^{s_c}(\mathbf{R} \times \mathbf{R}^d)}.$$

By (2.8) and (2.9),

$$\|F(w) - F(u)\|_{\dot{N}^{s_c}(\mathbf{R} \times \mathbf{R}^d)}$$
$$\lesssim \|v\|_{\dot{S}^{s_c}(\mathbf{R} \times \mathbf{R}^d)} \left(\|u\|^2_{L^{d+2}_{t,x}(\mathbf{R} \times \mathbf{R}^d)} + \|v\|^2_{L^{d+2}_{t,x}(\mathbf{R} \times \mathbf{R}^d)}\right) + \|v\|_{L^{d+2}_{t,x}(\mathbf{R} \times \mathbf{R}^d)}$$
$$\times \left(\|u\|_{\dot{S}^{s_c}(\mathbf{R} \times \mathbf{R}^d)} + \|v\|_{\dot{S}^{s_c}(\mathbf{R} \times \mathbf{R}^d)}\right) \left(\|u\|_{L^{d+2}_{t,x}(\mathbf{R} \times \mathbf{R}^d)} + \|v\|_{L^{d+2}_{t,x}(\mathbf{R} \times \mathbf{R}^d)}\right).$$
$$\tag{2.17}$$

2.1 3D and 4D Cubic NLS with Small Data

Therefore, by Strichartz estimates, the product rule, (2.13), (2.14), and (2.17),

$$\|v\|_{L^{d+2}_{t,x}(\mathbf{R}\times\mathbf{R}^d)}$$
$$\lesssim \varepsilon + \|v\|_{\dot{S}^{sc}(\mathbf{R}\times\mathbf{R}^d)} \left(\varepsilon_0 + \|v\|_{L^{d+2}_{t,x}(\mathbf{R}\times\mathbf{R}^d)}\right)^2$$
$$+ \|v\|_{L^{d+2}_{t,x}(\mathbf{R}\times\mathbf{R}^d)} \left(\|u\|_{\dot{S}^{sc}(\mathbf{R}\times\mathbf{R}^d)} + \|v\|_{\dot{S}^{sc}(\mathbf{R}\times\mathbf{R}^d)}\right) \left(\varepsilon_0 + \|v\|_{L^{d+2}_{t,x}(\mathbf{R}\times\mathbf{R}^d)}\right)$$

and

$$\|v\|_{\dot{S}^{sc}(\mathbf{R}\times\mathbf{R}^d)}$$
$$\lesssim E + E' + \|v\|_{\dot{S}^{sc}(\mathbf{R}\times\mathbf{R}^d)} \left(\varepsilon_0 + \|v\|_{L^{d+2}_{t,x}(\mathbf{R}\times\mathbf{R}^d)}\right)^2$$
$$+ \|v\|_{L^{d+2}_{t,x}(\mathbf{R}\times\mathbf{R}^d)} \left(\|u\|_{\dot{S}^{sc}(\mathbf{R}\times\mathbf{R}^d)} + \|v\|_{\dot{S}^{sc}(\mathbf{R}\times\mathbf{R}^d)}\right) \left(\varepsilon_0 + \|v\|_{L^{d+2}_{t,x}(\mathbf{R}\times\mathbf{R}^d)}\right).$$

Strichartz estimates, (2.11), (2.12), and (2.13) imply $\|u\|_{\dot{S}^{sc}(\mathbf{R}\times\mathbf{R}^d)} \lesssim E + E'$, so taking $\varepsilon_0(E,E')$ sufficiently small, say $\varepsilon_0 \ll \frac{1}{1+E+E'}$, proves (2.15) and (2.16). □

Setting $e = 0$, then (2.16) and Strichartz estimates imply that if u and w solve (2.4) with initial data

$$\|w(t_0)\|_{\dot{H}^{sc}_x(\mathbf{R}^d)}, \quad \|u(t_0)\|_{\dot{H}^{sc}_x(\mathbf{R}^d)} \leq \varepsilon_0,$$

then

$$\|u - w\|_{L^{d+2}_{t,x}(\mathbf{R}\times\mathbf{R}^d)} + \|u - w\|_{\dot{S}^{sc}(\mathbf{R}\times\mathbf{R}^d)} \lesssim \|u(t_0) - w(t_0)\|_{\dot{H}^{sc}(\mathbf{R}^d)}.$$

This proves continuous dependence on the initial data, which completes the proof of global well-posedness.

Setting

$$u_+ = u(t_0) - i \int_{t_0}^{\infty} e^{-i\tau\Delta} F(u(\tau)) d\tau$$

and

$$u_- = u(t_0) + i \int_{-\infty}^{t_0} e^{-i\tau\Delta} F(u(\tau)) d\tau,$$

implies scattering, since $u_+, u_- \in \dot{H}^{sc}(\mathbf{R}^d)$, and (2.6) and (2.7) imply that $u(t)$ scatters forward to $e^{it\Delta}u_+$ as $t \nearrow +\infty$, and $u(t)$ scatters backward in time to $e^{it\Delta}u_-$ as $t \searrow -\infty$. □

Taking $w = 0$, Lemma 2.3 implies global well-posedness and scattering for (2.4) when

$$\|e^{it\Delta}u_0\|_{L^{d+2}_{t,x}(\mathbf{R}\times\mathbf{R}^d)} \leq \varepsilon_0(\|u_0\|_{\dot{H}^{sc}}),$$

where in this case, $\varepsilon_0 \searrow 0$ as $\|u_0\|_{\dot H^{s_c}} \nearrow \infty$. Likewise, to prove global well-posedness on $[0,\infty)$ and that a solution scatters forward in time, it would be enough to show

$$\|e^{it\Delta}u_0\|_{L^{d+2}_{t,x}([0,\infty)\times\mathbf{R}^d)} \leq \varepsilon_0(\|u_0\|_{\dot H^{s_c}}). \tag{2.18}$$

Compare to (1.99), in which ε_0 is independent of the size of the initial data.

Following the arguments used to prove Theorems 1.21 and 1.22 we obtain the following theorems.

Theorem 2.4 (Local well-posedness)

1. *For any $u_0 \in \dot H^{s_c}_x(\mathbf{R}^d)$, there exists $T(u_0) > 0$ such that (2.1) is locally well posed on $[-T,T]$. The quantity $T(u_0)$ depends on the profile of the initial data as well as its size. Moreover, (2.1) is well posed on an open interval $I \subset \mathbf{R}$, $0 \in I$.*
2. *If $\sup(I) < +\infty$, where I is the maximal interval of existence of (2.1),*

$$\lim_{T \nearrow \sup(I)} \|u\|_{L^{d+2}_{t,x}([0,T]\times\mathbf{R}^d)} = \infty \quad \text{or} \quad \lim_{T \nearrow \sup(I)} \|u\|_{L^\infty_t \dot H^{s_c}([0,T]\times\mathbf{R}^d)} = \infty.$$

The corresponding result also holds for $\inf(T) > -\infty$.

Theorem 2.5 (Scattering) *A solution to (2.1) scatters forward in time to some $e^{it\Delta}u_+$, $u_+ \in \dot H^{s_c}_x(\mathbf{R}^d)$ if and only if for some $t_0 \in \mathbf{R}$,*

$$\|u\|_{L^{d+2}_{t,x}([t_0,\infty)\times\mathbf{R}^d)} < \infty \quad \text{and} \quad \|u\|_{L^\infty_t \dot H^{s_c}([t_0,\infty)\times\mathbf{R}^d)} < \infty.$$

Arguing as in the proof of Theorem 1.24 implies long-time perturbation results.

Theorem 2.6 (Long-time perturbations) *Let I be a compact time interval and let w be an approximate solution to (2.4) on $I \times \mathbf{R}^d$ in the sense that*

$$(i\partial_t + \Delta)w = F(w) + e.$$

Assume that u is a solution to (2.4) satisfying

$$\|u\|_{L^{d+2}_{t,x}(I\times\mathbf{R}^d)} \leq M$$

and

$$\|u\|_{L^\infty_t \dot H^{s_c}_x(I\times\mathbf{R}^d)} \leq E,$$

for some $M, E > 0$. Let $t_0 \in I$, and suppose that $u(t_0)$ is close to $w(t_0)$ in the sense that

$$\|u(t_0) - w(t_0)\|_{\dot H^{s_c}_x(\mathbf{R}^d)} \leq E'$$

2.2 Large Data Scattering for Radial, 3D Cubic NLS

and
$$\left\| e^{i(t-t_0)\Delta}\left(u(t_0) - w(t_0)\right) \right\|_{L_{t,x}^{d+2}(I \times \mathbf{R}^d)} \leq \varepsilon.$$

Also assume
$$\|e\|_{\dot{N}^{sc}(I \times \mathbf{R}^d)} \leq \varepsilon,$$

for some $0 < \varepsilon \leq \varepsilon_1(E, E', M)$. *Then*
$$\|u - w\|_{L_{t,x}^{d+2}(I \times \mathbf{R}^d)} \leq C(E, E', M, d)\varepsilon$$

and
$$\|w\|_{L_{t,x}^{d+2}(I \times \mathbf{R}^d)} \leq C(E, E', M, d).$$

Proof As in the proof of Theorem 1.24, Theorem 2.5 may be proved by partitioning I into subintervals and iterating the argument used to prove Lemma 2.3. □

2.2 Scattering for the Radial Cubic NLS in Three Dimensions with Large Data

Local conservation laws imply a large data scattering result for the defocusing, three-dimensional cubic nonlinear Schrödinger equation,

$$iu_t + \Delta u = |u|^2 u, \quad u(0,x) = u_0, \quad u: \mathbf{R} \times \mathbf{R}^3 \to \mathbf{C}. \tag{2.19}$$

Theorem 2.7 *The initial value problem* (2.19) *is globally well posed and scattering for radial initial data that lies in the inhomogeneous Sobolev space* $H_x^1(\mathbf{R}^3)$. *Furthermore,*

$$\|u\|_{L_{t,x}^5(\mathbf{R} \times \mathbf{R}^3)}^5 \lesssim \|u_0\|_{L_x^2(\mathbf{R}^3)}^{3/2} \|u_0\|_{\dot{H}_x^1(\mathbf{R}^3)}^{3/2} \left(1 + \|u_0\|_{L_x^2(\mathbf{R}^3)} \|u_0\|_{\dot{H}_x^1(\mathbf{R}^3)}\right)^{3/2}. \tag{2.20}$$

Both the left- and right-hand sides of (2.20) *are invariant under the scaling symmetry* (2.2), *when $p = 2$ and $d = 3$.*

Remark Compare Theorem 2.7 to Theorem 1.32.

Proof By Theorem 2.4, since $H_x^1(\mathbf{R}^3) \hookrightarrow \dot{H}_x^{1/2}(\mathbf{R}^3)$, (2.19) is locally well posed on some interval I with $[-T, T] \subset I$, $T(u_0) > 0$.

Proof of global well-posedness is a straightforward application of conservation of mass and energy. By the Sobolev embedding theorem, $\dot{H}_x^1(\mathbf{R}^3) \hookrightarrow$

$L_x^6(\mathbf{R}^3)$, so

$$E(u_0) \lesssim \|u_0\|_{\dot{H}_x^1(\mathbf{R}^3)}^2 \left(1 + \|u_0\|_{L_x^2(\mathbf{R}^3)} \|u_0\|_{\dot{H}_x^1(\mathbf{R}^3)}\right), \qquad (2.21)$$

and then by interpolation, for any $t \in I$,

$$\|u(t)\|_{\dot{H}_x^{1/2}(\mathbf{R}^3)}^2 \lesssim \|u(t)\|_{L_x^2(\mathbf{R}^3)} \|u(t)\|_{\dot{H}_x^1(\mathbf{R}^3)} \lesssim \|u(t)\|_{L_x^2(\mathbf{R}^3)} E(u(t))^{1/2}$$

$$\lesssim \|u_0\|_{L_x^2(\mathbf{R}^3)} \|u_0\|_{\dot{H}_x^1(\mathbf{R}^3)} \left(1 + \|u_0\|_{L_x^2(\mathbf{R}^3)} \|u_0\|_{\dot{H}_x^1(\mathbf{R}^3)}\right)^{1/2}, \qquad (2.22)$$

which implies the uniform bound $\|u\|_{L_t^\infty \dot{H}_x^{1/2}(I \times \mathbf{R}^3)} < \infty$. Moreover, by Hölder's inequality in time, the Sobolev embedding theorem, and conservation of energy, for any $t_0 \in I$,

$$\|e^{i(t-t_0)\Delta} u(t_0)\|_{L_{t,x}^5([t_0-T, t_0+T] \times \mathbf{R}^3)}$$

$$\lesssim T^{1/5} \|u(t_0)\|_{\dot{H}_x^{\frac{9}{10}}(\mathbf{R}^3)} \lesssim T^{1/5} \|u(t_0)\|_{L_x^2(\mathbf{R}^3)}^{1/10} \|u(t_0)\|_{\dot{H}_x^1(\mathbf{R}^3)}^{9/10}$$

$$\lesssim T^{1/5} \|u_0\|_{L_x^2(\mathbf{R}^3)}^{1/10} \|u_0\|_{\dot{H}_x^1(\mathbf{R}^3)}^{9/10} \left(1 + \|u_0\|_{L_x^2(\mathbf{R}^3)} \|u_0\|_{\dot{H}_x^1(\mathbf{R}^3)}\right)^{9/20}. \qquad (2.23)$$

The estimates (2.22) and (2.23) imply a uniform lower bound on $T(u(t_0))$, $t_0 \in I$. Iterating local well-posedness directly implies $I = \mathbf{R}$.

The proof of scattering uses the Morawetz estimate. Recall that the momentum

$$\int T_{0j}(t, x) \, dx$$

is a conserved quantity. Moreover, when $\mu = +1$, the time derivative of T_{0j} is the derivative of a negative definite quantity, plus three derivatives of a lower-order term. Indeed, by (1.115) and (1.116),

$$\partial_t T_{0j}(t, x) = -4 \partial_k \operatorname{Re}\left(\left(\overline{\partial_j u}\right)(\partial_k u)\right) - \partial_j |u|^4 + \partial_j \Delta |u|^2.$$

Integrating by parts,

$$\partial_t \int x_j T_{0j}(t, x) \, dx = 4 \int |\nabla u(t, x)|^2 \, dx + 3 \int |u(t, x)|^4 \, dx \sim E(u(t)).$$

This fact, combined with an appropriate choice of Y in (1.111), and the fundamental theorem of calculus, gives estimates of certain space-time integrals of $u(t, x)$.

Take (1.111) with $Y = h(x) \bar{u}(t, x)$, and let

$$M(t) = \int h(x) |u(t, x)|^2 \, dx. \qquad (2.24)$$

2.2 Large Data Scattering for Radial, 3D Cubic NLS

If $\mathcal{N} = |u|^p u$, (1.112) implies

$$\partial_t M(t) = -\int h(x) \partial_j T_{0j}(t,x) dx,$$

and (1.119) and (1.120) imply

$$\partial_{tt} M(t) = \int h(x) \left[\Delta|u|^4 + \partial_j \partial_k \left[4\operatorname{Re}\left(\left(\overline{\partial_j u}\right)(\partial_k u)\right) - \partial_j \partial_k |u|^2\right]\right] dx. \quad (2.25)$$

Integrating by parts,

$$\partial_{tt} M(t)$$
$$= \int_{\mathbf{R}^d} (-\Delta\Delta h(x))|u(t,x)|^2 + 4\partial_j \partial_k h(x) \cdot \operatorname{Re}(\partial_j u \partial_k \bar{u}) + (\Delta h(x)) \cdot |u(t,x)|^4 dx.$$
$$(2.26)$$

The choice $h(x) = |x|$ gives a Morawetz estimate for the nonlinear Schrödinger equation.

Theorem 2.8 (Morawetz estimate) *If u is a smooth solution to (2.4) on $I \subset \mathbf{R}$ when $d = 3, 4$,*

$$\int_I \int \frac{|u(t,x)|^4}{|x|} dx \, dt \lesssim \|u\|_{L_t^\infty L_x^2(I \times \mathbf{R}^d)} \|u\|_{L_t^\infty \dot{H}_x^1(I \times \mathbf{R}^d)} \quad (2.27)$$

and

$$\int_I \int_{|x| \lesssim |I|^{1/2}} \frac{|u(t,x)|^4}{|x|} dx \, dt \lesssim |I|^{1/2} E(u_0). \quad (2.28)$$

Remark Lin and Strauss (1978) proved (2.27) and Bourgain (1999) proved (2.28).

Proof When $h(x) = |x|$,

$$\Delta h(x) = \frac{d-1}{|x|}, \quad \partial_j \partial_k h(x) = \frac{\delta_{jk}}{|x|} - \frac{x_j x_k}{|x|^3}, \quad (2.29)$$

and when $d = 3$,

$$\Delta\Delta h(x) = -8\pi \delta(x), \quad (2.30)$$

where (2.30) holds in a distributional sense. Indeed,

$$-\Delta\Delta|x| = -\left(\partial_{rr} + \frac{2}{r}\partial_r\right)\left(\partial_{rr} + \frac{2}{r}\partial_r\right)r = -\left(\partial_{rr} + \frac{2}{r}\partial_r\right)\frac{2}{r}.$$

If f is a Schwartz function then switching to polar coordinates and integrating by parts,

$$\int f(x)(-\Delta\Delta|x|)\,dx = -\int_0^\infty \int_{S^2} f(r,\theta) r^2 \left(\partial_{rr} + \frac{2}{r}\partial_r\right) \frac{2}{r}\,dr\,d\theta$$

$$= \int_0^\infty \int_{S^2} f_r(r,\theta) r^2 \partial_r \left(\frac{2}{r}\right) dr\,d\theta$$

$$= -2\int_0^\infty \int_{S^2} f_r(r,\theta)\,dr\,d\theta = 8\pi f(0).$$

Therefore, by the fundamental theorem of calculus in time, (2.26), (2.29), and (2.30) imply

$$8\pi \int_I |u(t,0)|^2 dt + \int\int \frac{4}{|x|} \left[|\nabla u(t,x)|^2 - |\partial_r u(t,x)|^2\right] dx\,dt$$

$$+ \int_I \int \frac{2}{|x|} |u(t,x)|^4 dx\,dt$$

$$\leq 2\sup_{t\in I} |\partial_t M(t)| \leq 2\|u\|_{L_t^\infty L_x^2(I\times\mathbf{R}^3)} \|u\|_{L_t^\infty \dot{H}^1(I\times\mathbf{R}^3)}. \tag{2.31}$$

To make (2.31) completely rigorous, one could take a smooth approximation of the initial data and then show that (2.31) holds in the limit, using, by now standard, perturbative arguments. Since $|\partial_r u| \leq |\nabla u|$, all three terms on the left-hand side of (2.31) are positive, proving (2.27) when $d = 3$.

To prove (2.28) choose $h(x) = |x|\psi\left(\frac{x}{R}\right)$, where $\psi \in C^\infty(\mathbf{R}^3)$ is a smooth function, $\psi(x) = 1$ on $|x| \leq 1$, and $\psi(x) = 0$ for $|x| > 2$. For $|x| < R$, (2.29) and (2.30) hold. When $|x| > R$,

$$|\nabla h(x)| \lesssim 1, \quad |\partial_j \partial_k h(x)| \lesssim \frac{1}{R}, \quad |\Delta^2 h(x)| \lesssim \frac{1}{R^3}. \tag{2.32}$$

By the Sobolev embedding theorem and Hölder's inequality,

$$|\partial_t M(t)| \lesssim \int_{|x|\leq 2R} |u(t,x)||\nabla u(t,x)|\,dx \lesssim R\|\nabla u\|_{L^2(\mathbf{R}^d)}^2.$$

By (2.29) and (2.32),

$$4\int \mathrm{Re}\,(\partial_k u \partial_j \bar{u})(\partial_j \partial_k h(x))\,dx \gtrsim \frac{-1}{R}\|\nabla u\|_{L^2(\mathbf{R}^d)}^2.$$

By (2.30), (2.32), Hardy's inequality, and the support of $\psi\left(\frac{x}{R}\right)$,

$$-\int |u(t,x)|^2 (\Delta\Delta h(x))\,dx = -\int_{|x|\leq R} |u|^2 (\Delta\Delta h(x))\,dx - \int_{|x|>R} |u|^2 (\Delta\Delta h(x))\,dx$$

$$\gtrsim \frac{-1}{R}\|\nabla u\|_{L_x^2(\mathbf{R}^d)}^2.$$

2.2 Large Data Scattering for Radial, 3D Cubic NLS

Finally, by (2.29), (2.32), and the support of $\psi\left(\frac{x}{R}\right)$,

$$\int |u(t,x)|^4 \cdot \Delta h(x) \, dx \geq \int_{|x| \leq R} \frac{|u(t,x)|^4}{|x|} \, dx - C \int_{|x| \geq R} \frac{|u(t,x)|^4}{|x|} \, dx$$

$$\geq \int \frac{|u(t,x)|^4}{|x|} \, dx - \frac{C}{R} E(u_0).$$

Integrating $\partial_{tt} M(t)$ in time, by the fundamental theorem of calculus,

$$\int_I \int_{|x| \leq R} \frac{|u(t,x)|^4}{|x|} \, dx \, dt \lesssim \frac{|I|}{R} E(u_0) + R E(u_0).$$

Choosing $R \sim |I|^{1/2}$ proves (2.28) when $d = 3$.

When $d = 4$ the proofs of (2.27) and (2.28) are similar, except

$$-\Delta \Delta |x| = \frac{3}{|x|^3}.$$

$\square$

Next, by the fundamental theorem of calculus, if $f \in H^1(\mathbf{R}^3)$ is a radial function,

$$f^2(r) = -\int_r^\infty \partial_s (f^2(s)) \, ds = -2 \int_r^\infty f'(s) f(s) \, ds$$

$$\leq \frac{2}{r^2} \int_r^\infty s^2 |f'(s)| |f(s)| \, ds \lesssim \frac{1}{r^2} \|f\|_{\dot{H}^1(\mathbf{R}^3)} \|f\|_{L^2_x(\mathbf{R}^3)}. \quad (2.33)$$

This is the only time the radial symmetry is used in the proof of Theorem 2.7. Combining (2.27) and (2.33),

$$\|u\|^5_{L^5_{t,x}(\mathbf{R} \times \mathbf{R}^3)} \leq \||x|u\|_{L^\infty_{t,x}(\mathbf{R} \times \mathbf{R}^3)} \cdot \int_\mathbf{R} \int_{\mathbf{R}^3} \frac{|u(t,x)|^4}{|x|} \, dx \, dt$$

$$\lesssim \|u\|^{3/2}_{L^\infty_t \dot{H}^1_x(I \times \mathbf{R}^3)} \|u\|^{3/2}_{L^\infty_t L^2_x(I \times \mathbf{R}^3)}. \quad (2.34)$$

Then conservation of energy and (2.21) directly give (2.20). $\square$

Remark The interaction Morawetz estimate (Lemma 3.23), conservation of mass, and conservation of energy imply nonradial scattering for initial data satisfying the conditions of Theorem 2.7.

Remark Kenig and Merle (2010) proved global well-posedness and scattering for (2.19), assuming an a priori bound on $\|u(t)\|_{\dot{H}^{1/2}_x(\mathbf{R}^3)}$.

Remark See Duyckaerts et al. (2008) and Holmer and Roudenko (2008) for results in the focusing case.

2.3 The Radially Symmetric, Cubic Problem in Four Dimensions

Study of the four-dimensional cubic problem with large data represented an important development in the study of the scattering theory for critical, defocusing nonlinear Schrödinger equations. Observe that the small data results of Theorem 1.17 and Theorem 2.1 prove global well-posedness and scattering for the nonlinear Schrödinger equation with initial data lying in the critical Sobolev space, while the large data results in Theorem 1.32 and Theorem 2.7 hold only for initial data in a subspace of the critical Sobolev space. In other words, for (1.130) and (2.34), the right-hand sides are norms that bound the critical Sobolev spaces by interpolation, and are invariant under the critical scaling, but are strict subsets of the critical Sobolev space, which naturally raises the question of the long-term behavior of generic initial data in the critical Sobolev space.

Bourgain (1999) proved global well-posedness and scattering for the defocusing, energy-critical problem for radially symmetric data lying in $\dot{H}^1$ in dimensions three and four. In keeping with the theme of this chapter, only the proof of scattering for the cubic, four-dimensional nonlinear Schrödinger equation will be discussed here. Proof of the corresponding result for the three-dimensional, quintic problem is similar. This result proved to be quite foundational to the study of other nonlinear Schrödinger problems, and many of the same ideas were then incorporated into the concentration compactness method, which will be studied in the next chapter.

Therefore, there is great advantage in studying the argument of Bourgain (1999) in detail. The reader is encouraged to pay careful attention to how the Morawetz estimate and conservation laws are used in concert to bound the scattering size of a solution. To be sure, Grillakis (2000) independently proved global well-posedness for the radial, three-dimensional energy-critical problem. These results were subsequently extended to higher dimensions by Tao (2005). The upcoming proof contains many of the arguments found in Grillakis (2000) and Tao (2005) when such arguments simplify the exposition, without obscuring the ideas in Bourgain (1999).

Theorem 2.9 *The defocusing, cubic, initial value problem*

$$iu_t + \Delta u = F(u) = |u|^2 u, \qquad u(0,x) = u_0 \in \dot{H}^1\left(\mathbf{R}^4\right) \qquad (2.35)$$

is globally well posed and scattering for u_0 radial. Moreover, there exists a function $C\colon [0,\infty) \to [0,\infty)$, such that

$$\|u\|_{L^6_{t,x}(\mathbf{R}\times\mathbf{R}^4)} \leq C(E(u_0)), \qquad (2.36)$$

2.3 The Radially Symmetric, Cubic Problem in Four Dimensions

where $E(u(t))$ is the conserved energy.

Proof By the Sobolev embedding theorem,

$$\|u\|_{L^4_x(\mathbf{R}^4)} \lesssim \|u\|_{\dot H^1(\mathbf{R}^4)},$$

and therefore

$$E(u_0) \lesssim \|u_0\|^2_{\dot H^1(\mathbf{R}^4)} + \|u_0\|^4_{\dot H^1(\mathbf{R}^4)},$$

so Theorem 2.1 implies that for $E(u_0)$ small,

$$\|u\|_{L^6_{t,x}(\mathbf{R}\times\mathbf{R}^d)} \lesssim \|u_0\|_{\dot H^1(\mathbf{R}^4)} \lesssim E(u_0)^{1/2} \sim \|u_0\|_{\dot H^1(\mathbf{R}^4)}. \quad (2.37)$$

Starting with (2.37) as a base case, and then arguing in a manner analogous to the induction argument for natural numbers, to prove (2.36) it suffices to show that there exists some decreasing, continuous function $\varepsilon(E):[0,\infty)\to(0,\infty)$, such that

$$C(E) \lesssim_{C(E-\varepsilon(E))} 1. \quad (2.38)$$

Naively applying Theorem 2.6 does not suffice to prove (2.38), because Theorem 2.6 alone is not strong enough to disprove the possibility that $\varepsilon(E) \searrow 0$ as $E \nearrow E_0$, for some $E_0 < \infty$. The proof of (2.38) is considerably more complicated and involves considering a number of different cases.

The first case to consider is the case in which the initial data has small Besov norm.

Lemma 2.10 *For $u_0 \in \dot H^1(\mathbf{R}^4)$,*

$$\|e^{it\Delta}u_0\|_{L^6_{t,x}(\mathbf{R}\times\mathbf{R}^4)} \lesssim \|u_0\|^{1/3}_{\dot H^1(\mathbf{R}^4)} \left(\sup_j \|P_j u_0\|_{\dot H^1(\mathbf{R}^4)}\right)^{2/3}. \quad (2.39)$$

Then by (2.18), there exists an $\varepsilon(E) > 0$ such that (2.35) is globally well posed and scattering when

$$\sup_j \|P_j u_0\|_{\dot H^1(\mathbf{R}^4)} \leq \varepsilon(\|u_0\|_{\dot H^1}),$$

with scattering size bounded by

$$\|u\|_{L^6_{t,x}(\mathbf{R}\times\mathbf{R}^4)} \lesssim \|u_0\|^{1/3}_{\dot H^1(\mathbf{R}^4)} \left(\sup_j \|P_j u_0\|_{\dot H^1(\mathbf{R}^4)}\right)^{2/3}.$$

Proof Make a Littlewood–Paley decomposition,

$$|e^{it\Delta}u_0|^6 \lesssim \sum_{\substack{j_1\leq j_2\leq j_3 \\ \leq j_4\leq j_5\leq j_6}} |e^{it\Delta}P_{j_1}u_0||e^{it\Delta}P_{j_2}u_0||e^{it\Delta}P_{j_3}u_0|$$
$$\times |e^{it\Delta}P_{j_4}u_0||e^{it\Delta}P_{j_5}u_0||e^{it\Delta}P_{j_6}u_0|.$$

Then
$$\|e^{it\Delta}u_0\|_{L^6_{t,x}}^6$$
$$\lesssim \sum_{\substack{j_1 \leq j_2 \leq j_3 \\ \leq j_4 \leq j_5 \leq j_6}} \|e^{it\Delta}P_{j_1}u_0\|_{L^\infty_{t,x}} \|e^{it\Delta}P_{j_2}u_0\|_{L^\infty_{t,x}} \|e^{it\Delta}P_{j_3}u_0\|_{L^\infty_t L^4_x}$$
$$\times \|e^{it\Delta}P_{j_4}u_0\|_{L^\infty_t L^4_x} \|e^{it\Delta}P_{j_5}u_0\|_{L^2_t L^4_x} \|e^{it\Delta}P_{j_6}u_0\|_{L^2_t L^4_x}. \quad (2.40)$$

By Strichartz estimates, the Sobolev embedding theorem, and the Cauchy–Schwarz inequality,

$$(2.40) \lesssim \sum_{\substack{j_1 \leq j_2 \leq j_3 \\ \leq j_4 \leq j_5 \leq j_6}} 2^{j_1+j_2} \|P_{j_1}u_0\|_{\dot{H}^1(\mathbf{R}^4)} \|P_{j_2}u_0\|_{\dot{H}^1(\mathbf{R}^4)} \|e^{it\Delta}P_{j_3}u_0\|_{L^\infty_t L^4_x}$$
$$\times \|e^{it\Delta}P_{j_4}u_0\|_{L^\infty_t L^4_x} \|P_{j_5}u_0\|_{L^2_x(\mathbf{R}^4)} \|P_{j_6}u_0\|_{L^2_x(\mathbf{R}^4)}$$
$$\lesssim \|u_0\|_{\dot{H}^1(\mathbf{R}^4)}^2 \left(\sup_j \|P_j u_0\|_{\dot{H}^1(\mathbf{R}^4)}\right)^4, \quad (2.41)$$

which proves (2.39). The right-hand side of (2.41) can be improved to

$$\|e^{it\Delta}u_0\|_{L^6_{t,x}(\mathbf{R}\times\mathbf{R}^4)}^6$$
$$\lesssim \|u_0\|_{\dot{H}^1(\mathbf{R}^4)}^2 \left(\sup_j \|P_j u_0\|_{\dot{H}^1(\mathbf{R}^4)}\right)^2 \left(\sup_j \|e^{it\Delta}P_j u_0\|_{L^\infty_t L^4_x(\mathbf{R}\times\mathbf{R}^4)}\right)^2, \quad (2.42)$$

which will be relevant later. □

Now let $\varepsilon_0 = \varepsilon(\sqrt{2E})$, where $E = E(u_0)$. Suppose u is a solution to (2.35) on the maximal interval of existence I. For any $t \in I$, let

$$j(t) = \inf\{j \in \mathbf{Z} : \|P_j u(t)\|_{\dot{H}^1} \geq \varepsilon_0\}. \quad (2.43)$$

If $\|P_j u(t)\|_{\dot{H}^1} < \varepsilon_0$ for all $j \in \mathbf{Z}$ then set $j(t) = +\infty$.

If there exists $t_0 \in I$ such that $j(t_0) = \infty$ then by conservation of energy, $\|u(t_0)\|_{\dot{H}^1}^2 \leq 2E$, so Theorem 2.1 and (2.41) imply

$$\|u\|_{L^6_{t,x}(\mathbf{R}\times\mathbf{R}^4)}^6 \lesssim E\varepsilon_0^4.$$

If $j(t) < \infty$ for all $t \in I$ then consider two cases separately: one in which $u(t)$ is large in a Besov sense at frequencies far away from $j(t)$, and the other in which $u(t)$ is small in a Besov sense far away from $j(t)$.

In the former, the scattering size may be estimated using induction on energy.

2.3 The Radially Symmetric, Cubic Problem in Four Dimensions

Theorem 2.11 *There exists $L(E, C(E - \frac{\varepsilon_0^2}{8})) \in \mathbf{Z}$ such that if for some $t_0 \in I$ there exists $j_1 > j(t_0) + L(E, C(E - \frac{\varepsilon_0^2}{8}))$ or $j_1 < j(t_0) - L(E, C(E - \frac{\varepsilon_0^2}{8}))$ satisfying*

$$\|P_{j_1} u(t_0)\|_{\dot{H}^1(\mathbf{R}^4)} \geq \frac{\varepsilon_0}{2},$$

then $I = \mathbf{R}$ and

$$\|u\|_{L^6_{t,x}(\mathbf{R} \times \mathbf{R}^4)} \lesssim_{C(E - \frac{\varepsilon_0^2}{8})} 1.$$

Proof If $u(t_0)$ satisfies the conditions of the theorem, then $u(t_0)$ may be split into two pieces with energy strictly less than E. By induction, each piece will yield a global, scattering solution, and the two solutions will interact very weakly, and can be estimated using Theorem 2.6.

For any $l \in \mathbf{Z}$, $j(t_0) \leq l \leq j_1$,

$$\|u(t_0)\|^2_{\dot{H}^1(\mathbf{R}^4)} = \|u_{\leq l}(t_0)\|^2_{\dot{H}^1(\mathbf{R}^4)} + \|u_{>l}(t_0)\|^2_{\dot{H}^1(\mathbf{R}^4)} + 2\langle u_{\leq l}(t_0), u_{>l}(t_0)\rangle_{\dot{H}^1(\mathbf{R}^4)}, \tag{2.44}$$

and

$$\|u(t_0)\|^4_{L^4_x(\mathbf{R}^4)} = \|u_{\leq l}(t_0)\|^4_{L^4_x(\mathbf{R}^4)} + \|u_{>l}(t_0)\|^4_{L^4_x(\mathbf{R}^4)}$$
$$+ O\left(\||u_{>l}(t_0)||u_{\leq l}(t_0)||u(t_0)|^2\|_{L^1_x(\mathbf{R}^4)}\right).$$

By the Cauchy–Schwarz and Bernstein inequalities,

$$\frac{1}{L-1} \sum_{j(t_0) < l < j(t_0)+L} 2|\langle u_{>l}(t_0), u_{\leq l}(t_0)\rangle_{\dot{H}^1(\mathbf{R}^4)}| \lesssim \frac{1}{L-1} \|u(t_0)\|^2_{\dot{H}^1(\mathbf{R}^4)}$$

and

$$\frac{1}{L-1} \sum_{j(t_0) < l < j(t_0)+L} \|u(t_0)\|^2_{L^4_x} \|u_{>l}(t_0)\|_{L^2_x} \|u_{\leq l}(t_0)\|_{L^\infty_x}$$

$$\lesssim \frac{1}{L-1} \|u(t_0)\|^2_{\dot{H}^1} \sum_{j(t_0) < l < j(t_0)+L} \sum_{k \leq l < m} 2^{k-m} \|P_k u(t_0)\|_{\dot{H}^1} \|P_m u(t_0)\|_{\dot{H}^1}$$

$$\lesssim \frac{1}{L-1} \|u(t_0)\|^4_{\dot{H}^1(\mathbf{R}^4)}. \tag{2.45}$$

Therefore, for $L(E)$ sufficiently large, there exists $j(t_0) < l < j(t_0) + L$ such that

$$E(u_{\leq l}(t_0)), \; E(u_{\geq l}(t_0)) \leq E - \frac{\varepsilon_0^2}{8}.$$

Then by the inductive hypothesis, (2.35) with initial data $u_{\leq l}(t_0)$ and $u_{>l}(t_0)$ has a global solution with $L^6_{t,x}$-norm bounded by $C(E - \frac{\varepsilon^2}{8})$. Possibly after making L larger and combining this fact with Theorem 2.6 and a bilinear Strichartz estimate, the scattering norm is bounded by the right-hand side of (2.38). The argument in the case that $j_1 < j(t_0) - L$ is identical.

Theorem 2.12 (Bilinear Strichartz estimate) For $u_0, v_0 \in H^1(\mathbf{R}^4)$, $j \leq k$,

$$\left\| \left(P_j e^{it\Delta} u_0 \right) \left(P_k e^{it\Delta} v_0 \right) \right\|_{L^2_{t,x}(\mathbf{R} \times \mathbf{R}^4)} \lesssim \frac{2^{3j/2}}{2^{k/2}} \|P_j u_0\|_{L^2} \|P_k v_0\|_{L^2}.$$

This theorem will be proved later (in Theorem 4.6). □

Therefore, it only remains to consider the case when for any $t \in I$, $j(t) < \infty$,

$$\sup_{j \geq j(t)+L} \|P_j u(t)\|_{\dot{H}^1} < \frac{\varepsilon_0}{2} \qquad (2.46)$$

and

$$\sup_{j \leq j(t)-L} \|P_j u(t)\|_{\dot{H}^1} < \frac{\varepsilon_0}{2}. \qquad (2.47)$$

The fact that $j(t) < \infty$ for all $t \in I$, combined with (2.46) and (2.47), gives local control over the change of $j(t)$. Let ε_1 be a constant such that

$$0 < \varepsilon_1 \ll \varepsilon_0. \qquad (2.48)$$

Suppose J is an interval such that

$$\|u\|_{L^6_{t,x}(J \times \mathbf{R}^4)} \leq \varepsilon_1.$$

Then by Strichartz estimates,

$$\|\nabla u\|_{L^2_t L^4_x(J \times \mathbf{R}^4)} \lesssim \|u\|_{L^\infty_t \dot{H}^1(J \times \mathbf{R}^4)} + \|\nabla u\|_{L^2_t L^4_x(J \times \mathbf{R}^4)} \|u\|^2_{L^6_{t,x}(J \times \mathbf{R}^4)}, \qquad (2.49)$$

which implies

$$\|\nabla u\|_{L^2_t L^4_x(J \times \mathbf{R}^4)} \lesssim E^{1/2}.$$

Therefore, for any $t_1, t_2 \in J$,

$$\left\| \nabla \int_{t_1}^{t_2} e^{i(t_2-\tau)\Delta} F(u(\tau)) d\tau \right\|_{\dot{H}^1} \lesssim \varepsilon_1^2 E^{1/2} \ll \varepsilon_1 \ll \varepsilon_0,$$

which by (2.46) and (2.47) proves that for all $t_1, t_2 \in J$,

$$\sup_{j \geq j(t_1)+L} \|P_j u(t_2)\|_{\dot{H}^1(\mathbf{R}^4)} < \frac{3\varepsilon_0}{4} \quad \text{and} \quad \sup_{j \leq j(t_1)-L} \|P_j u(t_2)\|_{\dot{H}^1(\mathbf{R}^4)} < \frac{3\varepsilon_0}{4},$$

2.3 The Radially Symmetric, Cubic Problem in Four Dimensions

and therefore
$$|j(t_1) - j(t_2)| \leq L. \tag{2.50}$$

For each $t \in I$, by Duhamel's principle,
$$u(t) = e^{it\Delta}u_0 - i \int_0^t e^{i(t-\tau)\Delta} F(u(\tau)) d\tau.$$

By Strichartz estimates, $\|e^{it\Delta}u_0\|_{L^6_{t,x}(\mathbf{R}\times\mathbf{R}^4)} \lesssim E^{1/2}$, so when $\|u\|_{L^6_{t,x}(I\times\mathbf{R}^4)}$ is very large, the contribution of $e^{it\Delta}u_0$ to $\|u\|_{L^6_{t,x}}$ is small on average. To see this, partition I into $\sim \|u_0\|_{H^1}^6 / \varepsilon_1^6$ pieces I_k such that for each I_k,
$$\|e^{it\Delta}u_0\|_{L^6_{t,x}(I_k\times\mathbf{R}^4)} \leq \varepsilon_1.$$

To prove (2.38), it suffices to prove that there exists a bound uniform in k,
$$\|u\|_{L^6_{t,x}(I_k\times\mathbf{R}^4)} \lesssim C(E - \varepsilon(E)). \tag{2.51}$$

Next partition each I_k into subintervals $J_{k,l}$ such that on each $J_{k,l}$,
$$\|u\|_{L^6_{t,x}(J_{k,l}\times\mathbf{R}^4)} \leq \varepsilon_1. \tag{2.52}$$

For each $J_{k,l}$ define
$$N(J_{k,l}) = \inf_{t \in J_{k,l}} 2^{j(t)}. \tag{2.53}$$

Inequality (2.50) implies that (2.53) is well defined.

All but a finite number of the intervals $J_{k,l} \subset I_k$ have measure bounded by a constant times $N(J_{k,l})^{-2}$.

Lemma 2.13 *Suppose $J_{k,l} = [a_{k,l}, b_{k,l}]$. If $|J_{k,l}| \geq \frac{1}{N(J_{k,l})^2 \varepsilon_1}$, where $|J_{k,l}| = b_{k,l} - a_{k,l}$, then*
$$\left\| e^{i(t-b_{k,l})\Delta} u(b_{k,l}) \right\|_{L^6_{t,x}(I_k \cap [b_{k,l},\infty))} \ll 1.$$

Proof By (2.47) and Lemma 2.10,
$$\left\| e^{i(t-b_{k,l})\Delta} P_{\leq j(b_{k,l})-L} u(b_{k,l}) \right\|_{L^6_{t,x}(\mathbf{R}\times\mathbf{R}^4)} \ll 1. \tag{2.54}$$

By Duhamel's principle,
$$e^{i(t-b_{k,l})\Delta} P_{\geq j(b_{k,l})-L} u(b_{k,l}) = e^{it\Delta} P_{\geq j(b_{k,l})-L} u_0$$
$$- i \int_0^{a_{k,l}} e^{i(t-\tau)\Delta} P_{\geq j(b_{k,l})-L} F(u(\tau)) d\tau$$
$$- i \int_{a_{k,l}}^{b_{k,l}} e^{i(t-\tau)\Delta} P_{\geq j(b_{k,l})-L} F(u(\tau)) d\tau.$$

By definition of I_k and the Littlewood–Paley theorem,

$$\left\| e^{it\Delta} P_{\geq j(b_{k,l}) - L} u_0 \right\|_{L^6_{t,x}(I_k \times \mathbf{R}^3)} \lesssim \varepsilon_1. \tag{2.55}$$

By Strichartz estimates and (2.52),

$$\left\| \int_{a_{k,l}}^{b_{k,l}} e^{i(t-\tau)\Delta} P_{\geq j(b_{k,l}) - L} F(u(\tau)) d\tau \right\|_{L^6_{t,x}(\mathbf{R} \times \mathbf{R}^4)}$$
$$\lesssim \|\nabla u\|_{L^2_t L^4_x (J_{k,l} \times \mathbf{R}^4)} \|u\|^2_{L^6_{t,x}(J_{k,l} \times \mathbf{R}^4)} \lesssim E^{1/2} \varepsilon_1^2. \tag{2.56}$$

By dispersive estimates and the product rule,

$$\left\| -i\nabla \int_0^{a_{k,l}} e^{i(t-\tau)\Delta} P_{\geq j(b_{k,l}) - L} F(u(\tau)) d\tau \right\|_{L^\infty_x}$$
$$\lesssim \|\nabla u\|_{L^\infty_t L^2_x} \|u\|^2_{L^\infty_t L^4_x} \left(\int_0^{a_{k,l}} \frac{1}{(t-\tau)^2} d\tau \right) \lesssim \frac{E}{(t-a_{k,l})}, \tag{2.57}$$

and by conservation of energy,

$$\left\| -i\nabla \int_0^{a_{k,l}} e^{i(t-\tau)\Delta} P_{\geq j(b_{k,l}) - L} F(u(\tau)) d\tau \right\|_{L^2_x} \lesssim \|u(a_{k,l})\|_{\dot{H}^1} + \|u_0\|_{\dot{H}^1}$$
$$\lesssim E^{1/2}. \tag{2.58}$$

Interpolating (2.57) and (2.58),

$$\left\| -i\nabla \int_0^{a_{k,l}} e^{i(t-\tau)\Delta} P_{\geq j(b_{k,l}) - L} F(u(\tau)) d\tau \right\|_{L^6_{t,x}(I_k \cap [b_{k,l},\infty))} \lesssim \frac{E^{5/6}}{(b_{k,l} - a_{k,l})^{1/2}}$$
$$\lesssim E^{5/6} 2^{j(b_{k,l})} \varepsilon_1^{1/2}.$$

Therefore, by Bernstein's inequality,

$$\left\| -i \int_0^{a_{k,l}} e^{i(t-\tau)\Delta} P_{\geq j(b_{k,l}) - L} F(u(\tau)) d\tau \right\|_{L^6_{t,x}(I_k \cap [b_{k,l},\infty))} \lesssim E^{5/6} 2^L \varepsilon_1^{1/2}. \tag{2.59}$$

Combining (2.54), (2.55), (2.56), and (2.59) proves the lemma. $\square$

Then by Lemma 2.3,

$$\|u\|_{L^6_{t,x}(I_k \cap [a_{k,l},\infty) \times \mathbf{R}^3)} \leq 1.$$

Thus, to prove (2.51), it suffices to prove that (2.51) holds on the subinterval $I_k^{(0)} \subset I_k$, where $I_k^{(0)}$ is the union of consecutive intervals $J_{k,l} \subset I_k$ that satisfy

$$|J_{k,l}| \leq \frac{1}{\varepsilon_1 N(J_{k,l})^2}. \tag{2.60}$$

2.3 The Radially Symmetric, Cubic Problem in Four Dimensions

Partition $I_k^{(0)}$ into a union of consecutive intervals $I_{k,l}$ for which

$$\frac{1}{2} \leq \|u\|_{L_{t,x}^6(I_{k,l} \times \mathbf{R}^4)} \leq 1.$$

By (2.50),

$$|j(t_1) - j(t_2)| \leq L\varepsilon_1^{-6} \quad \text{for all } t_1, t_2 \in I_{k,l}. \tag{2.61}$$

Set $N(I_{k,l}) = \sup_{t \in I_{k,l}} 2^{j(t)}$. By (2.60) and (2.61),

$$|I_{k,l}| \lesssim_{\varepsilon_1} \frac{1}{N(I_{k,l})^2}. \tag{2.62}$$

Meanwhile, by the Sobolev embedding theorem, for any $t_0 \in I_{k,l}$,

$$\left\|e^{i(t-t_0)\Delta} P_{\leq j(t_0)+L} u(t_0)\right\|_{L_x^6(\mathbf{R}^4)} \lesssim 2^{\frac{j(t_0)+L}{3}} \|u(t_0)\|_{\dot{H}^1(\mathbf{R}^4)}. \tag{2.63}$$

Therefore, combining Hölder's inequality in time, Theorem 2.4, (2.46), (2.47), and (2.63) implies $|I_{k,l}| \gtrsim N(I_{k,l})^{-2}$, and therefore

$$|I_{k,l}| \sim N(I_{k,l})^{-2}. \tag{2.64}$$

The bound (2.64) necessitates a positive lower bound on

$$\frac{1}{|I_{k,l}|} \int_{I_{k,l}} \|u(t)\|_{L^4}^4 \, dt. \tag{2.65}$$

By (2.48), (2.49), and (2.62),

$$\|\nabla u\|_{L_t^2 L_x^4(I_{k,l} \times \mathbf{R}^4)} \lesssim E^{1/2}, \tag{2.66}$$

so by interpolation and the Sobolev embedding theorem,

$$1 \lesssim \|u\|_{L_{t,x}^6(I_{k,l} \times \mathbf{R}^4)} \lesssim \||\nabla|^{1/3} u\|_{L_t^6 L_x^4(I_{k,l} \times \mathbf{R}^4)}$$
$$\lesssim \|\nabla u\|_{L_t^2 L_x^4(I_{k,l} \times \mathbf{R}^4)}^{1/3} \|u\|_{L_t^\infty L_x^4(I_{k,l} \times \mathbf{R}^4)}^{2/3} \lesssim E^{1/6} \|u\|_{L_t^\infty L_x^4(I_{k,l} \times \mathbf{R}^4)}^{2/3}.$$

Therefore, by interpolation, $\|u\|_{L_t^\infty L_x^4(I_{k,l} \times \mathbf{R}^4)} \gtrsim_E 1$. Next, as in (2.41) and (2.42),

$$\|u(t_0)\|_{L_x^4(\mathbf{R}^4)}^4 \lesssim \sum_{\substack{j_1 \leq j_2 \\ \leq j_3 \leq j_4}} \|P_{j_1} u(t_0)\|_{L_x^\infty} \|P_{j_2} u(t_0)\|_{L_x^4} \|P_{j_3} u(t_0)\|_{L_x^4} \|P_{j_4} u(t_0)\|_{L_x^2}$$

$$\lesssim \|u(t_0)\|_{\dot{H}^1}^2 \left(\sup_j \|P_j u(t_0)\|_{L_x^4}^2\right). \tag{2.67}$$

Combining (2.46), (2.47), and (2.67), there exists $t_0 \in I_{k,l}$ such that

$$\sup_{j(t_0)-L \leq j \leq j(t_0)+L} \|P_j u(t_0)\|_{L_x^4} \gtrsim 1.$$

Furthermore, inserting (2.35) and integrating by parts,

$$\left|\frac{d}{dt}\|P_ju(t)\|_{L^4_x}^4\right| \lesssim \int |P_ju(t,x)|^2 |\nabla(P_ju(t,x))|^2 dx$$
$$+ \int |P_ju(t,x)|^3 |P_j(|u|^2 u)(t,x)| dx \lesssim_E 2^{2j}. \quad (2.68)$$

Therefore, there exists some $c(E) > 0$ and some $j(t_0) - L \le j \le j(t_0) + L$ such that

$$\|P_ju(t)\|_{L^4_x} \gtrsim 1 \quad \text{for all } t \in [t_0 - c2^{-2j}, t_0 + c2^{-2j}]. \quad (2.69)$$

By (2.60), (2.69) represents a positive fraction of each subinterval $I_{k,l}$.

For any $t \in [t_0 - c2^{-2j}, t_0 + c2^{-2j}]$, $u(t)$ must be localized near the origin in an L^4 sense. By the fundamental theorem of calculus and Hölder's inequality,

$$|f(x)| \lesssim \int_{|x|}^\infty |f'(r)| dr$$
$$\le \sum_{j \ge 0} \int_{2^j|x|}^{2^{j+1}|x|} |f'(r)| dr \le \sum_{j \ge 0} \left(\int_{2^j|x|}^{2^{j+1}|x|} |f'(r)|^2 r^3 dr\right)^{1/2} \cdot 2^{-j}|x|^{-1}$$
$$\lesssim \frac{1}{|x|} \|f\|_{\dot{H}^1(\mathbf{R}^4)}. \quad (2.70)$$

Therefore, for $C(E) > 0$ sufficiently large, by (2.70) and Bernstein's inequality,

$$\left\|\chi\left(\frac{2^j x}{C}\right) P_ju(t)\right\|_{L^4_x(\mathbf{R}^4)}$$
$$\ge \|P_ju(t)\|_{L^4_x(\mathbf{R}^4)} - \frac{2^{j/2}}{C^{1/2}} \|P_ju(t)\|_{L^2_x(\mathbf{R}^4)}^{1/2} \||x|P_ju(t)\|_{L^\infty_x(\mathbf{R}^4)}^{1/2}$$
$$\gtrsim_E 1.$$

Furthermore,

$$\left\|\left[\chi\left(\frac{2^j x}{C}\right), P_j\right] u\right\|_{L^4_x}$$
$$= \left\|2^{4j} \int K(2^j(x-y)) \left[\chi\left(\frac{2^j x}{C}\right) - \chi\left(\frac{2^j y}{C}\right)\right] u(y) dy\right\|_{L^4_x}$$
$$\lesssim \left\|2^{4j} \int K(2^j(x-y)) \frac{2^j|x-y|}{C} u(y) dy\right\|_{L^4_x} \lesssim \frac{1}{C}\|u\|_{L^4_x}.$$

Therefore, for $C(E)$ sufficiently large,

$$\left\|P_j\left(\chi\left(\frac{2^j x}{C}\right) u\right)\right\|_{L^4_x} \gtrsim_E 1,$$

2.3 The Radially Symmetric, Cubic Problem in Four Dimensions

so by the Littlewood–Paley theorem,

$$\int \frac{1}{|x|} |u(t,x)|^4 dx \gtrsim \frac{2^j}{C}. \tag{2.71}$$

Then (2.62), (2.69), and (2.71) imply

$$\int_{I_{k,l}} \int \frac{1}{|x|} |u(t,x)|^4 dx dt \gtrsim \frac{1}{C} 2^{-j} \sim_{\varepsilon_1, E} |I_{k,l}|^{1/2}. \tag{2.72}$$

Meanwhile, by the Morawetz estimate in (2.28),

$$\int_{I_k^{(0)}} \int \frac{|u(t,x)|^4}{|x|} dx dt \lesssim E |I_k^{(0)}|^{1/2},$$

which by (2.72) implies

$$\sum_{I_{k,l} \subset I_k^{(0)}} |I_{k,l}|^{1/2} \lesssim_{\varepsilon_1, E} |I_k^{(0)}|^{1/2}. \tag{2.73}$$

Observe that (2.73) gives an upper bound on the number of consecutive intervals $I_{k,l} \subset I_k^{(0)}$ of the same length. For example, if $|I_{k,l}| = 1$ for each l, then the number of $I_{k,l}$ intervals would be bounded by a constant that depends on E and $\varepsilon_1(E)$, which would then imply the desired (2.51) bound.

However, (2.73) does not automatically imply a uniform bound on the total number of $I_{k,l}$ subintervals of $I_k^{(0)}$ under the general constraint of (2.50). For example, one could take $|I_{k,l}| = 2^{-l}$, $l \geq 0$ and then (2.73) would not give any kind of upper bound on the number of $I_{k,l}$ intervals.

However, in the second case, conservation of energy would give a bound. Indeed, by (2.73), the general fact that $l^1 \subset l^2 \subset l^\infty$, and Hölder's inequality,

$$\sum_{I_{k,l} \subset I_k^{(0)}} |I_{k,l}|^{1/2} \sim_{\varepsilon_1, E} |I_k^{(0)}|^{1/2} \quad \text{and} \quad \sup_{I_{k,l} \subset I_k^{(0)}} |I_{k,l}|^{1/2} \sim_{\varepsilon_1, E} |I_k^{(0)}|^{1/2}. \tag{2.74}$$

Therefore, (2.74) implies that there are $\leq N(\varepsilon_1, E)$ intervals $I_{k,l}$ such that

$$|I_{k,l}| \sim |I_k^{(0)}|, \tag{2.75}$$

where $N(\varepsilon_1, E)$ is some fixed quantity. If

$$\|u\|_{L^6_{t,x}\left(I_k^{(0)} \times \mathbf{R}^4\right)} = M,$$

with $M \gg N$, then by the pigeonhole principle, after removing the intervals $I_{k,l}$ that satisfy (2.75), there exists an interval $I_k^{(1)} \subset I_k^{(0)}$ such that

$$\|u\|_{L^6_{t,x}\left(I_k^{(1)} \times \mathbf{R}^4\right)} \geq \frac{M}{N}.$$

Moreover, by (2.74) there exists $\varepsilon_2(E) > 0$, $\varepsilon_2 \ll \varepsilon_1$ such that

$$|I_k^{(1)}| \leq (1 - \varepsilon_2)|I_k^{(0)}|.$$

Now if $\frac{M}{N} \gg N$ the same procedure may be applied to $I_k^{(1)}$. In fact, if $M \gg N^n$, there is a nested sequence of intervals

$$I_k^{(n)} \subset I_k^{(n-1)} \subset \cdots \subset I_k^{(0)},$$

for each $1 \leq m \leq n$,

$$|I_k^{(m)}| \leq (1 - \varepsilon_2)|I_k^{(m-1)}|, \tag{2.76}$$

and there exists

$$I_{k,l}^{(m)} \subset I_k^{(m)} \quad \text{such that} \quad |I_{k,l}^{(m)}| \sim_{\varepsilon_1, E} |I_k^{(m)}|. \tag{2.77}$$

Fix $t_0 \in I_k^{(n)}$. Relation (2.77) implies that for all $0 \leq m \leq n$, there exists $I_{k,l}^{(m)} \subset I_k^{(m)}$ such that

$$|I_{k,l}^{(m)}| \gtrsim_{\varepsilon_1, E} \text{dist}\left(t_0, I_{k,l}^{(m)}\right). \tag{2.78}$$

Inequality (2.78) implies that some of the energy at level $j(t)$ (see (2.43)) remains at t_0. By (1.112), for any $j \in \mathbf{Z}$,

$$\begin{aligned}
\frac{d}{dt}\|P_{>j}u(t)\|_{L_x^2}^2 &= \int \text{Im}\left(\overline{P_{>j}u}P_{>j}F(u)\right)dx \\
&= \int \text{Im}\left(\overline{P_{>j}u}(P_{>j}F(u) - F(P_{>j}u))\right)dx \\
&\lesssim \|P_{>j}u\|_{L_x^2}\|P_{\leq j}u\|_{L_x^\infty}\|u\|_{L_x^4}^2
\end{aligned} \tag{2.79}$$

and

$$\begin{aligned}
\frac{d}{dt}\|P_{\leq j}u(t)\|_{L_x^2}^2 &= \int \text{Im}\left(\overline{P_{\leq j}u}P_{\leq j}F(u)\right)dx \\
&= \int \text{Im}\left(\overline{P_{\leq j}u}(P_{\leq j}F(u) - F(P_{\leq j}u))\right)dx \\
&\lesssim \|P_{\leq j}u\|_{L_x^\infty}\|P_{>j}u\|_{L_x^2}\|u\|_{L_x^4}^2.
\end{aligned} \tag{2.80}$$

Choose $t_m \in I_{k,l}^{(m)}$. Assume without loss of generality that $t_m > t_0$. Now for $L_1 \geq 2L$, (2.45) and (2.62) imply

$$\frac{1}{L_1 - L}\int_{t_0}^{t_m} \sum_{j(t_m)-L_1 \leq j \leq j(t_m)-L} \|u(t)\|_{L_x^4(\mathbf{R}^4)}^2 \|P_{\leq j}u(t)\|_{L_x^\infty(\mathbf{R}^4)} \|P_{>j}u(t)\|_{L_x^2(\mathbf{R}^4)} dt$$

$$\lesssim_{\varepsilon_1, E} \frac{1}{L_1 - L}|I_{j,k}^{(m)}| \lesssim_{\varepsilon_1, E} \frac{1}{L_1 - L}2^{-2j(t_m)}. \tag{2.81}$$

2.3 The Radially Symmetric, Cubic Problem in Four Dimensions

Therefore, the pigeonhole principle implies that there exists some $j(t_m) - L_1 \leq j \leq j(t_m) - L$ such that the integral

$$\int_{t_0}^{t_m} \|P_{>j}u(t)\|_{L_x^2} \|P_{\leq j}u(t)\|_{L_x^\infty} \|u(t)\|_{L_x^4}^2 \, dt$$

is bounded by the right-hand side of (2.81). Then by definition of $j(t_m)$ (see (2.43)) and Bernstein's inequality, for $L_1(\varepsilon_1, E, L)$ sufficiently large, (2.79) and (2.80) imply

$$\|P_{j(t_m)-L_1 \leq \cdot \leq j(t_m)+L_1} u(t_0)\|_{L_x^2(\mathbf{R}^4)} \gtrsim \varepsilon_0 2^{-2j(t_m)-L},$$

and therefore

$$\|P_{j(t_m)-L_1 \leq \cdot \leq j(t_m)+L_1} u(t_0)\|_{\dot{H}_x^1(\mathbf{R}^4)} \gtrsim \varepsilon_0 2^{-L-L_1}. \tag{2.82}$$

Inequalities (2.66) and (2.76) also imply that there exists some constant $J(L_1, \varepsilon_2)$ such that if $M \gg N^n$, then there exist at least $\frac{n}{J(L_1, \varepsilon_2)}$ integers $0 \leq m' \leq n$ such that the intervals $j(t_{m'}) - L_1 \leq \cdot \leq j(t_{m'}) + L_1$ are disjoint.

Conservation of energy and (2.82) then give an upper bound on the number of such m', which in turn gives an upper bound on $\frac{n}{J(L_1, \varepsilon_2)}$, which finally gives an upper bound on M, such that

$$\|u\|_{L_{t,x}^6(I_k^{(0)} \times \mathbf{R}^4)} = M.$$

This finally concludes the proof of Theorem 2.9. $\square$

3

The Energy-Critical Problem in Higher Dimensions

3.1 Small Data Energy-Critical Problem

Plugging $s_c = 1$ into (2.3) shows that the general, energy-critical problem is given by (2.1) with $p = \frac{4}{d-2}$. In dimensions $d \geq 5$, this means that the energy-critical problem has the added difficulty that the nonlinearity is no longer smooth as a function of u. Rather, the function $F(x) = |x|^{\frac{4}{d-2}} x$ has a derivative that is only Hölder continuous of order $\frac{4}{d-2}$. This fact complicates the computation of $\|F(u) - F(v)\|_{\dot{N}^1}$, a necessary part of proving scattering for small data.

Remark The small data problem when $d = 3$ ($p = 4$) may be analyzed in a manner very similar to the analysis in dimension $d = 4$.

Theorem 3.1 (Global well-posedness and scattering for small critical data) *There exists $\varepsilon_0(d) > 0$ such that if $\|u_0\|_{\dot{H}^1(\mathbf{R}^d)} = \varepsilon < \varepsilon_0$, then*

$$\begin{aligned} iu_t + \Delta u = F(u) = \mu |u|^{\frac{4}{d-2}} u, \quad \mu = \pm 1, \\ u(0) = u_0 \in \dot{H}^1(\mathbf{R}^d) \end{aligned} \quad (3.1)$$

is globally well posed and scattering in $\dot{H}^1(\mathbf{R}^d)$. In fact, this result holds for

$$\left\| e^{it\Delta} u_0 \right\|_{L_{t,x}^{\frac{2(d+2)}{d-2}}(\mathbf{R} \times \mathbf{R}^d)} = \varepsilon \leq \varepsilon_0 \left(\|u_0\|_{\dot{H}^1(\mathbf{R}^d)} \right).$$

Proof This result was proved in Cazenave and Weissler (1988). The proof presented here is the proof in Tao and Visan (2005), which also contained a perturbative result for the energy-critical problem.

When $d = 3, 4, 5$, and denoting $\frac{2(d+2)}{d-2}$ by B, define the set of functions

$$X = \left\{ u : \|u\|_{\dot{S}^1(\mathbf{R} \times \mathbf{R}^d)} \leq C(d) \|u_0\|_{\dot{H}^1(\mathbf{R}^d)}; \quad \|u\|_{L_{t,x}^B(\mathbf{R} \times \mathbf{R}^d)} \leq 2\varepsilon_0 \right\}.$$

3.1 Small Data Energy-Critical Problem

By the chain rule and Strichartz estimates,

$$\|\Phi(u)\|_{\dot{S}^1(\mathbf{R}\times\mathbf{R}^d)} \lesssim \|u_0\|_{\dot{H}^1(\mathbf{R}^d)} + \|\nabla u\|_{L_{t,x}^{\frac{2(d+2)}{d}}(\mathbf{R}\times\mathbf{R}^d)} \|u\|_{L_{t,x}^B(\mathbf{R}\times\mathbf{R}^d)}^{\frac{4}{d-2}}$$

and

$$\|\Phi(u)\|_{L_{t,x}^B(\mathbf{R}\times\mathbf{R}^d)} \le \|e^{it\Delta}u_0\|_{L_{t,x}^B(\mathbf{R}\times\mathbf{R}^d)} + C(d)\|u\|_{L_{t,x}^B(\mathbf{R}\times\mathbf{R}^d)}^{\frac{4}{d-2}} \|u\|_{\dot{S}^1(\mathbf{R}\times\mathbf{R}^d)}.$$

Therefore, when $d = 3, 4, 5$, for $\varepsilon_0(d, \|u_0\|_{\dot{H}^1}) > 0$ sufficiently small, $\frac{4}{d-2} > 1$, $\varepsilon < \varepsilon_0$, and $u \in V$ imply that $\Phi(u) \in V$.

When $d \ge 6$, choose q satisfying

$$\frac{2}{q} + \frac{4}{d-2}\left(\frac{1}{q} - \frac{4}{d(d+2)}\right) = 1.$$

Then let q_1 satisfy

$$\frac{2}{q_1} = d\left(\frac{1}{2} - \frac{1}{q}\right) - \frac{d-2}{d+2},$$

and define the set of functions

$$X = \left\{u : \|u\|_{\dot{S}^1(\mathbf{R}\times\mathbf{R}^d)} \le C(d)\|u_0\|_{\dot{H}^1(\mathbf{R}^d)}; \quad \||\nabla|^{\frac{4}{d+2}} u\|_{L_t^{q_1} L_x^q(\mathbf{R}\times\mathbf{R}^d)} \le 2\varepsilon_1\right\}.$$

By the fractional chain rule, Duhamel's principle, the Sobolev embedding theorem, $\frac{1}{q} > \frac{1}{2} - \frac{1}{d}$, and Lemma 1.10,

$$\||\nabla|^{\frac{4}{d+2}} \Phi(u)\|_{L_t^{q_1} L_x^q(\mathbf{R}\times\mathbf{R}^d)}$$
$$\lesssim \||\nabla|^{\frac{4}{d+2}} e^{it\Delta} u_0\|_{L_t^{q_1} L_x^q(\mathbf{R}\times\mathbf{R}^d)} + \||\nabla|^{\frac{4}{d+2}} u\|_{L_t^{q_1} L_x^q(\mathbf{R}\times\mathbf{R}^d)}^{\frac{d+2}{d-2}}. \quad (3.2)$$

By the product rule, the Sobolev embedding theorem, and Strichartz estimates,

$$\|\Phi(u)\|_{\dot{S}^1(\mathbf{R}\times\mathbf{R}^d)} \lesssim \|u_0\|_{\dot{H}^1(\mathbf{R}^d)} + \|\nabla u\|_{L_t^2 L_x^{\frac{2d}{d-2}}(\mathbf{R}\times\mathbf{R}^d)} \||\nabla|^{\frac{4}{d+2}} u\|_{L_t^{q_1} L_x^q(\mathbf{R}\times\mathbf{R}^d)}^{\frac{4}{d-2}}. \quad (3.3)$$

By interpolation and Strichartz estimates, there exists $0 < \alpha(d) < 1$ such that

$$\||\nabla|^{\frac{4}{d+2}} e^{it\Delta} u_0\|_{L_t^{q_1} L_x^q(\mathbf{R}\times\mathbf{R}^d)} \lesssim \|u_0\|_{\dot{H}^1(\mathbf{R}^d)}^{1-\alpha} \|e^{it\Delta} u_0\|_{L_{t,x}^{\frac{2(d+2)}{d-2}}(\mathbf{R}\times\mathbf{R}^d)}^{\alpha} \quad (3.4)$$

and

$$\||\nabla|^{\frac{4}{d+2}} u\|_{L_t^{q_1} L_x^q(\mathbf{R}\times\mathbf{R}^d)} \lesssim \|u\|_{\dot{S}^1(\mathbf{R}\times\mathbf{R}^d)}^{1-\alpha} \|u\|_{L_{t,x}^{\frac{2(d+2)}{d-2}}(\mathbf{R}\times\mathbf{R}^d)}^{\alpha}, \quad (3.5)$$

and therefore $\varepsilon_1 \lesssim_{d, \|u_0\|_{\dot{H}^1}} \varepsilon_0^\alpha$. So, for $\varepsilon_0(d, \|u_0\|_{\dot{H}^1})$ small enough, $\Phi : X \to X$.

To prove a contraction, by Taylor's theorem,

$$F(u) - F(v)$$
$$= (v-u) \int_0^1 F_z(u + \tau(v-u)) d\tau + (\overline{v-u}) \int_0^1 F_{\bar{z}}(u + \tau(v-u)) d\tau, \quad (3.6)$$

where F_z and $F_{\bar{z}}$ are the complex derivatives

$$F_z = \frac{1}{2}\left(\frac{\partial F}{\partial x} - i\frac{\partial F}{\partial y}\right) \quad \text{and} \quad F_{\bar{z}} = \frac{1}{2}\left(\frac{\partial F}{\partial x} + i\frac{\partial F}{\partial y}\right).$$

By the chain rule,

$$\nabla F(u(x)) = F_z(u(x)) \nabla u(x) + F_{\bar{z}}(u(x)) \nabla \bar{u}(x).$$

For $3 \leq d \leq 5$, F_z and $F_{\bar{z}}$ are differentiable functions, so by the chain rule and product rule (and again using the notation B for $\frac{2(d+2)}{d-2}$),

$$\|\Phi(u) - \Phi(v)\|_{\dot{S}^1(\mathbf{R} \times \mathbf{R}^d)}$$
$$\lesssim_d \|u-v\|_{\dot{S}^1(\mathbf{R} \times \mathbf{R}^d)} \left(\|u\|_{\dot{S}^1(\mathbf{R} \times \mathbf{R}^d)} + \|v\|_{\dot{S}^1(\mathbf{R} \times \mathbf{R}^d)}\right) \||u| + |v|\|_{L_{t,x}^B(\mathbf{R} \times \mathbf{R}^d)}^{\frac{6-d}{d-2}}$$
$$\lesssim \varepsilon_0^{\frac{6-d}{d-2}} \left(\|u\|_{\dot{S}^1(\mathbf{R} \times \mathbf{R}^d)} + \|v\|_{\dot{S}^1(\mathbf{R} \times \mathbf{R}^d)}\right) \|u-v\|_{\dot{S}^1(\mathbf{R} \times \mathbf{R}^d)}.$$

This gives a contraction in $\dot{S}^1(\mathbf{R} \times \mathbf{R}^d)$ for $\varepsilon_0(d, \|u_0\|_{\dot{H}^1(\mathbf{R}^d)})$ sufficiently small and $d = 3, 4, 5$.

For dimensions $d \geq 6$ the proof will utilize (3.2). To simplify notation, write

$$\|u\|_{\mathscr{S}(I \times \mathbf{R}^d)} = \left\||\nabla|^{\frac{4}{d+2}} u\right\|_{L_t^{q_1} L_x^q(I \times \mathbf{R}^d)}.$$

By the fractional product rule, fractional chain rule, (3.2), and (3.6),

$$\|\Phi(u) - \Phi(v)\|_{\mathscr{S}(\mathbf{R} \times \mathbf{R}^d)}$$
$$\lesssim_d \|u-v\|_{\mathscr{S}(\mathbf{R} \times \mathbf{R}^d)} \left(\|u\|_{\mathscr{S}(\mathbf{R} \times \mathbf{R}^d)} + \|v\|_{\mathscr{S}(\mathbf{R} \times \mathbf{R}^d)}\right)^{\frac{4}{d-2}}$$
$$\lesssim \varepsilon_1^{\frac{4}{d-2}} \|u-v\|_{\mathscr{S}(\mathbf{R} \times \mathbf{R}^d)}.$$

This implies a contraction in $\mathscr{S}(\mathbf{R} \times \mathbf{R}^d)$ for $\varepsilon_1 > 0$ sufficiently small, which by (3.4) and (3.5) follows from $\varepsilon_0(d, \|u_0\|_{\dot{H}^1(\mathbf{R}^d)}) > 0$ sufficiently small. Thus there exists a unique $u \in \mathscr{S}(\mathbf{R} \times \mathbf{R}^d)$ satisfying

$$u(t) = e^{it\Delta} u_0 - i \int_0^t e^{i(t-\tau)\Delta} F(u(\tau)) d\tau.$$

By (3.3), $\|F(u)\|_{\dot{N}^1(\mathbf{R} \times \mathbf{R}^d)} \lesssim \|u_0\|_{\dot{H}^1(\mathbf{R}^d)}$, proving $u(t) \in C_t^0(\mathbf{R}, \dot{H}_x^1(\mathbf{R}^d))$.

3.1 Small Data Energy-Critical Problem

Furthermore, by the chain rule,

$$\|F(u) - F(v)\|_{\dot{N}^1(\mathbf{R} \times \mathbf{R}^d)}$$
$$\lesssim \|u - v\|_{\dot{S}^1(\mathbf{R} \times \mathbf{R}^d)} \left(\|u\|_{\mathscr{S}(\mathbf{R} \times \mathbf{R}^d)} + \|v\|_{\mathscr{S}(\mathbf{R} \times \mathbf{R}^d)} \right)^{\frac{4}{d-2}}$$
$$+ \|u - v\|_{\mathscr{S}(\mathbf{R} \times \mathbf{R}^d)}^{\frac{4}{d-2}} \left(\|u\|_{\dot{S}^1(\mathbf{R} \times \mathbf{R}^d)} + \|v\|_{\dot{S}^1(\mathbf{R} \times \mathbf{R}^d)} \right),$$

which will be useful later for perturbative results, since then,

$$\|u - v\|_{\dot{S}^1(\mathbf{R} \times \mathbf{R}^d)} \lesssim \|u_0\|_{\dot{H}^1(\mathbf{R}^d)} \|u - v\|_{\mathscr{S}(\mathbf{R} \times \mathbf{R}^d)}^{\frac{4}{d-2}} + \varepsilon_1^{\frac{4}{d-2}} \|u - v\|_{\dot{S}^1(\mathbf{R} \times \mathbf{R}^d)}.$$

Thus $\varepsilon_1 > 0$ is small and $\|u - v\|_{\mathscr{S}(\mathbf{R} \times \mathbf{R}^d)} \to 0$ implies

$$\|\Phi(u) - \Phi(v)\|_{\dot{S}^1(\mathbf{R} \times \mathbf{R}^d)} \to 0.$$

Continuous dependence on initial data again follows from a perturbation lemma.

Lemma 3.2 (Perturbation lemma) *Let I be a compact time interval and let w be an approximate solution to* (3.1),

$$(i\partial_t + \Delta)w = F(w) + e,$$

for some e. Suppose also that

$$\|w\|_{L_t^\infty \dot{H}^1(I \times \mathbf{R}^d)} \leq E.$$

Let u solve (3.1) *on* $I \times \mathbf{R}^d$, *and for* $t_0 \in I$ *let* $u(t_0) \in \dot{H}^1(\mathbf{R}^d)$ *be close to* $w(t_0)$,

$$\|u(t_0) - w(t_0)\|_{\dot{H}^1(\mathbf{R}^d)} \leq E',$$

for some $E' > 0$. *Moreover, assume the smallness conditions*

$$\|u\|_{\dot{W}(I \times \mathbf{R}^d)} = \|\nabla u\|_{L_{t,x}^{\frac{2(d+2)}{d}}(I \times \mathbf{R}^d)} \leq \varepsilon_0,$$

$$\|e\|_{\dot{N}^1(I \times \mathbf{R}^d)} \leq \varepsilon,$$

and

$$\|e^{i(t-t_0)\Delta}(u(t_0) - w(t_0))\|_{\dot{W}(\mathbf{R} \times \mathbf{R}^d)} \leq \varepsilon, \quad (3.7)$$

for some $0 < \varepsilon \leq \varepsilon_0(E, E')$. *Then*

$$\|u - w\|_{\dot{W}(I \times \mathbf{R}^d)} \lesssim \varepsilon + \varepsilon^{\frac{d}{d-2}\left(\frac{4}{d-2} - \frac{2}{d}\right)}, \quad (3.8)$$

$$\|u-w\|_{\dot{S}^1(I\times\mathbf{R}^d)} \lesssim E' + \varepsilon + \varepsilon^{\frac{d}{d-2}\left(\frac{4}{d-2}-\frac{2}{d}\right)}, \tag{3.9}$$

$$\|w\|_{\dot{S}^1(I\times\mathbf{R}^d)} \lesssim E + E' + \varepsilon + \varepsilon^{\frac{d}{d-2}\left(\frac{4}{d-2}-\frac{2}{d}\right)}, \tag{3.10}$$

and

$$\|(i\partial_t + \Delta)(u-w) + e\|_{\dot{N}^1(I\times\mathbf{R}^d)} \lesssim \varepsilon + \varepsilon^{\frac{d}{d-2}\left(\frac{4}{d-2}-\frac{2}{d}\right)}. \tag{3.11}$$

Taking $e = 0$ implies

$$\|u-w\|_{\dot{S}^1(\mathbf{R}\times\mathbf{R}^d)} \lesssim \|u(t_0) - w(t_0)\|_{\dot{H}^1(\mathbf{R}^d)} + \|u(t_0) - w(t_0)\|_{\dot{H}^1(\mathbf{R}^d)}^{\frac{d}{d-2}\left(\frac{4}{d-2}-\frac{2}{d}\right)}, \tag{3.12}$$

which proves continuous dependence on the initial data.

To prove scattering forward in time, since $F(u) \in \dot{N}^1(\mathbf{R} \times \mathbf{R}^d)$, set

$$u_+ = u(t_0) - i\int_{t_0}^{\infty} e^{-it\Delta} F(u(t))\,dt.$$

Then

$$\lim_{T\to\infty} \|F(u)\|_{\dot{N}^1([T,\infty)\times\mathbf{R}^d)} = 0,$$

so

$$\|e^{iT\Delta}u_+ - u(T)\|_{\dot{H}^1(\mathbf{R}^d)} = \left\|\int_T^{\infty} e^{i(t-\tau)\Delta} F(u(\tau))\,d\tau\right\|_{\dot{H}^1(\mathbf{R}^d)} \to 0.$$

The argument proving scattering backward in time is identical. This completes the proof Theorem 3.1, assuming Lemma 3.2 is true. □

Proof of Lemma 3.2 The proof of Lemma 3.2 is also simpler for $3 \leq d \leq 5$ than for $d \geq 6$. By Strichartz estimates, if $v = u - w$,

$$\|v\|_{\dot{W}(I\times\mathbf{R}^d)}$$
$$\lesssim \|e^{i(t-t_0)\Delta}(w(t_0) - u(t_0))\|_{\dot{W}(I\times\mathbf{R}^d)} + \|F(w) - F(u) + e\|_{\dot{N}^1(I\times\mathbf{R}^d)}.$$

When $3 \leq d \leq 5$, applying the product rule to $F(w) - F(u)$,

$$\|F(w) - F(u)\|_{\dot{N}^1(I\times\mathbf{R}^d)} \lesssim \|v\|_{\dot{W}(I\times\mathbf{R}^d)} \left(\|u\|_{\dot{W}(I\times\mathbf{R}^d)}^{\frac{4}{d-2}} + \|w\|_{\dot{W}(I\times\mathbf{R}^d)}^{\frac{4}{d-2}}\right).$$

This implies

$$\|v\|_{\dot{W}(I\times\mathbf{R}^d)} \lesssim \varepsilon + \|v\|_{\dot{W}(I\times\mathbf{R}^d)} \left(\varepsilon_0^{\frac{4}{d-2}} + \|v\|_{\dot{W}(I\times\mathbf{R}^d)}^{\frac{4}{d-2}}\right),$$

3.1 Small Data Energy-Critical Problem

which proves (3.8) and (3.11). Next,

$$\|u-w\|_{\dot{S}^1(I\times\mathbf{R}^d)} \lesssim \|w(t_0)-u(t_0)\|_{\dot{H}^1(\mathbf{R}^d)} + \|F(w)-F(u)+e\|_{\dot{N}^1(I\times\mathbf{R}^d)}$$
$$\lesssim E' + \varepsilon.$$

This takes care of (3.9). Finally, since

$$\|u\|_{\dot{S}^1(I\times\mathbf{R}^d)} \lesssim \|u(t_0)-w(t_0)\|_{\dot{H}^1(\mathbf{R}^d)} + \|w(t_0)\|_{\dot{H}^1(\mathbf{R}^d)}$$
$$+ \|\nabla u\|_{L_{t,x}^{\frac{2(d+2)}{d}}(I\times\mathbf{R}^d)} \|u\|_{\dot{S}^1(I\times\mathbf{R}^d)}^{\frac{4}{d-2}},$$

we have

$$\|u\|_{\dot{S}^1(I\times\mathbf{R}^d)} \lesssim E + E' + \varepsilon,$$

which proves (3.10).

For $d \geq 6$, since $\partial_t e^{it\Delta} = i\Delta e^{it\Delta}$, the Sobolev embedding theorem in time and (3.7) imply

$$\|e^{i(t-t_0)\Delta}v(t_0)\|_{\mathscr{S}(I\times\mathbf{R}^d)} \lesssim \varepsilon.$$

This implies

$$\|v\|_{\mathscr{S}(I\times\mathbf{R}^d)} \lesssim \|e^{i(t-t_0)\Delta}v(t_0)\|_{\dot{W}(I\times\mathbf{R}^d)} + \|w-u\|_{\mathscr{S}(I\times\mathbf{R}^d)}$$
$$\times \left(\|w\|_{\mathscr{S}(I\times\mathbf{R}^d)}^{\frac{4}{d-2}} + \|u\|_{\mathscr{S}(I\times\mathbf{R}^d)}^{\frac{4}{d-2}} \right) + \|e\|_{\dot{N}^1(I\times\mathbf{R}^d)},$$

which implies

$$\|v\|_{\mathscr{S}(I\times\mathbf{R}^d)} \lesssim \varepsilon + \|v\|_{\mathscr{S}(I\times\mathbf{R}^d)} \left(\|u\|_{\dot{W}(I\times\mathbf{R}^d)}^{\frac{4}{d-2}} + \|v\|_{\dot{W}(I\times\mathbf{R}^d)}^{\frac{4}{d-2}} \right)$$
$$\lesssim \varepsilon + \varepsilon_0^{\frac{4}{d-2}} \|v\|_{\mathscr{S}(I\times\mathbf{R}^d)} + \|v\|_{\mathscr{S}(I\times\mathbf{R}^d)} \|v\|_{\dot{W}(I\times\mathbf{R}^d)}^{\frac{4}{d-2}}. \quad (3.13)$$

By Strichartz estimates and the product rule,

$$\|v\|_{\dot{W}(I\times\mathbf{R}^d)} \lesssim \varepsilon + \|v\|_{\dot{W}(I\times\mathbf{R}^d)} \|u\|_{\dot{W}(I\times\mathbf{R}^d)}^{\frac{4}{d-2}}$$
$$+ \|v\|_{\mathscr{S}(I\times\mathbf{R}^d)}^{\frac{4}{d-2}-\frac{2}{d}} \|v\|_{\dot{W}(I\times\mathbf{R}^d)}^{\frac{2}{d}} \left(\|u\|_{\dot{W}(I\times\mathbf{R}^d)} + \|v\|_{\dot{W}(I\times\mathbf{R}^d)} \right)$$
$$\lesssim \varepsilon + \varepsilon_0^{\frac{4}{d-2}} \|v\|_{\dot{W}(I\times\mathbf{R}^d)} + \varepsilon_0 \|v\|_{\dot{W}(I\times\mathbf{R}^d)}^{\frac{2}{d}} \|v\|_{\mathscr{S}(I\times\mathbf{R}^d)}^{\frac{4}{d-2}-\frac{2}{d}}$$
$$+ \|v\|_{\mathscr{S}(I\times\mathbf{R}^d)}^{\frac{4}{d-2}-\frac{2}{d}} \|v\|_{\dot{W}(I\times\mathbf{R}^d)}^{1+\frac{2}{d}}. \quad (3.14)$$

Therefore,

$$\|v\|_{\dot W} + \|v\|_{\mathscr{S}} \lesssim \varepsilon + \varepsilon_0 \left(\|v\|_{\mathscr{S}} + \|v\|_{\dot W}\right)^{\frac{4}{d-2}} + \left(\|v\|_{\mathscr{S}} + \|v\|_{\dot W}\right)^{\frac{d+2}{d-2}},$$

which implies

$$\|v\|_{\dot W} + \|v\|_{\mathscr{S}} \lesssim \varepsilon + \varepsilon_0^{\frac{d-2}{d-6}}.$$

Then, since $\varepsilon < \varepsilon_0$ and ε_0 is small,

$$\|v\|_{\mathscr{S}(I\times\mathbf{R}^d)} + \|v\|_{\dot W} \lesssim \varepsilon_0. \tag{3.15}$$

Plugging (3.15) into (3.13) implies

$$\|v\|_{\mathscr{S}(I\times\mathbf{R}^d)} \lesssim \varepsilon. \tag{3.16}$$

Then plugging (3.16) into (3.14) implies

$$\|v\|_{\dot W(I\times\mathbf{R}^d)} \lesssim \varepsilon^{\frac{d}{d-2}\cdot\left(\frac{4}{d-2}-\frac{2}{d}\right)}.$$

This completes the proof of Lemma 3.2. $\square$

Combining the modifications in higher dimensions used in the above theorem, and arguing as in Theorems 2.4 and 2.5, we obtain the following theorems.

Theorem 3.3 (Energy-critical local well-posedness)

1. For $u_0 \in \dot H^1(\mathbf{R}^d)$, there exists $T(u_0) > 0$ such that (3.1) is locally well posed in $\dot H^1(\mathbf{R}^d)$ on $[0,T]$. In particular, (3.1) is well posed on an open interval $I \subset \mathbf{R}$, $0 \in I$.
2. If $\sup(I) < +\infty$, where I is the maximal interval of existence of (3.1),

$$\lim_{T \nearrow \sup(I)} \|u\|_{L_{t,x}^{\frac{2(d+2)}{d-2}}([0,T]\times\mathbf{R}^d)} = \infty.$$

The corresponding result holds for $\inf(T) > -\infty$.

Theorem 3.4 (Scattering) A solution to (3.1) scatters forward in time to some $e^{it\Delta}u_+$, $u_+ \in \dot H_x^1(\mathbf{R}^d)$, if and only if, for some $t_0 \in \mathbf{R}$,

$$\|u\|_{L_{t,x}^{\frac{2(d+2)}{d-2}}([t_0,\infty)\times\mathbf{R}^d)} < \infty \quad \text{and} \quad \|u\|_{L_t^\infty \dot H^1([t_0,\infty)\times\mathbf{R}^d)} < \infty.$$

The backward-in-time scattering result is similar.

Proof of long-time perturbative results is also more difficult in dimensions $d \geq 5$.

3.1 Small Data Energy-Critical Problem

Theorem 3.5 (Long-time perturbations) *Let I be a compact time interval and let w be an approximate solution to (3.1) on $I \times \mathbf{R}^d$ in the sense that*

$$(i\partial_t + \Delta)w = F(w) + e.$$

Assume that u solves (3.1),

$$\|u\|_{L_{t,x}^{\frac{2(d+2)}{d-2}}(I \times \mathbf{R}^d)} \leq M,$$

and

$$\|u\|_{L_t^\infty \dot H^1(I \times \mathbf{R}^d)} \leq E,$$

for some $M, E > 0$. Let $t_0 \in I$, and suppose that $u(t_0)$ is close to $w(t_0)$,

$$\|u(t_0) - w(t_0)\|_{\dot H^1(\mathbf{R}^d)} \leq E'. \tag{3.17}$$

Also assume that

$$\|e^{i(t-t_0)\Delta}(u(t_0) - w(t_0))\|_{\dot W(I \times \mathbf{R}^d)} \leq \varepsilon$$

and

$$\|e\|_{\dot N^1(I \times \mathbf{R}^d)} \leq \varepsilon,$$

for some $0 < \varepsilon \leq \varepsilon_1(E, E', M, d)$. Then

$$\|u - w\|_{\dot W(I \times \mathbf{R}^d)} \leq C(E, E', M, d)\left(\varepsilon + \varepsilon^{\frac{d}{d-2}\left(\frac{4}{d-2} - \frac{2}{d}\right)}\right),$$

$$\|u - w\|_{\dot S^1(I \times \mathbf{R}^d)} \leq C(E, E', M, d)\left(E' + \varepsilon + \varepsilon^{\frac{d}{d-2}\left(\frac{4}{d-2} - \frac{2}{d}\right)}\right),$$

and

$$\|w\|_{\dot S^1(I \times \mathbf{R}^d)} \leq C(E, E', M, d).$$

Proof This result was also proved in Tao and Visan (2005). Divide I into $\sim \left(1 + \frac{M}{\varepsilon_0}\right)^{\frac{2(d+2)}{d-2}}$ subintervals J_k such that on each J_k,

$$\|u\|_{L_{t,x}^{\frac{2(d+2)}{d-2}}(J_k \times \mathbf{R}^d)} \leq \varepsilon_0,$$

where $\varepsilon_0(d)$ is a small fixed constant. For each $J_k = [a_k, b_k]$,

$$\|u\|_{\dot S^1(J_k \times \mathbf{R}^d)} \lesssim \|u(a_k)\|_{\dot H^1(\mathbf{R}^d)} + \|F(u)\|_{\dot N^1(J_k \times \mathbf{R}^d)} + \|e\|_{\dot N^1(J_k \times \mathbf{R}^d)}$$

$$\lesssim E + \varepsilon_0^{\frac{4}{d-2}} \|u\|_{\dot S^1(J_k \times \mathbf{R}^d)} + \varepsilon.$$

This implies

$$\|u\|_{\dot S^1(J_k \times \mathbf{R}^d)} \lesssim_{M,d} E,$$

and thus
$$\|u\|_{\dot{S}^1(I\times\mathbf{R}^d)} \leq C(M,d)E. \tag{3.18}$$

Therefore, it is possible to partition I into $C(E,M,d)$ subintervals I_j such that
$$\|u\|_{\dot{W}(I_j\times\mathbf{R}^d)} \leq \varepsilon_0.$$

Suppose
$$\left\|e^{i(t-t_j)\Delta}(u(t_j)-w(t_j))\right\|_{\mathscr{S}(I_j\times\mathbf{R}^d)} \leq \varepsilon, \tag{3.19}$$

for $\varepsilon < \varepsilon_0$ sufficiently small. Then
$$\|u-w\|_{\mathscr{S}(I_j\times\mathbf{R}^d)} \lesssim \left\|e^{i(t-t_j)\Delta}(u(t_j)-w(t_j))\right\|_{\mathscr{S}(I_j\times\mathbf{R}^d)} + \varepsilon.$$

Moreover,
$$\left\|e^{i(t-t_{j+1})\Delta}(u(t_{j+1})-w(t_{j+1}))\right\|_{\mathscr{S}(I\times\mathbf{R}^d)}$$
$$\lesssim_{E,E'} \left\|e^{i(t-t_j)\Delta}(u(t_j)-w(t_j))\right\|_{\mathscr{S}(I\times\mathbf{R}^d)} + \varepsilon.$$

Therefore, there exists $\varepsilon_1(M,E,E',d)$ sufficiently small such that if $0 < \varepsilon \leq \varepsilon_1$,
$$\|u-w\|_{\mathscr{S}(I\times\mathbf{R}^d)} \leq C(M,E,E',d)\varepsilon. \tag{3.20}$$

Plugging (3.20) into (3.14) implies
$$\|u-w\|_{\dot{W}(I\times\mathbf{R}^d)} \leq C(M,E,E',d)\left(\varepsilon + \varepsilon^{\frac{d}{d-2}\left(\frac{4}{d-2}-\frac{2}{d}\right)}\right). \tag{3.21}$$

Next, by (3.17), (3.21), and Strichartz estimates,
$$\|u-w\|_{\dot{S}^1(I\times\mathbf{R}^d)} \leq C(M,E,E',d)\left(E' + \varepsilon + \varepsilon^{\frac{d}{d-2}\left(\frac{4}{d-2}-\frac{2}{d}\right)}\right).$$

Then by (3.18) the proof of Theorem 3.5 is complete. $\square$

3.2 Profile Decomposition for the Energy-Critical Problem

To prove scattering for the defocusing, energy-critical problem
$$iu_t + \Delta u = |u|^{\frac{4}{d-2}}u, \quad u(0,x) = u_0, \quad u : \mathbf{R}\times\mathbf{R}^d \to \mathbf{C}, \tag{3.22}$$
with large data, define the quantity corresponding to (2.36).

3.2 Profile Decomposition for the Energy-Critical Problem

Definition 3.6 (Scattering size function) For any energy $0 \leq E < \infty$, let

$$C(E) = \sup\left\{\|u\|_{L_{t,x}^{\frac{2(d+2)}{d-2}}(I \times \mathbf{R}^d)} : I \text{ is the maximal interval} \right.$$

$$\left. \text{of existence of } u, \text{ and } E(u(t)) = E\right\}.$$

It is clear that $C(0) = 0$, and by perturbation theory, $C(E)$ is a continuous function of E, so if $C(E) < \infty$ for all $E < \infty$ does not hold, there must exist a minimum E_0 for which $C(E) < \infty$ for all $E < E_0$, and $C(E_0) = \infty$.

Consider a sequence of initial data $u_{0,n} \in \dot{H}^1(\mathbf{R}^d)$ such that u_n is the corresponding solution to (3.22), with $E(u_{0,n}) \nearrow E_0$, and

$$\|u_n\|_{L_{t,x}^{\frac{2(d+2)}{d-2}}(\mathbf{R} \times \mathbf{R}^d)} \nearrow \infty.$$

Keraani's (2001) showed that $u_{0,n}$ must converge weakly to some nonzero element of $\dot{H}^1(\mathbf{R}^d)$, modulo symmetries of the equation (3.22).

Theorem 3.7 *Suppose $\{f_n\} \subset \dot{H}^1(\mathbf{R}^d)$ is a bounded sequence of functions,*

$$\|f_n\|_{\dot{H}^1(\mathbf{R}^d)} \leq A,$$

and

$$\lim_{n \to \infty} \|e^{it\Delta} f_n\|_{L_{t,x}^{\frac{2(d+2)}{d-2}}(\mathbf{R} \times \mathbf{R}^d)} = \varepsilon > 0.$$

Then after passing to a subsequence there exists $\phi \in \dot{H}^1(\mathbf{R}^d)$, and sequences $\lambda_n \in (0, \infty)$, $(t_n, x_n) \in \mathbf{R} \times \mathbf{R}^d$ such that

$$\lambda_n^{\frac{d-2}{2}} \left[e^{it_n \Delta} f_n\right](\lambda_n x + x_n) \rightharpoonup \phi(x), \tag{3.23}$$

weakly in $\dot{H}^1$, and taking

$$\phi_n(x) = \lambda_n^{-\left(\frac{d-2}{2}\right)} e^{-it_n \Delta} \phi\left(\frac{x - x_n}{\lambda_n}\right) = \lambda_n^{-\left(\frac{d-2}{2}\right)} \left[e^{-i\frac{t_n}{\lambda_n^2}\Delta} \phi\right]\left(\frac{x - x_n}{\lambda_n}\right),$$

$$\lim_{n \to \infty} \|f_n\|_{\dot{H}^1(\mathbf{R}^d)}^2 - \|f_n - \phi_n\|_{\dot{H}^1(\mathbf{R}^d)}^2 = \|\phi\|_{\dot{H}^1(\mathbf{R}^d)}^2 \gtrsim \varepsilon^2 \left(\frac{\varepsilon}{A}\right)^{2\alpha},$$

and

$$\limsup_{n \to \infty} \|e^{it\Delta}(f_n - \phi_n)\|_{L_{t,x}^B(\mathbf{R} \times \mathbf{R}^d)}^B \leq \varepsilon^B \left[1 - c\left(\frac{\varepsilon}{A}\right)^\beta\right], \tag{3.24}$$

where $B = \frac{2(d+2)}{d-2}$ and c, α, and β are constants that depend on dimension.

Both the energy and (3.22) are preserved under spatial translation

$$T_{x_0} : u(x) \mapsto u(x - x_0). \tag{3.25}$$

Also by (2.2), the energy and (3.22) are preserved under the scaling symmetry

$$R_\lambda : u(x) \mapsto \lambda^{\frac{d-2}{2}} u(\lambda x). \tag{3.26}$$

The invariances (3.25) and (3.26) also preserve scattering size. The proof is straightforward for (3.25). For (3.26) make a change of variables $x' = \lambda x$, $t' = \lambda t$. Then

$$\int_0^{\frac{T_0}{\lambda^2}} \int_{\mathbf{R}^d} |\lambda^{\frac{d-2}{2}} u(\lambda^2 t, \lambda x)|^{\frac{2(d+2)}{d-2}} dx\, dt = \int_0^{T_0} \int_{\mathbf{R}^d} |u(t,x)|^{\frac{2(d+2)}{d-2}} dx\, dt.$$

Also, if (p,q) is an admissible pair, then

$$\|\nabla u\|_{L_t^p L_x^q([0,T_0]\times \mathbf{R}^d)} = \lambda^{\frac{d}{2} - \frac{2}{p} - \frac{d}{q}} \left(\int_0^{T_0} \left(\int_{\mathbf{R}^d} |(\nabla u)(t',x')|^q dx' \right)^{p/q} dt' \right)^{1/p}.$$

Translation in time is a more delicate issue, although (1.15) implies that the $\dot{H}^1$-norm is preserved under the action of $e^{it_0 \Delta}$. The Strichartz norm of a free solution is also preserved under translation in time. More will be said later about translation in time.

Proof of Theorem 3.7 The proof is slightly different for dimensions $d \geq 6$ than for dimensions $d = 3,4,5$, so only the argument for $d \geq 6$ will be presented here. Make a Littlewood–Paley decomposition. By Hölder's inequality,

$$|e^{it\Delta} f_n|^{\frac{2d}{d-2}} \lesssim \sum_{j_1 \leq j_2} |e^{it\Delta} P_{j_1} f_n| |e^{it\Delta} P_{j_2} f_n| |e^{it\Delta} P_{\leq j_1} f_n|^{\frac{4}{d-2}}$$

$$+ \sum_{j_1 \leq j_2} |e^{it\Delta} P_{j_1} f_n| |e^{it\Delta} P_{j_2} f_n| |e^{it\Delta} P_{\geq j_2} f_n|^{\frac{4}{d-2}}$$

$$+ \sum_{j_1 \leq j_2} |e^{it\Delta} P_{j_1} f_n| |e^{it\Delta} P_{j_2} f_n| |e^{it\Delta} P_{j_1 \leq \cdot \leq j_2} f_n|^{\frac{4}{d-2}}.$$

By Bernstein's inequality, Hölder's inequality, the Sobolev embedding

3.2 Profile Decomposition for the Energy-Critical Problem

theorem, and rearranging the order of summation,

$$
\begin{aligned}
\left\| e^{it\Delta} f_n \right\|_{L_t^\infty L_x^{\frac{2d}{d-2}}(\mathbf{R}\times\mathbf{R}^d)}^{\frac{2d}{d-2}}
&\lesssim \sum_{j_1 \leq j_2} \left\| e^{it\Delta} P_{\leq j_1} f_n \right\|_{L_t^\infty L_x^{\frac{2d}{d-4}}(\mathbf{R}\times\mathbf{R}^d)}^{\frac{2d}{d-4}} \left\| P_{\geq j_2} f_n \right\|_{L_t^\infty L_x^2(\mathbf{R}\times\mathbf{R}^d)} \\
&\quad \times \sum_{j_1 \leq k \leq j_2} \left\| e^{it\Delta} P_k f_n \right\|_{L_t^\infty L_x^{\frac{2d}{d-2}}(\mathbf{R}\times\mathbf{R}^d)}^{\frac{4}{d-2}} \\
&\lesssim \sum_{j_1 \leq j_2} \left\| P_{\leq j_1} f_n \right\|_{\dot H^2} \left\| P_{\geq j_2} f_n \right\|_{L^2} \sum_{j_1 \leq k \leq j_2} \left\| e^{it\Delta} P_k f_n \right\|_{L_t^\infty L_x^{\frac{2d}{d-2}}(\mathbf{R}\times\mathbf{R}^d)}^{\frac{4}{d-2}} \\
&\lesssim \|f_n\|_{\dot H^1}^2 \left(\sup_j \left\| e^{it\Delta} P_j f_n \right\|_{L_t^\infty L_x^{\frac{2d}{d-2}}(\mathbf{R}\times\mathbf{R}^d)}^{\frac{4}{d-2}} \right).
\end{aligned}
\tag{3.27}
$$

Remark If $d = 3, 4, 5$ then $\frac{4}{d-2} > 1$, which would necessitate handling the sums in (3.27) in a slightly different manner. The Sobolev embedding theorem $\dot H^2(\mathbf{R}^d) \hookrightarrow L^{\frac{2d}{d-4}}(\mathbf{R}^d)$ would also fail to hold in dimensions $d = 3, 4$, again necessitating a modification of (3.27).

Therefore, for some $\alpha(d) > 0$,

$$\sup_j \left\| e^{it\Delta} P_j f_n \right\|_{L_t^\infty L_x^{\frac{2d}{d-2}}(\mathbf{R}\times\mathbf{R}^d)} \gtrsim \varepsilon \left(\frac{\varepsilon}{A}\right)^\alpha. \tag{3.28}$$

Remark The term $\alpha(d)$ may change from line to line during the proof.

By Bernstein's inequality,

$$\left\| e^{it\Delta} P_j f_n \right\|_{L_t^\infty L_x^2(\mathbf{R}\times\mathbf{R}^d)} \lesssim 2^{-j} \|f_n\|_{\dot H^1} \leq 2^{-j} A,$$

so by interpolation, (3.28) implies

$$\sup_j 2^{-j\frac{(d-2)}{2}} \left\| P_j e^{it\Delta} P_j f_n \right\|_{L_{t,x}^\infty(\mathbf{R}\times\mathbf{R}^d)} \gtrsim \varepsilon \left(\frac{\varepsilon}{A}\right)^\alpha.$$

Thus there exist some x_n, t_n, j_n such that

$$\left| \left(e^{it_n \Delta} P_{j_n} f_n \right)(x_n) \right| \gtrsim \varepsilon \left(\frac{\varepsilon}{A}\right)^\alpha 2^{j_n \frac{(d-2)}{2}}.$$

Taking $\lambda_n = 2^{-j_n}$ and letting

$$g_n(x) = \lambda_n^{\frac{d-2}{2}} \left(e^{it_n \Delta} f_n \right)(\lambda_n x + x_n),$$

we get

$$|P_1 g_n(0)| \gtrsim \varepsilon \left(\frac{\varepsilon}{A}\right)^\alpha. \tag{3.29}$$

By Alaoglu's theorem, a bounded subset of $\dot H^1(\mathbf{R}^d)$ is weakly compact, so there exists $\phi \in \dot H^1(\mathbf{R}^d)$ such that after passing to a subsequence,

$$\lambda_n^{\frac{d-2}{2}}\left[e^{it_n\Delta}f_n\right](\lambda_n x + x_n) = g_n(x) \rightharpoonup \phi(x), \tag{3.30}$$

weakly in $\dot H^1$, which proves (3.23). Moreover, (3.29) implies that there exists $\psi \in L^{\frac{2d}{d+2}}(\mathbf{R}^d) \subset \dot H^{-1}(\mathbf{R}^d)$, $\|\psi\|_{\dot H^{-1}} \sim \left(\frac{\varepsilon}{A}\right)^\alpha$ with Fourier support on $|\xi| \sim 1$, such that for all n,

$$|\langle \psi, g_n\rangle| \gtrsim \varepsilon\left(\frac{\varepsilon}{A}\right)^\alpha,$$

which implies

$$\|\phi\|_{\dot H^1(\mathbf{R}^d)}^2 \gtrsim \varepsilon^2\left(\frac{\varepsilon}{A}\right)^\alpha.$$

Furthermore, by (3.30) and the invariance of $\dot H^1$ under (3.25), (3.26), and the operator $e^{it_0\Delta}$,

$$\langle f_n, f_n\rangle_{\dot H^1} - \langle f_n - \phi_n, f_n - \phi_n\rangle_{\dot H^1} - \langle \phi_n, \phi_n\rangle_{\dot H^1} = \langle \phi_n, 2(f_n - \phi_n)\rangle_{\dot H^1}$$
$$= 2\langle \phi, g_n - \phi\rangle_{\dot H^1} \to 0.$$

Next let $\chi \in C_0^\infty(\mathbf{R}^d)$ be a cutoff function: $\chi = 1$ on $|x| \leq 1$ and χ is supported on $|x| \leq 2$. By Rellich's theorem, (3.30) implies that for any $R < \infty$ and $t \in \mathbf{R}$,

$$\chi\left(\frac{x}{R}\right)e^{it\Delta}g_n \to \chi\left(\frac{x}{R}\right)e^{it\Delta}\phi,$$

in $L^2(\mathbf{R}^d)$, so therefore

$$e^{it\Delta}g_n \to e^{it\Delta}\phi,$$

pointwise almost everywhere in $\mathbf{R}\times\mathbf{R}^d$. Then by Fatou's lemma,

$$\lim_{n\to\infty}\left[\|e^{it\Delta}g_n\|_{L^B_{t,x}(\mathbf{R}\times\mathbf{R}^d)}^B - \|e^{it\Delta}\phi\|_{L^B_{t,x}(\mathbf{R}\times\mathbf{R}^d)}^B - \|e^{it\Delta}(\phi - g_n)\|_{L^B_{t,x}(\mathbf{R}\times\mathbf{R}^d)}^B\right] = 0,$$

where here and below we write $B = \frac{2(d+2)}{d-2}$ for clarity. Therefore, (3.24) follows from

$$\|e^{it\Delta}\phi\|_{L^B_{t,x}(\mathbf{R}\times\mathbf{R}^d)} \gtrsim \varepsilon\left(\frac{\varepsilon}{A}\right)^\beta,$$

which in turn follows directly from interpolation and the fact that

$$\|e^{it\Delta}P_1 g_n\|_{L^\infty_t L^{\frac{2d}{d-2}}_x(\mathbf{R}\times\mathbf{R}^d)} \gtrsim \varepsilon\left(\frac{\varepsilon}{A}\right)^\alpha. \tag{3.31}$$

Making a computation similar to (2.68), (3.31) implies that for some $t'_n \in \mathbf{R}$,

$$\|e^{it\Delta}P_1 g_n\|_{L^{\frac{2d}{d-2}}(\mathbf{R}^d)} \gtrsim \varepsilon\left(\frac{\varepsilon}{A}\right)^\alpha, \tag{3.32}$$

3.2 Profile Decomposition for the Energy-Critical Problem

for all $t \in [t'_n - c\left(\frac{\varepsilon}{A}\right)^\beta, t'_n + c\left(\frac{\varepsilon}{A}\right)^\beta]$, where $c > 0$ is a small constant. Interpolating (3.32) with

$$\left\| e^{it\Delta} P_1 g_n \right\|_{L_t^\infty L_x^2(\mathbf{R} \times \mathbf{R}^d)} \leq A, \tag{3.33}$$

and the Littlewood–Paley theorem proves

$$\left\| e^{it\Delta} \phi \right\|_{L_{t,x}^{\frac{2(d+2)}{d-2}}(\mathbf{R} \times \mathbf{R}^d)} \gtrsim \varepsilon^{\frac{2(d+2)}{d-2}} \left(\frac{\varepsilon}{A}\right)^\beta. \tag{3.34}$$

This completes the proof of Theorem 3.7. □

Armed with Theorem 3.7, Keraani's (2001) proved a profile decomposition result. We introduce the following terminology from that paper.

Definition 3.8 (Asymptotic orthogonality) A sequence λ_n, for $\lambda_n > 0$, is called a scale. Call the sequence $(z_n) = (t_n, x_n)$, $t_n \in \mathbf{R}$, $x_n \in \mathbf{R}^d$ a core. Two pairs (λ, z) and (λ', z') are called asymptotically orthogonal if

$$\lim_{n \to \infty} \left[\frac{\lambda_n}{\lambda'_n} + \frac{\lambda'_n}{\lambda_n} + \frac{|t_n - t'_n|}{\lambda_n \lambda'_n} + \frac{|x_n - x'_n|}{(\lambda_n \lambda'_n)^{1/2}} \right] = +\infty.$$

Theorem 3.9 (Profile decomposition) *Let $u_n \in \dot{H}^1(\mathbf{R}^d)$ be a bounded sequence of functions. Then there exists a subsequence u_n and a sequence of scales λ_n^j, cores (t_n^j, x_n^j), and functions $\phi_j \in \dot{H}^1(\mathbf{R}^d)$, with ϕ^j possibly equal to zero, such that for any $J < \infty$,*

$$u_n = \sum_{j=1}^{J} (\lambda_n^j)^{-\left(\frac{d-2}{2}\right)} e^{-it_n^j \Delta} \phi^j \left(\frac{x - x_n^j}{\lambda_n^j}\right) + w_n^J, \tag{3.35}$$

with the remainder estimate

$$\lim_{J \to \infty} \limsup_{n \to \infty} \left\| e^{it\Delta} w_n^J \right\|_{L_{t,x}^{\frac{2(d+2)}{d-2}} \cap L_t^\infty L_x^{\frac{2d}{d-2}}(\mathbf{R} \times \mathbf{R}^d)} = 0. \tag{3.36}$$

Moreover, for any $1 \leq j \leq J$,

$$(\lambda_n^j)^{\frac{d-2}{2}} \left[e^{it_n^j \Delta} w_n^J \right] (\lambda_n^j x + x_n^j) \rightharpoonup 0, \text{ weakly} \tag{3.37}$$

and

$$\lim_{n \to \infty} \left[\|u_n\|_{\dot{H}^1(\mathbf{R}^d)}^2 - \sum_{j=1}^{J} \|\phi^j\|_{\dot{H}^1(\mathbf{R}^d)}^2 - \|w_n^J\|_{\dot{H}^1(\mathbf{R}^d)}^2 \right] = 0. \tag{3.38}$$

Finally, (λ_n^j, z_n^j) and (λ_n^k, z_n^k) are asymptotically orthogonal when $j \neq k$, and ϕ^j and ϕ^k are nonzero.

Proof This proof may be used for any dimension $d \geq 3$. If

$$\liminf_{n \to \infty} \|e^{it\Delta} u_n\|_{L_{t,x}^{\frac{2(d+2)}{d-2}}(\mathbf{R} \times \mathbf{R}^d)} = 0, \tag{3.39}$$

then simply take $J = 0$ and $w_n^0 = u_n$, after passing to a subsequence such that (3.39) holds with liminf replaced by lim. In this case

$$\lim_{n \to \infty} \|e^{it\Delta} u_n\|_{L_t^\infty L_x^{\frac{2d}{d-2}}(\mathbf{R} \times \mathbf{R}^d)} = 0. \tag{3.40}$$

Otherwise, (3.31)–(3.33) would imply the existence of a ϕ satisfying (3.34).

Now suppose

$$\liminf_{n \to \infty} \|e^{it\Delta} u_n\|_{L_{t,x}^{\frac{2(d+2)}{d-2}}(\mathbf{R} \times \mathbf{R}^d)} = \varepsilon > 0,$$

and pass to a subsequence satisfying

$$\lim_{n \to \infty} \|\nabla u_n\|_{L^2(\mathbf{R}^d)} \leq A,$$

$$\lim_{n \to \infty} \|e^{it\Delta} u_n\|_{L_{t,x}^{\frac{2(d+2)}{d-2}}(\mathbf{R} \times \mathbf{R}^d)} = \varepsilon.$$

By Theorem 3.7, there exists $\phi^1 \in \dot{H}^1$ such that after passing to a further subsequence,

$$\|\phi^1\|_{\dot{H}^1(\mathbf{R}^d)} \gtrsim \varepsilon \left(\frac{\varepsilon}{A}\right)^\alpha$$

and

$$(\lambda_n^1)^{\frac{d-2}{2}} \left[e^{it_n^1 \Delta} u_n\right](\lambda_n^1 x + x_n^1) \rightharpoonup \phi^1(x) \in \dot{H}^1(\mathbf{R}^d), \text{ weakly.}$$

Let

$$w_n^1 = u_n - (\lambda_n^1)^{-\frac{d-2}{2}} e^{-it_n^1 \Delta} \phi^1 \left(\frac{x - x_n^1}{\lambda_n^1}\right).$$

Again by Theorem 3.7, possibly after passing to a subsequence,

$$\lim_{n \to \infty} \left[\|\nabla u_n\|_{L^2(\mathbf{R}^d)}^2 - \|\nabla \phi^1\|_{L^2(\mathbf{R}^d)}^2 - \|\nabla w_n^1\|_{L^2(\mathbf{R}^2)}^2\right] = 0$$

and

$$(\lambda_n^1)^{\frac{d-2}{2}} \left[e^{it_n^1 \Delta} w_n^1\right](\lambda_n^1 x + x_n^1) \rightharpoonup 0, \text{ weakly.}$$

Now suppose (3.35), (3.37), and (3.38) hold for $1 \leq j \leq J_0 - 1$, and (λ_n^j, z_n^j) are pairwise asymptotically orthogonal. If

$$\liminf_{n \to \infty} \|e^{it\Delta} w_n^{J_0-1}\|_{L_{t,x}^{\frac{2(d+2)}{d-2}}(\mathbf{R} \times \mathbf{R}^d)} = 0,$$

3.2 Profile Decomposition for the Energy-Critical Problem

then by (3.40), the proof is complete. Otherwise, after passing to a further subsequence,

$$\lim_{n\to\infty} \left\| e^{it\Delta} w_n^{J_0-1} \right\|_{L_{t,x}^{\frac{2(d+2)}{d-2}}(\mathbf{R}\times\mathbf{R}^d)} = \varepsilon(J_0) > 0$$

and

$$\left(\lambda_n^{J_0}\right)^{\frac{d-2}{2}} \left[e^{it_n^{J_0}\Delta} w_n^{J_0-1} \right] \left(\lambda_n^{J_0} x + x_n^{J_0}\right) \rightharpoonup \phi^{J_0}(x), \tag{3.41}$$

weakly. By Theorem 3.7,

$$\left\| \phi^{J_0} \right\|^2_{\dot{H}^1(\mathbf{R}^d)} \gtrsim \varepsilon(J_0)^2 \left(\frac{\varepsilon(J_0)}{A} \right)^\alpha$$

and

$$\lim_{n\to\infty} \left[\left\| w_n^{J_0-1} \right\|^2_{\dot{H}^1} - \left\| \phi^{J_0} \right\|^2_{\dot{H}^1(\mathbf{R}^d)} - \left\| w_n^{J_0} \right\|^2_{\dot{H}^1(\mathbf{R}^d)} \right] = 0.$$

Therefore, (3.38) continues to hold.

Pairwise asymptotic orthogonality also continues to hold. By contradiction, suppose that for some $1 \le j < J_0$,

$$\frac{\lambda_n^j}{\lambda_n^{J_0}} + \frac{\lambda_n^{J_0}}{\lambda_n^j} + \frac{|x_n^j - x_n^{J_0}|}{(\lambda_n^j \lambda_n^{J_0})^{1/2}} + \frac{|t_n^j - t_n^{J_0}|}{\lambda_n^j \lambda_n^{J_0}}$$

has a bounded subsequence. Then there exists $\lambda \in (0,\infty)$, $x_0 \in \mathbf{R}^d$, and $t \in \mathbf{R}$ such that along some further subsequence,

$$\frac{\lambda_n^{J_0}}{\lambda_n^j} \to \lambda, \quad \frac{x_n^j - x_n^{J_0}}{\lambda_n^{J_0}} \to x_0, \quad \frac{t_n^j - t_n^{J_0}}{(\lambda_n^{J_0})^2} \to t.$$

By (3.37),

$$\left(\lambda_n^j\right)^{\frac{d-2}{2}} \left[e^{it_n^j \Delta} w_n^{J_0-1} \right] \left(\lambda_n^j x + x_n^j\right) \rightharpoonup 0, \text{ weakly,}$$

while by (3.41),

$$\left(\lambda_n^{J_0}\right)^{\frac{d-2}{2}} \left[e^{it_n^{J_0}\Delta} w_n^{J_0-1} \right] \left(\lambda_n^{J_0} x + x_n^{J_0}\right) \rightharpoonup \phi^{J_0}(x), \text{ weakly.} \tag{3.42}$$

Now by (1.42),

$$(3.42) = \left(\lambda_n^{J_0}\right)^{\frac{d-2}{2}} \left[e^{i\left(t_n^{J_0} - t_n^j\right)\Delta} e^{it_n^j \Delta} w_n^{J_0-1} \right] \left(\lambda_n^{J_0} x + x_n^{J_0}\right)$$

$$= \exp\left(i \frac{(t_n^{J_0} - t_n^j)}{(\lambda_n^{J_0})^2} \Delta \right) \left(\lambda_n^{J_0}\right)^{\frac{d-2}{2}} \left[e^{it_n^j \Delta} w_n^{J_0-1} \right] \left(\lambda_n^{J_0} x + x_n^{J_0}\right)$$

$$= \exp\left(i\frac{(t_n^{J_0} - t_n^j)}{(\lambda_n^{J_0})^2}\Delta\right) T_{\frac{x_n^j - x_n^{J_0}}{\lambda_n^{J_0}}} (\lambda_n^{J_0})^{\frac{d-2}{2}} \left[e^{it_n^j\Delta} w_n^{J_0-1}\right] (\lambda_n^{J_0}x + x_n^j)$$

$$= \exp\left(i\frac{(t_n^{J_0} - t_n^j)}{(\lambda_n^{J_0})^2}\Delta\right) T_{\frac{x_n^j - x_n^{J_0}}{\lambda_n^{J_0}}} R_{\frac{\lambda_n^{J_0}}{\lambda_n^j}} (\lambda_n^j)^{\frac{d-2}{2}} \left[e^{it_n^j\Delta} w_n^{J_0-1}\right] (\lambda_n^j x + x_n^j)$$

$$\rightharpoonup e^{it\Delta} T_{x_0} R_\lambda 0 = 0,$$

weakly in $\dot{H}^1$, where we have used (3.25) and (3.26). This gives a contradiction with (3.42) since $\phi^{J_0} \neq 0$. Therefore, pairwise asymptotic orthogonality of (λ_n^j, z_n^j) must also hold for $1 \leq j \leq J_0$.

Now verify (3.37). Let

$$w_n^{J_0} = w_n^{J_0-1} - (\lambda_n^{J_0})^{-\frac{d-2}{2}} \exp\left(-it_n^{J_0}\Delta\right) \phi^{J_0}\left(\frac{x - x_n^{J_0}}{\lambda_n^{J_0}}\right).$$

It is clear from Theorem 3.7 that

$$(\lambda_n^{J_0})^{\frac{d-2}{2}} \left[\exp\left(it_n^{J_0}\Delta\right) w_n^{J_0}\right] (\lambda_n^{J_0} x + x_n^{J_0}) \rightharpoonup 0.$$

Also, by induction, for $1 \leq j < J_0$,

$$\lim_{n \to \infty} (\lambda_n^j)^{\frac{d-2}{2}} \left[\exp\left(it_n^j\Delta\right) w_n^{J_0-1}\right] (\lambda_n^j x + x_n^j) \rightharpoonup 0.$$

Next, by pairwise asymptotic orthogonality up to J_0,

$$\left(\frac{\lambda_n^j}{\lambda_n^{J_0}}\right)^{\frac{d-2}{2}} \exp\left(i\frac{(t_n^j - t_n^{J_0})}{(\lambda_n^j)^2}\Delta\right) \phi^{J_0}\left(\frac{\lambda_n^j x + x_n^j - x_n^{J_0}}{\lambda_n^{J_0}}\right) \rightharpoonup 0,$$

weakly. Then (3.37) follows from the definition of $w_n^{J_0}$.

Finally, by (3.24), there exists some $\delta(J_0) > 0$ such that

$$\limsup_{n \to \infty} \|e^{it\Delta} w_n^{J_0}\|_{L_{t,x}^{\frac{2(d+2)}{d-2}}(\mathbf{R} \times \mathbf{R}^d)}$$
$$< \limsup_{n \to \infty} (1 - \delta(J_0)) \|e^{it\Delta} w_n^{J_0-1}\|_{L_{t,x}^{\frac{2(d+2)}{d-2}}(\mathbf{R} \times \mathbf{R}^d)}. \tag{3.43}$$

Moreover, $\delta(J) \searrow 0$ can only occur as $J \to \infty$ if

$$\lim_{J \to \infty} \limsup_{n \to \infty} \|e^{it\Delta} w_n^J\|_{L_{t,x}^{\frac{2(d+2)}{d-2}}(\mathbf{R} \times \mathbf{R}^d)} = 0. \tag{3.44}$$

3.2 Profile Decomposition for the Energy-Critical Problem

Thus, combining (3.43), (3.44), and the discussion at the beginning of the proof immediately following (3.39) gives (3.36). □

Asymptotic orthogonality of individual profiles implies energy decoupling as $n \to \infty$.

Theorem 3.10 *Suppose $u_{0,n}$ is a sequence of functions bounded in $\dot{H}^1(\mathbf{R}^d)$ and $E(u_{0,n}) \to E_0 \in \mathbf{R}$. Then after making the profile decomposition of Theorem 3.9 and setting*

$$u_n^j(0) = \left(\lambda_n^j\right)^{-\frac{d-2}{2}} e^{-it_n^j \Delta} \phi^j \left(\frac{x - x_n^j}{\lambda_n^j}\right)$$

$$= \left(\lambda_n^j\right)^{-\frac{d-2}{2}} \left[\exp\left(-i \frac{t_n^j}{(\lambda_n^j)^2} \Delta\right) \phi^j\right] \left(\frac{x - x_n^j}{\lambda_n^j}\right),$$

$$\lim_{J \to \infty} \lim_{n \to \infty} \sum_{j=1}^{J} E\left(u_n^j(0)\right) + E\left(w_n^J\right) = E_0. \tag{3.45}$$

Proof From the definition of the energy,

$$E(u(t)) = \frac{1}{2} \int |\nabla u(t,x)|^2 dx + \frac{d-2}{2d} \int |u(t,x)|^{\frac{2d}{d-2}} dx,$$

and (3.36), (3.38), it only remains to prove that

$$\lim_{J \to \infty} \lim_{n \to \infty} \left[\|u_n(0)\|_{L^{2d/(d-2)}(\mathbf{R}^d)}^{2d/(d-2)} - \sum_{j=1}^{J} \|u_n^j(0)\|_{L^{2d/(d-2)}(\mathbf{R}^d)}^{2d/(d-2)} \right] = 0. \tag{3.46}$$

This fact follows from the pairwise asymptotic orthogonality of (λ_n^j, z_n^j). If $|t_n^j/(\lambda_n^j)^2| \to +\infty$, then the dispersive estimates combined with the dominated convergence theorem imply that $\left\|\exp\left(-i \frac{t_n^j}{(\lambda_n^j)^2}\Delta\right) \phi^j\right\|_{L_x^{\frac{2d}{d-2}}(\mathbf{R}^d)} \to 0$. Now suppose that $t_n^j/(\lambda_n^j)^2$ and $t_n^k/(\lambda_n^k)^2$ converge to constants in $\mathbf{R}$. In this case,

$$\frac{\lambda_n^j}{\lambda_n^k} + \frac{\lambda_n^k}{\lambda_n^j} + \frac{|x_n^j - x_n^k|}{(\lambda_n^j)^{1/2}(\lambda_n^k)^{1/2}} \to +\infty,$$

and then the dominated convergence theorem implies that for fixed j, k,

$$\int_{\mathbf{R}^d} \left| \frac{1}{(\lambda_n^j)^{\frac{d-2}{2}}} u_n^j\left(0, \frac{x - x_n^j}{(\lambda_n^j)}\right) \right| \times \left| \frac{1}{(\lambda_n^k)^{\frac{d-2}{2}}} u_n^k\left(0, \frac{x - x_n^k}{(\lambda_n^k)}\right) \right|^{\frac{d+2}{d-2}} dx \to 0.$$

Also, (3.38) combined with the Sobolev embedding theorem implies that all the quantities in (3.46) are finite, so by definition of w_n^J,

$$\lim_{J \to \infty} \limsup_{n \to \infty} \left[\|u_n(0)\|_{L_x^{2d/(d-2)}(\mathbf{R}^d)}^{2d/(d-2)} - \sum_{j=1}^{J} \|u_n^j(0)\|_{L_x^{2d/(d-2)}(\mathbf{R}^d)}^{2d/(d-2)} \right]$$

$$\lesssim \lim_{J \to \infty} \limsup_{n \to \infty} \|w_n^J\|_{L_x^{\frac{2d}{d-2}}(\mathbf{R}^d)} \left[\|u_n(0)\|_{L_x^{\frac{2d}{d-2}}(\mathbf{R}^d)} + \|w_n^J\|_{L_x^{\frac{2d}{d-2}}(\mathbf{R}^d)} \right]^{\frac{d+2}{d-2}} = 0.$$

This proves (3.45). $\square$

Now let $u_{0,n} \in \dot{H}^1$ be a sequence of initial data, and u_n the solution to (3.22) with initial data $u(0) = u_{0,n}$ on $\mathbf{R}$, with $E(u_{0,n}) \nearrow E_0$ and $\|u_n\|_{L_{t,x}^{\frac{2(d+2)}{d-2}}(\mathbf{R} \times \mathbf{R}^d)} \geq 2n$. Moreover, there exists some $t_n \in \mathbf{R}$ such that, after translating in time $u(t) \mapsto u(t - t_n)$,

$$\|u_n\|_{L_{t,x}^{\frac{2(d+2)}{d-2}}([0,\infty) \times \mathbf{R}^d)} \geq n \qquad (3.47)$$

and

$$\|u_n\|_{L_{t,x}^{\frac{2(d+2)}{d-2}}((-\infty,0] \times \mathbf{R}^d)} \geq n. \qquad (3.48)$$

Because E_0 is the minimal energy such that $C(E_0) = \infty$, if $E_0 < \infty$,

$$\lim_{J \to \infty} \limsup_{n \to \infty} E(w_n^J) = 0.$$

Otherwise, by (3.45),

$$\tilde{u}_{n,0}^J = \sum_{j=1}^{J} u_n^j(0),$$

with $E(\tilde{u}_{n,0}^J) < E_0 - \delta$ for some fixed $\delta > 0$, independent of J. Then if $\tilde{u}_n^J$ is a solution to the initial value problem with initial data $\tilde{u}_{0,n}^J$,

$$\|\tilde{u}_n^J\|_{L_{t,x}^{\frac{2(d+2)}{d-2}}(\mathbf{R} \times \mathbf{R}^d)} \leq C(E_0 - \delta) < \infty.$$

Then by (3.36) and the proof of Theorem 3.5, in particular (3.19)–(3.20), for n and J sufficiently large,

$$\|u_n\|_{L_{t,x}^{\frac{2(d+2)}{d-2}}(\mathbf{R} \times \mathbf{R}^d)} \leq C(E_0 - \delta) + \varepsilon(J, n), \qquad (3.49)$$

with

$$\lim_{J \to \infty} \limsup_{n \to \infty} \varepsilon(J, n) = 0. \qquad (3.50)$$

3.2 Profile Decomposition for the Energy-Critical Problem

This contradicts (3.47) and (3.48).

Next, if $E_0 < \infty$, then for each j such that ϕ^j is nonzero, $t_n^j/(\lambda_n^j)^2$ must remain uniformly bounded. Without loss of generality suppose that it is not true when $j = 1$. Take $J = 1$ and let

$$\tilde{u}_{n,0} = w_n^1.$$

Then by (3.45), since ϕ^1 is nonzero,

$$\limsup_{n \to \infty} E(w_n^1) < E_0 - \delta, \tag{3.51}$$

for some $\delta > 0$. Therefore, in this case, if $\tilde{u}_n$ is the solution to (3.22) with initial data $\tilde{u}_{n,0}$, then

$$\|\tilde{u}_n\|_{L_{t,x}^{\frac{2(d+2)}{d-2}}(\mathbb{R} \times \mathbb{R}^d)} \leq C(E_0 - \delta), \tag{3.52}$$

and by the dominated convergence theorem either

$$\liminf_{n \to \infty} \|e^{it\Delta}\phi^1\|_{L_{t,x}^{\frac{2(d+2)}{d-2}}([0,\infty) \times \mathbb{R}^d)} = 0 \tag{3.53}$$

or

$$\liminf_{n \to \infty} \|e^{it\Delta}\phi^1\|_{L_{t,x}^{\frac{2(d+2)}{d-2}}((-\infty,0] \times \mathbb{R}^d)} = 0. \tag{3.54}$$

By Theorem 3.5 and (3.52), (3.53) contradicts (3.47) and (3.54) contradicts (3.48).

Therefore, after passing to a subsequence,

$$\frac{t_n^j}{(\lambda_n^j)^2} \to t_j \in \mathbb{R}. \tag{3.55}$$

It is possible to replace ϕ^j with $e^{-it_j\Delta}\phi^j$ and set $t_n^j \equiv 0$, absorbing the error into w_n^J, since by Strichartz estimates, $t_n^j \to 0$ implies that for any j,

$$\lim_{n \to \infty} \|e^{it\Delta}\phi^j - e^{i(t-t_n^j)\Delta}\phi^j\|_{L_{t,x}^{\frac{2(d+2)}{d-2}}(\mathbb{R} \times \mathbb{R}^d)} = 0. \tag{3.56}$$

Finally, $\phi^j = 0$ for all $j \geq 2$, possibly after an initial relabeling. Relabel so that the energy of ϕ^j is decreasing as $j \nearrow \infty$ and suppose that this was not true, and that in fact $E(\phi^2) = \delta > 0$. Let u^j be the solution to the initial value problem with initial data ϕ^j. For each j, by (3.45),

$$\|u^j\|_{L_{t,x}^{\frac{2(d+2)}{d-2}}(\mathbb{R} \times \mathbb{R}^d)} \leq C(E_0 - \delta). \tag{3.57}$$

Also, by (3.45), $E(\phi^j) \geq \varepsilon_0$ for only finitely many j, so by Theorem 3.1, for j sufficiently large,

$$\|u^j\|^2_{L^{\frac{2(d+2)}{d-2}}_{t,x}(\mathbf{R}\times\mathbf{R}^d)} \lesssim E(\phi^j).$$

Now by asymptotic orthogonality and the bilinear Strichartz estimate, when $j \neq k$,

$$\limsup_{n\to\infty} \left\| \left|\frac{1}{(\lambda_n^j)^{\frac{d-2}{2}}} u^j\left(\frac{t}{(\lambda_n^j)^2}, \frac{x-x_n^j}{\lambda_n^j}\right)\right| \right.$$
$$\left. \times \left|\frac{1}{(\lambda_n^k)^{\frac{d-2}{2}}} u^k\left(\frac{t}{(\lambda_n^k)^2}, \frac{x-x_n^k}{\lambda_n^k}\right)\right|^{\frac{d+2}{d-2}} \right\|_{\dot N^1(\mathbf{R}\times\mathbf{R}^d)} \to 0. \quad (3.58)$$

Therefore, (3.57)–(3.58) combined with Theorem 3.5 imply

$$\lim_{n\to\infty} \|u_n\|_{L^{\frac{2(d+2)}{d-2}}_{t,x}(\mathbf{R}\times\mathbf{R}^d)} \lesssim C(E_0-\delta).$$

Putting all of this together, we have the following result.

Theorem 3.11 *If $C(E) < \infty$ for all $E < E_0 < \infty$ and $C(E_0) = \infty$, then there exists a solution $u(t)$ on a maximal interval of existence with initial data $E(\phi^1) = E_0$, such that*

$$\|u\|_{L^{\frac{2(d+2)}{d-2}}_{t,x}(I\cap[0,\infty)\times\mathbf{R}^d)} = \|u\|_{L^{\frac{2(d+2)}{d-2}}_{t,x}(I\cap(-\infty,0]\times\mathbf{R}^d)} = \infty.$$

The proof of Theorem 3.11 could also be applied to $u(t_n)$, for some sequence $t_n \in I$. Because $E(u(t_n)) = E(\phi^1)$, a subsequence weakly converging in $\dot H^1$ must converge in $\dot H^1$-norm.

Remark To completely mirror the proof of Theorem 3.11, this argument could be applied to a sequence $(1-\varepsilon_n)u(t_n)$ for some $\varepsilon_n \searrow 0$.

In conclusion, $u(t)$ lies in a compact subset of $\dot H^1$, modulo scaling and translation symmetries, so there exists $\lambda(t): I \to (0,\infty)$ and $x(t): I \to \mathbf{R}^d$ such that

$$(\lambda(t))^{\frac{d-2}{2}} u(t, \lambda(t)x + x(t)) \quad (3.59)$$

lies in some compact subset of $\dot H^1(\mathbf{R}^d)$ for all $t \in I$. Setting $N(t) = \frac{1}{\lambda(t)}$, by the Arzelà–Ascoli theorem, this implies that for any $\eta > 0$, there exists $C(\eta) < \infty$ such that

$$\int_{|x-x(t)| \geq C(\eta)\lambda(t)} |\nabla u(t,x)|^2 + \int_{|x-x(t)| \geq C(\eta)\lambda(t)} |u(t,x)|^{\frac{2d}{d-2}} dx < \eta \quad (3.60)$$

and

$$\int_{|\xi|\geq C(\eta)N(t)} |\xi|^2 |\hat{u}(t,\xi)|^2 d\xi + \int_{|\xi|\leq \frac{N(t)}{C(\eta)}} |\xi|^2 |\hat{u}(t,\xi)|^2 d\xi < \eta. \quad (3.61)$$

Definition 3.12 (Almost-periodic solution) A solution for which (3.59) lies in a compact set for all $t \in I$ is called an almost-periodic solution.

Remark Compare the derivation of (3.61) to the derivation of (2.69) in Section 2.3. The relation (2.69) shows that either a nonzero fraction of the solution lies in a compact set modulo scaling symmetries (but not translation in space due to the radial symmetry of the solution) for a nonzero fraction of the time of its existence, or it is possible to apply induction on energy to prove scattering.

The derivation of (3.61) carries this argument to its natural limit. Either, at the minimal energy, the entire solution must lie in a compact subset of $\dot{H}^1\left(\mathbf{R}^d\right)$, modulo symmetries, for the entire time of its existence, or it is possible to apply an induction on energy-type argument, as in (3.49).

3.3 Global Well-Posedness and Scattering for the Energy-Critical Problem When $d \geq 5$

The quantities $N(t)$ and $x(t)$ in (3.60) and (3.61) may be chosen to be differentiable in time.

Theorem 3.13 *We can choose $N(t)$ and $x(t)$ in (3.59) to satisfy*

$$|N'(t)| \lesssim N(t)^3 \quad (3.62)$$

and

$$|x'(t)| \lesssim \frac{1}{N(t)} = \lambda(t). \quad (3.63)$$

Moreover, for any $t \in I$,

$$\|u(t)\|_{L_x^{\frac{2d}{d-2}}(\mathbf{R}^d)} \gtrsim 1. \quad (3.64)$$

Proof Let $K \subset \dot{H}^1$ be a compact subset of $\dot{H}^1$ such that (3.59) lies in K for all $t \in I$. By Theorem 3.3, each $u_0 \in K$ is the initial data function for a local solution on $[-T(u_0), T(u_0)]$, with

$$\|u\|_{L_{t,x}^{\frac{2(d+2)}{d-2}}([-T(u_0),T(u_0)]\times\mathbf{R}^d)} = \varepsilon, \quad (3.65)$$

where $\varepsilon > 0$ is a small, fixed number. Furthermore, by Theorem 3.5, $T(u_0)$ is continuous on $\dot{H}^1$. Therefore, by compactness there must exist some $T_0 > 0$

such that $T(u_0) \geq T_0$ for all $u_0 \in K$. Equation (3.65) also implies that the linear solution dominates the Duhamel expression for a solution on $[-T_0, T_0]$. Since $e^{it\Delta}$ is a Fourier multiplier with absolute value 1, this implies

$$N(t) \sim 1 \tag{3.66}$$

for all $t \in [-T_0, T_0]$ and $u_0 \in K$. Moreover, by (1.28) and $N(t) \sim 1$, for any $u_0 \in K$, $|x(t)| \lesssim 1$ for all $t \in [-T_0, T_0]$. Then by rescaling, (3.66) implies (3.62), $|x(t)| \lesssim 1$ and rescaling gives (3.63), and

$$\|u\|_{L_{t,x}^{\frac{2(d+2)}{d-2}}\left[t_0 - \frac{T_0}{N(t_0)^2}, t_0 + \frac{T_0}{N(t_0)^2}\right]} \leq \varepsilon. \tag{3.67}$$

□

Let G be the group of symmetries generated by translation and scaling (see (3.59)). To prove that (3.22) is globally well posed and scattering, it suffices to show that the only almost-periodic solution to (3.22) is $u \equiv 0$. To prove this, it is enough to exclude the existence of a finite set of solutions which may be obtained by taking the limit of a generic almost-periodic solution. Indeed, if u is a generic almost-periodic solution, $u(t_n)$, $t_n \in I$ has a subsequence that converges in $\dot{H}^1/G$ to some $u_0 \in \dot{H}^1(\mathbf{R}^d)$. Moreover, by Theorem 3.5, u_0 is the initial condition to a blowup solution.

Theorem 3.14 (Energy-critical scenarios) *If there is a nonzero almost-periodic solution to the defocusing, energy-critical problem, then there exists a nonzero almost-periodic solution on an interval I such that one of two scenarios holds:*

$$I = \mathbf{R}, \quad N(t) \geq 1 \quad \text{for all } t \in \mathbf{R} \tag{3.68}$$

or

$$\sup(I) < \infty \quad \text{and} \quad \lim_{t \to \sup(I)} N(t) = +\infty. \tag{3.69}$$

Remark One may obtain (3.69) for a generic $I \neq \mathbf{R}$ by time reversal symmetry.

Proof This argument follows the reduction in Killip et al. (2009). Define the quantity

$$\operatorname{osc}(T) = \inf_{t_0 \in I} \frac{\sup\{N(t) : t \in I \text{ and } |t - t_0| \leq TN(t_0)^{-2}\}}{\inf\{N(t) : t \in I \text{ and } |t - t_0| \leq TN(t_0)^{-2}\}}.$$

First suppose

$$\lim_{T \to \infty} \operatorname{osc}(T) < \infty. \tag{3.70}$$

3.3 Global Well-Posedness and Scattering When $d \geq 5$

In this case, there exists a sequence $t_n \in I$ such that

$$\lim_{n \to \infty} \frac{\sup\{N(t) : t \in I \text{ and } |t - t_n| \leq nN(t_n)^{-2}\}}{\inf\{N(t) : t \in I \text{ and } |t - t_n| \leq nN(t_n)^{-2}\}} < \infty.$$

Moreover, after rescaling, (3.70) implies that $N(t) \sim 1$ for all $t \in I$, and by (3.67), $I = \mathbf{R}$.

Now suppose

$$\lim_{T \to \infty} \operatorname{osc}(T) = \infty.$$

For any $t_0 \in I$, define

$$a(t_0) = \frac{N(t_0)}{\sup\{N(t) : t \in I \text{ and } t \geq t_0\}} + \frac{N(t_0)}{\sup\{N(t) : t \in I \text{ and } t \leq t_0\}}.$$

Suppose

$$\inf_{t_0 \in I} a(t_0) = 0.$$

In that case, for any n, there exist $t_n^-, t_n, t_n^+ \in I$, $t_n^- < t_n < t_n^+$, such that

$$N(t_n^+) = N(t_n^-) = nN(t_n).$$

Choose $t_{0,n}$ satisfying

$$t_n^- < t_{0,n} < t_n^+, \quad N(t_{0,n}) = \inf_{t \in [t_n^-, t_n^+]} N(t).$$

Then after rescaling and translating in space, $u(t_{0,n})$ converges in $\dot{H}^1(\mathbf{R}^d)/G$ to some u_0. Moreover, u_0 is the initial condition for an almost-periodic solution satisfying $N(t) \geq 1$ for all $t \in I$. Either $I = \mathbf{R}$, in which case (3.68) holds, or $I \neq \mathbf{R}$, and after time reversal symmetry (3.69) holds.

Finally, suppose

$$\inf_{t_0 \in I} a(t_0) = 2\varepsilon > 0.$$

If there exist sequences $t_n \nearrow \sup(I)$ and $t_n' \searrow \inf(I)$ such that for all n,

$$\frac{N(t_n)}{\sup\{N(t) : t \in I, t \leq t_n\}} \geq \varepsilon, \quad \frac{N(t_n')}{\sup\{N(t) : t \in I, t \leq t_n'\}} \geq \varepsilon, \quad (3.71)$$

then (3.71), $t_n \nearrow \sup(I)$, $t_n' \searrow \inf(I)$ would imply $N(t_1) \sim N(t_2)$ for any $t_1, t_2 \in I$, which would imply that, after rescaling, $N(t) \sim 1$ for all $t \in I$, $I = \mathbf{R}$. On the other hand, by time reversal symmetry, suppose without loss of generality that there exists $t_0 \in I$, such that for all $t_n \geq t_0$, $t_n \in I$,

$$\frac{N(t_n)}{\sup\{N(t) : t \in I, t \geq t_n\}} \geq \varepsilon.$$

Since $\lim_{T \to \infty} \operatorname{osc}(T) = \infty$, there exists T sufficiently large so that $\operatorname{osc}(T) > \frac{2}{\varepsilon}$. Now define a sequence $t_n \nearrow \sup(I)$ in the following way. Start the sequence with t_0 and rescale so that $N(t_0) = 1$. For any n, define $t'_n = t_n + 4TN(t_n)^{-2}$. If $N(t'_n) \leq \frac{1}{2} N(t_n)$ then take $t_{n+1} = t'_n$. If $N(t'_n) \geq \frac{1}{2} N(t_n)$ then

$$\left[t'_n - TN(t'_n)^{-2}, t'_n + TN(t'_n)^{-2} \right] \subset \left[t_n, t_n + 8TN(t_n)^{-2} \right]. \tag{3.72}$$

Combining (3.72) with $\operatorname{osc}(T) > \frac{2}{\varepsilon}$ implies that there exists $t_{n+1} \in [t_n, t_n + 8TN(t_n)^{-2}]$ such that $N(t_{n+1}) \leq \frac{1}{2} N(t_n)$. Arguing by induction, for every n, there exists $t_n \in I$, $t_n > t_0$, such that

$$\frac{N(t_n)}{\sup\{N(t) : t \in I, t \geq t_n\}} \geq \varepsilon, \qquad t_0 < t_n < t_0 + T 2^{2n}, \tag{3.73}$$

and

$$N(t_n) \leq 2^{-n} N(t_0). \tag{3.74}$$

Then by (3.62) and (3.73) and (3.74), we have $N(t) \sim (t - t_0)^{-1/2}$ for all $t - t_0 > 1$.

By (3.59), the sequence $u(t_n)$ has a subsequence that converges in $\dot H^1(\mathbf{R}^d)/G$ to $u_0 \in \dot H^1(\mathbf{R}^d)$, which is the initial value of a blowup solution. Moreover, (3.74) implies $N(t) \sim t^{-1/2}$ for all $t > 0$, which is clearly a finite time blowup solution to (3.22). This completes the proof of Theorem 3.14. $\square$

It is straightforward to rule out the existence of a finite time blowup solution to the energy-critical problem.

Theorem 3.15 *There does not exist an almost-periodic solution to the energy-critical problem that blows up in finite time.*

Proof Let $\psi \in C_0^\infty(\mathbf{R}^d)$ be a cutoff function satisfying $\psi(x) = 1$ for $|x| \leq 1$ and $\psi(x) = 0$ for $|x| > 2$. Without loss of generality suppose $\sup(I) < \infty$ and let $T = \sup(I)$. For any $R > 0$, (3.60), Hölder's inequality, and $N(t) \nearrow \infty$ imply

$$\lim_{t \nearrow T} \int_{\mathbf{R}^d} \left| \psi\left(\frac{x}{R}\right) u(t,x) \right|^2 dx = 0. \tag{3.75}$$

By (1.112) and integrating by parts,

$$\frac{d}{dt} \int_{\mathbf{R}^d} \psi\left(\frac{x}{R}\right)^2 |u(t,x)|^2 dx$$

$$\lesssim \frac{1}{R} \int_{|x| \leq 2R} |\nabla u(t,x)| |u(t,x)| \psi\left(\frac{x}{R}\right) dx$$

$$\leq \frac{1}{R} \left(\int_{\mathbf{R}^d} \left| \psi\left(\frac{x}{R}\right) u(t,x) \right|^2 dx \right)^{1/2} \left(\int_{\mathbf{R}^d} |\nabla u(t,x)|^2 dx \right)^{1/2}.$$

3.3 Global Well-Posedness and Scattering When $d \geq 5$

In particular, by the fundamental theorem of calculus and (3.75),

$$\left(\int_{\mathbf{R}^d} \left|\psi\left(\frac{x}{R}\right) u(t,x)\right|^2 dx\right)^{1/2} \lesssim \frac{(T-t)E_0}{R}.$$

Taking $R \to \infty$ proves $u(0,x) \equiv 0$, which certainly can be continued to a global solution. $\square$

Therefore, to prove scattering for (3.22), it remains to show that the only almost-periodic solution satisfying (3.68) is $u \equiv 0$. For radially symmetric data, it is enough to use a generalization of the Morawetz estimate (2.28).

Theorem 3.16 *The defocusing, energy-critical initial value problem*

$$iu_t + \Delta u = F(u) = |u|^{\frac{4}{d-2}} u, \quad u(0,x) = u_0 \in \dot{H}^1(\mathbf{R}^d)$$

is globally well posed and scattering when u_0 is radial.

Proof Radial symmetry combined with (3.60) implies $x(t) \equiv 0$. Generalizing (2.28), for any T, d,

$$\int_{-T}^{T} \int \frac{|u(t,x)|^{\frac{2d}{d-2}}}{|x|} dx\, dt \lesssim E(u_0) T^{1/2}. \qquad (3.76)$$

However, by (3.60), (3.61), $x(t) \equiv 0$, $N(t) \geq 1$, and (3.64),

$$1 \lesssim \int \frac{|u(t,x)|^{\frac{2d}{d-2}}}{|x|} dx \qquad (3.77)$$

with a bound that is uniform in t. Since $I = \mathbf{R}$, T may be chosen to be sufficiently large so that (3.77) contradicts (3.76) when u is not the identically zero solution. $\square$

Remark A variation of this argument was actually introduced in Kenig and Merle (2006) to prove scattering for the radial, focusing, energy-critical problem.

Remark The proof of Theorem 3.16 could also be applied to a solution that is symmetric across d linearly independent hyperplanes.

For a general nonradial solution, the argument will require a few more deductions concerning an almost-periodic solution. Suppose u is an almost-periodic solution to

$$iu_t + \Delta u = F(u),$$

on a maximal interval I. By Duhamel's principle, for any $t_0, t_1 \in I$,

$$u(t_1) = e^{i(t_1-t_0)\Delta} u(t_0) - i \int_{t_0}^{t_1} e^{i(t_1-\tau)\Delta} F(u(\tau))\, d\tau.$$

Then if $t_0 \nearrow \sup(I)$ or $t_0 \searrow \inf(I)$, almost-periodicity of u implies

$$e^{i(t_1-t_0)\Delta} u(t_0) \rightharpoonup 0, \tag{3.78}$$

weakly in $\dot{H}^1(\mathbf{R}^d)$.

Indeed, if $N(t_0) \nearrow +\infty$ or $N(t_0) \searrow 0$ then (3.78) follows from (3.60). On the other hand, if $N(t_0) \sim N(t_1)$ as $t_0 \nearrow \sup(I)$ or $t_0 \searrow \inf(I)$, Theorem 3.13 implies $\sup(I) = +\infty$ or $\inf(I) = -\infty$ respectively. Then by (1.19) and (3.60), it follows that (3.78) holds.

Therefore, by Rellich's theorem, for any $2 \le p < \infty$, $j \in \mathbf{Z}$, and $t \in I$,

$$\|P_j u(t)\|_{L_x^p(\mathbf{R}^d)} \lesssim \lim_{t_- \searrow \inf(I)} \left\| \int_{t_-}^{t} e^{i(t-\tau)\Delta} P_j F(u(\tau))\, d\tau \right\|_{L_x^p(\mathbf{R}^d)}$$

$$\text{and} \quad \|P_j u(t)\|_{L_x^p(\mathbf{R}^d)} \lesssim \lim_{t_+ \nearrow \sup(I)} \left\| \int_{t}^{t_+} e^{i(t-\tau)\Delta} P_j F(u(\tau))\, d\tau \right\|_{L_x^p(\mathbf{R}^d)}. \tag{3.79}$$

This fact shows that $u(t) \in L_t^\infty L_x^p$ for some $p < \frac{2d}{d-2}$, which gives an important improvement over the Sobolev embedding theorem.

Theorem 3.17 *Suppose u is an almost-periodic solution satisfying (3.68). Then when $d \ge 5$, for any q satisfying*

$$\frac{2(d+2)}{d} < q \le \frac{2d}{d-2},$$

$$\|u(t)\|_{L_t^\infty L_x^q(\mathbf{R} \times \mathbf{R}^d)} \lesssim_q 1. \tag{3.80}$$

Proof Inequality (3.61) and $N(t) \ge 1$ imply that for any $\eta > 0$, there exists $j_0(\eta) \in \mathbf{Z}$, $j_0(\eta) < 0$, such that

$$\|P_{\le j_0} u(t)\|_{L_t^\infty \dot{H}^1(\mathbf{R} \times \mathbf{R}^d)} \le \eta. \tag{3.81}$$

Now take $\frac{1}{p} = \frac{1}{2} - \frac{1}{d-2}$. By the Sobolev embedding theorem, for any $t \in \mathbf{R}$,

$$\left\| \int_{t-2^{-2j}}^{t+2^{-2j}} e^{i(t-\tau)\Delta} P_j F(u(\tau))\, d\tau \right\|_{L_x^p}$$

$$\lesssim 2^{\frac{jd}{d-2}} \left\| \int_{t-2^{-2j}}^{t+2^{-2j}} e^{i(t-\tau)\Delta} P_j F(u(\tau))\, d\tau \right\|_{L_x^2}$$

$$\lesssim 2^{\frac{2jd}{d-2}} \int_{t-2^{-2j}}^{t+2^{-2j}} \|P_j F(u(\tau))\|_{L_x^{p'}}\, d\tau$$

$$\lesssim 2^{\frac{4j}{d-2}} \|F(u(t))\|_{L_t^\infty L_x^{p'}(\mathbf{R} \times \mathbf{R}^d)}.$$

3.3 Global Well-Posedness and Scattering When $d \geq 5$

Meanwhile, by (1.20),

$$\left\| \int_{t+2^{-2j}}^{\infty} e^{i(t-\tau)\Delta} P_j F(u(\tau)) d\tau \right\|_{L_x^p}$$
$$\lesssim \int_{t+2^{-2j}}^{\infty} \frac{1}{|t-\tau|^{\frac{d}{d-2}}} \| P_j F(u(\tau)) \|_{L_x^{p'}} d\tau$$
$$\lesssim 2^{\frac{4j}{d-2}} \| F(u(t)) \|_{L_t^\infty L_x^{p'}(\mathbf{R} \times \mathbf{R}^d)}.$$

Therefore, by (3.79),

$$\| P_j u(t) \|_{L_t^\infty L_x^p(\mathbf{R} \times \mathbf{R}^d)} \lesssim 2^{\frac{4j}{d-2}} \| P_j F(u(t)) \|_{L_t^\infty L_x^{p'}(\mathbf{R} \times \mathbf{R}^d)}. \tag{3.82}$$

Split F as follows:

$$F(u) = F(P_{\leq j} u) + O\left(|P_{>j} u| |P_{\leq j} u|^{\frac{4}{d-2}} \right) + O\left(|P_{\geq j} u|^{\frac{d+2}{d-2}} \right).$$

When $d = 5$, Bernstein's inequality implies

$$\| P_j F(P_{\leq j} u) \|_{L_x^{6/5}} \lesssim 2^{-j} \| \nabla F(P_{\leq j} u) \|_{L_x^{6/5}} \lesssim 2^{-j} \| \nabla u \|_{L_x^2} \| P_{\leq j} u \|_{L_x^4}^{4/3}. \tag{3.83}$$

Then if $j \leq j_0$, for any small, fixed $\varepsilon > 0$,

$$\sum_{k \leq j} \| P_k u \|_{L_x^4} \lesssim \sum_{k \leq j} \| P_k u \|_{L_x^6}^{3/4} \| P_k u \|_{L_x^2}^{1/4}$$
$$\lesssim_\varepsilon \eta^{1/4} 2^{-\frac{3\varepsilon j}{4}} 2^{\frac{3j}{4}} \left(\sup_{k \leq j_0} 2^{\varepsilon k} 2^{-\frac{4k}{3}} \| P_k u \|_{L_x^6} \right)^{3/4}, \tag{3.84}$$

and therefore by (3.81), (3.82), and (3.83),

$$\sup_{j \leq j_0} 2^{\varepsilon j} \| P_j F(P_{\leq j} u) \|_{L_t^\infty L_x^{6/5}(\mathbf{R} \times \mathbf{R}^d)} \lesssim_\varepsilon \eta^{1/3} \left(\sup_{k \leq j_0} 2^{\varepsilon k} 2^{-\frac{4k}{3}} \| P_k u \|_{L_t^\infty L_x^6} \right).$$

Also by Bernstein's inequality and (3.84),

$$\sup_{j \leq j_0} 2^{\varepsilon j} \| P_{>j} u \|_{L_x^2(\mathbf{R}^d)} \| P_{\leq j} u \|_{L_x^4(\mathbf{R}^d)}^{4/3} \lesssim 2^{(\varepsilon-1)j} \| \nabla u \|_{L_x^2} \| P_{\leq j} u \|_{L_x^4}^{4/3}$$
$$\lesssim_\varepsilon \eta^{1/3} \left(\sup_{k \leq j_0} 2^{\varepsilon k} 2^{-\frac{4k}{3}} \| P_k u \|_{L_t^\infty L_x^6} \right).$$

Finally, by Bernstein's inequality,

$$2^{\varepsilon j} \| F(P_{>j} u) \|_{L_x^{6/5}} \lesssim 2^{\varepsilon j} \left(\sum_{j \leq k \leq j_0} \| P_k u \|_{L_x^2}^{4/7} \| P_k u \|_{L_x^6}^{3/7} + \sum_{k \geq j_0} \| P_k u \|_{L_x^{14/5}} \right)^{7/3}$$
$$\lesssim C_\varepsilon \eta^{4/3} \left(\sup_{k \leq j_0} 2^{\varepsilon k} 2^{-\frac{4k}{3}} \| P_k u \|_{L_x^6} \right) + C_\eta.$$

Therefore, by (3.82),

$$\sup_{j \le j_0} 2^{\varepsilon j} 2^{-\frac{4j}{3}} \|P_j u\|_{L_t^\infty L_x^6(\mathbf{R} \times \mathbf{R}^d)}$$
$$\lesssim C_\eta + C_\varepsilon \eta^{1/3} \left(\sup_{j \le j_0} 2^{\varepsilon j} 2^{-\frac{4j}{3}} \|P_j u\|_{L_t^\infty L_x^6(\mathbf{R} \times \mathbf{R}^d)} \right). \tag{3.85}$$

Choosing $\eta(\varepsilon) > 0$ sufficiently small,

$$\sup_{j \le j_0} 2^{\varepsilon j} 2^{-\frac{4j}{3}} \|P_j u\|_{L_t^\infty L_x^6(\mathbf{R} \times \mathbf{R}^d)} \lesssim_\varepsilon 1.$$

Since this estimate holds for any $\varepsilon > 0$, interpolating (3.85) with Bernstein's inequality $\|P_j u\|_{L_x^2} \lesssim 2^{-j}$ and the Sobolev embedding theorem proves (3.80) when $d = 5$.

The computations are similar when $d \ge 6$. By Bernstein's inequality,

$$2^{\varepsilon j} \|P_j F(P_{\le j} u)\|_{L_x^{p'}} \lesssim 2^{\varepsilon j} 2^{-\frac{4j}{d-2}} \||\nabla|^{\frac{4}{d-2}} F(P_{\le j} u)\|_{L_x^{p'}}$$
$$\lesssim 2^{\varepsilon j} 2^{-\frac{4j}{d-2}} \|\nabla P_{\le j} u\|_{L_x^2}^{\frac{4}{d-2}} \|P_{\le j} u\|_{L_x^p}$$
$$\lesssim_\varepsilon \eta^{\frac{4}{d-2}} \left(\sup_{k \le j_0} 2^{\varepsilon k} 2^{-\frac{4k}{d-2}} \|P_k u\|_{L_x^p} \right). \tag{3.86}$$

Furthermore, using Bernstein's inequality once again,

$$\left\| |P_{>j} u| |P_{\le j} u|^{\frac{4}{d-2}} \right\|_{L_x^{p'}}$$
$$\lesssim \|P_{\le j} u\|_{L_x^p}^{\frac{4}{d-2}} \left(\sum_{j \le k \le j_0} \|P_k u\|_{L_x^p}^{1-\frac{4}{d-2}} \|P_k u\|_{L_x^2}^{\frac{4}{d-2}} + \sum_{k \ge j_0} \|P_k u\|_{L_x^p}^{1-\frac{4}{d-2}} \|P_k u\|_{L_x^2}^{\frac{4}{d-2}} \right)$$
$$\lesssim_\varepsilon \eta^{\frac{4}{d-2}} \left(\sup_{k \le j_0} 2^{\varepsilon k} 2^{-\frac{4k}{d-2}} \|P_k u\|_{L_x^p} \right) + \left(\sup_{k \le j_0} 2^{\varepsilon k} 2^{-\frac{4k}{d-2}} \|P_k u\|_{L_x^p} \right)^{\frac{4}{d-2}}. \tag{3.87}$$

Finally, by Bernstein's inequality,

$$\left(\sum_{j \le k \le j_0} \|P_k u\|_{L_x^2}^{\frac{4}{d+2}} \|P_k u\|_{L_x^p}^{\frac{d-2}{d+2}} + \sum_{j \ge j_0} \|P_k u\|_{L_x^2}^{\frac{4}{d+2}} \|P_k u\|_{L_x^p}^{\frac{d-2}{d+2}} \right)^{\frac{d+2}{d-2}}$$
$$\lesssim C_\varepsilon \eta^{\frac{4}{d-2}} \left(\sup_{k \le j_0} 2^{\varepsilon k} 2^{-\frac{4k}{d-2}} \|P_k u\|_{L_x^p} \right) + C_\eta. \tag{3.88}$$

Relations (3.86)–(3.88) imply that for any $\varepsilon > 0$,

$$\left(\sup_{k \le j_0} 2^{\varepsilon k} 2^{-\frac{4k}{d-2}} \|P_k u\|_{L_x^p} \right) \lesssim_\varepsilon 1,$$

3.3 Global Well-Posedness and Scattering When $d \geq 5$

which proves (3.80) for $d \geq 6$. □

Arguing by induction, and using Theorem 3.17 as a base case, it is possible to show that a solution satisfying (3.60) with $N(t) \geq 1$ for all $t \in \mathbf{R}$ must be uniformly bounded in L^p for some $p < 2$, which by duality and the Sobolev embedding theorem implies $u \in L_t^\infty \dot{H}^{-\varepsilon}$ for some $\varepsilon > 0$.

Theorem 3.18 *Let $d \geq 5$ and let u be a global solution to the energy-critical problem that is almost-periodic modulo symmetries. Suppose, in addition, that $\inf_{t \in \mathbf{R}} N(t) \geq 1$. Then $u \in L_t^\infty \dot{H}_x^{-\varepsilon(d)} (\mathbf{R} \times \mathbf{R}^d)$ for some $\varepsilon = \varepsilon(d) > 0$. In particular, this means that $u \in L_t^\infty L_x^2$, and indeed lies in a compact subset of $L_x^2(\mathbf{R}^d)$ modulo the translation symmetry.*

Proof Theorem 3.18 follows directly from an inductive proposition.

Proposition 3.19 (Inductive proposition) *Let $d \geq 5$ and let u be as in Theorem 3.18. Assume further that $|\nabla|^s F(u) \in L_t^\infty L_x^q$ for $\frac{2(d-2)(d+1)}{d^2+3d-6} = q$ and some $0 \leq s \leq 1$. Then there exists some $s_0(q,d) > 0$ such that $u \in L_t^\infty \dot{H}_x^{s-s_0}$.*

By the fractional product rule,

$$\left\| |\nabla|^s F(u) \right\|_{L_x^q(\mathbf{R}^d)} \lesssim \left\| |\nabla|^s u \right\|_{L_x^2(\mathbf{R}^d)} \|u\|_{L_x^{\frac{2(d+1)}{d-1}}(\mathbf{R}^d)}^{\frac{4}{d-2}}.$$

Starting with Theorem 3.17 and iterating Proposition 3.19 proves Theorem 3.18. □

Proof of Proposition 3.19 Proposition 3.19 is equivalent to showing that for some $s_0 > 0$, say $s_0 = \frac{d}{q} - \frac{d+4}{2} > 0$, and all $j \leq 0$,

$$\left\| |\nabla|^s P_j u \right\|_{L_t^\infty L_x^2} \lesssim 2^{js_0}.$$

By time translation symmetry it suffices to show that

$$\left\| |\nabla|^s P_j u(0) \right\|_{L_x^2} \lesssim 2^{js_0}.$$

The proof of this fact uses the double Duhamel argument.

Lemma 3.20 (Double Duhamel lemma) *If I is the maximal interval of existence and u is an almost-periodic solution, then for any $t_1 \in I$,*

$\langle u(t_1), u(t_1) \rangle_{\dot{H}^1}$

$$= -\lim_{t_0 \searrow \inf(I)} \lim_{t_2 \nearrow \sup(I)} \left\langle \int_{t_0}^{t_1} e^{i(t_1-\tau)\Delta} F(u(\tau)) d\tau, \int_{t_1}^{t_2} e^{i(t_1-t)\Delta} F(u(t)) dt \right\rangle_{\dot{H}^1}.$$
(3.89)

Proof By (3.78), for any $t_0 < t_1 < t_2$,

$$\langle u(t_1), u(t_1) \rangle_{\dot H^1}$$
$$= \left\langle e^{i(t_1-t_0)\Delta} u(t_0) - i \int_{t_0}^{t_1} e^{i(t_1-\tau)\Delta} F(u(\tau)) d\tau, \right.$$
$$\left. e^{i(t_1-t_2)\Delta} u(t_2) - i \int_{t_2}^{t_1} e^{i(t_1-t)\Delta} F(u(t)) dt \right\rangle_{\dot H^1}$$
$$= \lim_{t_2 \nearrow \sup(I)} \left\langle e^{i(t_1-t_0)\Delta} u(t_0) - i \int_{t_0}^{t_1} e^{i(t_1-\tau)\Delta} F(u(\tau)) d\tau, \right.$$
$$\left. i \int_{t_1}^{t_2} e^{i(t_1-t)\Delta} F(u(t)) dt \right\rangle_{\dot H^1}$$
$$= - \lim_{t_0 \searrow \inf(I)} \lim_{t_2 \nearrow \sup(I)} \left\langle \int_{t_0}^{t_1} e^{i(t_1-\tau)\Delta} F(u(\tau)) d\tau, \int_{t_1}^{t_2} e^{i(t_1-t)\Delta} F(u(t)) dt \right\rangle_{\dot H^1}.$$
$\square$

Remark Equation (3.89) also holds for the inner product $\langle P_j \cdot, P_j \cdot \rangle_{\dot H^1}$.

Integrating one term in (3.89) forward in time and the other term backward in time takes full advantage of the dispersive estimates, since in that case $|t - \tau| = 0$ if and only if $t = \tau = t_1$.

By (3.89),

$$\||\nabla|^s P_j u(0)\|_{L_x^2}^2 \leq \int_0^\infty \int_{-\infty}^0 |\langle P_j |\nabla|^s F(u(t)), e^{i(t-\tau)\Delta} P_j |\nabla|^s F(u(\tau)) \rangle| \, dt \, d\tau.$$

By (1.19),

$$\left| (P_j |\nabla|^s F(u(t)), e^{i(t-\tau)\Delta} P_j |\nabla|^s F(u(\tau)))_{L^2} \right|$$
$$\lesssim |t-\tau|^{d(\frac{1}{2}-\frac{1}{q})} \||\nabla|^s F(u)\|_{L_T^\infty L_x^q}^2.$$

Also, by the Sobolev embedding theorem,

$$\left| (P_j |\nabla|^s F(u(t)), e^{i(t-\tau)\Delta} P_j |\nabla|^s F(u(\tau)))_{L^2} \right| \lesssim 2^{2j(\frac{d}{q}-\frac{d}{2})} \||\nabla|^s F(u)\|_{L_T^\infty L_x^q}^2.$$

Since $\frac{d}{q} - \frac{d}{2} > 2$, there exists $s_0 > 0$ such that

$$\||\nabla|^s P_j u(0)\|_{L_x^2}^2 \lesssim \||\nabla|^s F(u)\|_{L_T^\infty L_x^q}^2 \int_0^\infty \int_{-\infty}^0 \min(|t-\tau|^{-1}, 2^{2j})^{\frac{d}{q}-\frac{d}{2}}$$
$$\lesssim 2^{4js_0} \||\nabla|^s F(u)\|_{L_T^\infty L_x^q}^2.$$

This proves the proposition. $\square$

Armed with Theorem 3.18, scattering follows directly.

3.3 Global Well-Posedness and Scattering When $d \geq 5$

Theorem 3.21 *The defocusing, energy-critical initial value problem*

$$iu_t + \Delta u = F(u) = |u|^{\frac{4}{d-2}} u, \quad u(0,x) = u_0 \in \dot{H}^1\left(\mathbf{R}^d\right) \quad (3.90)$$

is globally well posed and scattering for any $u_0 \in \dot{H}^1\left(\mathbf{R}^d\right)$, $d \geq 5$.

Proof By Theorem 3.18, if u is an almost-periodic solution satisfying (3.68) then there exists some $\varepsilon(d) > 0$ such that

$$\sup_{t \in \mathbf{R}} \|u(t)\|_{H_x^{-\varepsilon(d)}(\mathbf{R}^d) \cap \dot{H}_x^1(\mathbf{R}^d)} < \infty. \quad (3.91)$$

For any $\eta > 0$,

$$\|P_{\leq c(\eta)N(t)} u\|_{\dot{H}_x^1(\mathbf{R}^d)} < \eta, \quad (3.92)$$

and interpolating (3.91) with (3.92), along with Bernstein's inequality, then

$$\|P_{>c(\eta)N(t)} u\|_{L_x^2(\mathbf{R}^d)} \lesssim \frac{1}{N(t) c(\eta)}$$

implies

$$\|u(t)\|_{L_x^2(\mathbf{R}^d)} \lesssim \frac{1}{N(t) c(\eta)} + \eta^{\frac{\varepsilon(d)}{1+\varepsilon(d)}}.$$

If $\limsup_{t \nearrow +\infty} N(t) = +\infty$, since $\eta > 0$ is arbitrary, conservation of mass implies that $\|u(t)\|_{L^2(\mathbf{R}^d)} = 0$, and therefore $u \equiv 0$.

Thus, if u is a nonzero, almost-periodic solution to (3.90), $N(t) \sim 1$.

Lemma 3.22 *When $d \geq 3$, for any interval $I \subset \mathbf{R}$,*

$$\left\| |\nabla|^{\frac{3-d}{4}} u \right\|_{L_{t,x}^4(I \times \mathbf{R}^d)} \lesssim \|\nabla u\|_{L_t^\infty L_x^2(I \times \mathbf{R}^d)}^{1/4} \|u\|_{L_t^\infty L_x^2(I \times \mathbf{R}^d)}^{3/4}.$$

Postponing the proof of Lemma 3.22 and interpolating

$$\left\| |\nabla|^{\frac{3-d}{4}} u \right\|_{L_{t,x}^4(\mathbf{R} \times \mathbf{R}^d)} < \infty$$

with the energy bound $\|\nabla u\|_{L_t^\infty L_x^2(\mathbf{R} \times \mathbf{R}^d)} < \infty$ implies

$$\|u\|_{L_t^{d+1} L_x^{\frac{2(d+1)}{d-1}}(\mathbf{R} \times \mathbf{R}^d)} < \infty.$$

Because $N(t) \sim 1$ for all t, (3.60), (3.64), and Hölder's inequality imply that for any $j \in \mathbf{Z}$,

$$\|u(t)\|_{L_t^{d+1} L_x^{\frac{2(d+1)}{d-1}}([j,j+1] \times \mathbf{R}^d)} \gtrsim 1.$$

Therefore, there do not exist any nonzero almost-periodic solutions to (3.90), which proves scattering. $\square$

3.4 Interaction Morawetz Estimate

Lemma 3.22 is part of a family of estimates known as interaction Morawetz estimates.

Lemma 3.23 (Interaction Morawetz estimate) *If u solves*

$$iu_t + \Delta u = \mathcal{N} = |u|^p u$$

on $I \times \mathbf{R}^d$ for some $p > 0$, then for $d \geq 3$,

$$\left\| |\nabla|^{\frac{3-d}{4}} u \right\|^4_{L^4_{t,x}(I \times \mathbf{R}^d)} \lesssim \|u\|^3_{L^\infty_t L^2_x(I \times \mathbf{R}^d)} \|u\|_{L^\infty_t \dot{H}^1(I \times \mathbf{R}^d)}. \tag{3.93}$$

Proof Recall that the Morawetz estimate of (2.27) was quite useful in the study of radially symmetric problems, due to the presence of the weight $1/|x|$ combined with the fact that radiality implies that u must remain centered at the origin. However, for general nonradial problems there is no a priori reason for choosing one element of $\mathbf{R}^d$ as the origin. (Although see Kenig and Merle (2010) for (2.27) applied to a nonradial problem.)

Instead, the approach of Colliander *et al.* (2004) was to replace (2.27) with an estimate that is centered around the solution u. Recall from (2.24)–(2.25) that if

$$M_h(t) = \int h(x) |u(t,x)|^2 dx,$$

then

$$\partial_t M_h(t) = \int h_j(x) T_{0j}(t,x) dx \tag{3.94}$$

and

$$\partial_{tt} M_h(t) = \int (-\Delta \Delta h(x)) |u|^2 + 4h_{jk} \operatorname{Re}(\partial_j \bar{u} \partial_k u) + 2h_j \{\mathcal{N}, u\}^j_p dx. \tag{3.95}$$

Translating $h(x)$, let

$$M^y_h(t) = \int h(x-y) |u(t,x)|^2 dx. \tag{3.96}$$

For any weight function $g(t,y) \in L^1(\mathbf{R}^n)$, take

$$\int \int g(t,y) h(x-y) |u(t,x)|^2 dx dy. \tag{3.97}$$

Then integrating by parts in time and following the argument in the proof of

3.4 Interaction Morawetz Estimate

Theorem 2.8, if u solves the cubic problem in three dimensions, then by (2.31),

$$8\pi \int_0^T \int \int g(t,y) \delta(x-y) |u(t,x)|^2 dx dy dt$$
$$+ 4 \int_0^T \int \int g(t,y) \frac{1}{|x-y|} \left[|\nabla u(t,x)|^2 - |\partial_{r(y)} u(t,x)|^2 \right] dx dy dt$$
$$+ 2 \int_0^T \int \int g(t,y) \frac{|u(t,x)|^4}{|x-y|} dx dy dt$$
$$\leq \int_0^T \int g(t,y) \partial_{tt} \int h(x-y) |u(t,x)|^2 dx dy dt$$
$$= - \int_0^T \int \partial_t g(t,y) \left(\partial_t \int h(x-y) |u(t,x)|^2 dx \right) dy dt$$
$$+ \int g(T,y) \left(\partial_t \int h(x-y) |u(t,x)|^2 dx|_{t=T} \right) dy$$
$$- \int g(0,y) \left(\partial_t \int h(x-y) |u(t,x)|^2 dx|_{t=0} \right) dy, \qquad (3.98)$$

where δ is the usual delta function, $r(y)$ is the radial unit vector centered at $y \in \mathbf{R}^d$, and

$$\partial_{r(y)} = \frac{(x-y)}{|x-y|} \left(\frac{(x-y)}{|x-y|} \cdot \nabla \right). \qquad (3.99)$$

The weight $g(t,y)$ should depend on the size of u in some sense. In that case the Morawetz quantity will have the largest weight precisely where the solution u is located. Moreover, the term

$$\int_0^T \int \partial_t g(t,y) \left(\partial_t \int h(x-y) |u(t,x)|^2 dx \right) dy dt \qquad (3.100)$$

is the only one in (3.98) that does not have a counterpart in the proof of Theorem 2.8. The presence of $\partial_t g(t,y)$ means that this term may be much more effectively estimated when $g(t,y)$ is a conserved quantity.

Momentum does not control the size of u; in fact, momentum can often be zero for nonzero u. This leaves two options for $g(t,y)$: mass or energy. Colliander *et al.* (2004) chose $g(t,y) = |u(t,y)|^2$, proving Lemma 3.22 when $d = 3$. Subsequent work by Colliander *et al.* (2009), Planchon and Vega (2009), and Tao *et al.* (2007a) extended the interaction Morawetz estimates to other dimensions. There are a number of good reasons to choose $g(t,y) = |u(t,y)|^2$, not the least of which is because then in three dimensions the first term on the left-hand side of (3.98) is the L^4-norm of u in space and time.

Remark It is useful to think of the standard Morawetz estimate in Theorem 2.8 as showing that a solution to a defocusing problem must spread out and move away from the origin. The appropriately named interaction Morawetz estimate shows that a solution must move away from itself.

Let
$$M^{\text{int}}(t) = \int |u(t,y)|^2 M_h^y(t)\,dy. \tag{3.101}$$

By (1.112), (3.94), (3.95), (3.96), and the fact that $h(x-y)$ is an even function,
$$\frac{1}{2}\partial_t M_h^{\text{int}}(t) = \int |u(t,y)|^2 \partial_t M_h^y(t)\,dy \tag{3.102}$$

and
$$\frac{1}{2}\partial_{tt} M_h^{\text{int}}(t) = \int |u(t,y)|^2 \partial_{tt} M_h^y(t)\,dy - 2\int \partial_k T_{0k}(t,y)\,\partial_t M_h^y(t)\,dy. \tag{3.103}$$

Integrating by parts,
$$(3.103) = \int (-\Delta\Delta h(x-y))|u(t,x)|^2 |u(t,y)|^2 dx\,dy \tag{3.104}$$

$$+ 4\int h_{jk}(x-y)\,\text{Re}\,(\partial_j \bar{u}\partial_k u)(t,x)|u(t,y)|^2 dx\,dy \tag{3.105}$$

$$- 4\int h_{jk}(x-y)\,\text{Im}\,(\bar{u}\partial_j u)(t,x)\,\text{Im}\,(\bar{u}\partial_k u)(t,y)\,dx\,dy \tag{3.106}$$

$$+ 2\int h_j(x-y)\{\mathcal{N},u\}_p^j(t,x)|u(t,y)|^2 dx\,dy \tag{3.107}$$

$$+ 4\int h_j(x-y)\,\text{Im}\,(\bar{u}\partial_j u)(t,x)\{\mathcal{N},u\}_m(t,y)\,dx\,dy. \tag{3.108}$$

Observe that (3.106) and (3.108) are the terms arising from (3.100). Decompose ∇ as follows:
$$\nabla = \mathbf{\nabla}_y + \partial_{r(y)},$$

where $\partial_{r(y)}$ is given by (3.99), and refers to a radial derivative with origin shifted to y, and $\mathbf{\nabla}$ to angular derivatives with origin shifted to y. Since $\mathbf{\nabla}_y$ and $\partial_{r(y)}$ are orthogonal vectors, if $h(x) = |x|$, then $h_{jk}(x) = \frac{\delta_{jk}}{|x|} - \frac{x_j x_k}{|x|^3}$, and

$$h_{jk}(x-y)\,\text{Re}\,(\partial_j \bar{u}\partial_k u)(t,x)|u(t,y)|^2$$
$$- h_{jk}(x-y)\,\text{Im}\,(\bar{u}\partial_j u)(t,x)\,\text{Im}\,(\bar{u}\partial_k u)(t,y)$$
$$= \frac{1}{|x-y|}|\mathbf{\nabla} u(t,x)|^2 |u(t,y)|^2 - \frac{1}{|x-y|}|\partial_{r(y)} u(t,x)|^2 |u(t,y)|^2$$

3.4 Interaction Morawetz Estimate

$$-\frac{1}{|x-y|}\operatorname{Im}(\bar{u}\nabla u)(t,x)\cdot\operatorname{Im}(\bar{u}\nabla u)(t,y)$$

$$+\frac{1}{|x-y|}\operatorname{Im}\left(\bar{u}\partial_{r(y)}u\right)(t,x)\cdot\operatorname{Im}\left(\bar{u}\partial_{r(x)}u\right)(t,y)$$

$$=|\not{\nabla}_y u(t,x)|^2|u(t,y)|^2 - \operatorname{Im}(\bar{u}\not{\nabla}_y u)(t,x)\cdot\operatorname{Im}(\bar{u}\not{\nabla}_x u)(t,y). \quad (3.109)$$

By the Cauchy–Schwarz inequality,

$$\left|\overline{u(t,x)}\not{\nabla}_y u(t,x)\right|\left|\overline{u(t,y)}\not{\nabla}_x u(t,y)\right|$$
$$\leq \frac{1}{2}|\not{\nabla}_y u(t,x)|^2|u(t,y)|^2 + \frac{1}{2}|u(t,x)|^2|\not{\nabla}_x u(t,y)|^2, \quad (3.110)$$

so $(3.105)+(3.106) \geq 0$. Therefore,

$$\int_I\!\!\int\!\!\int (-\Delta\Delta h(x-y))|u(t,x)|^2|u(t,y)|^2 dx\,dy\,dt$$

$$+2\int_I\!\!\int\!\!\int h_j(x-y)\{\mathcal{N},u\}_p^j(t,x)|u(t,y)|^2 dx\,dy\,dt \quad (3.111)$$

$$\leq \partial_t M^{\text{int}}(T) - \partial_t M^{\text{int}}(0) \quad (3.112)$$

$$-4\int_I\!\!\int\!\!\int h_j(x-y)\operatorname{Im}(\bar{u}\partial_j u)(t,x)\{\mathcal{N},u\}_m(t,y)\,dx\,dy\,dt, \quad (3.113)$$

and we may proceed as in the proof of Theorem 2.8. Since $|h_j(x)| \leq 1$, we have

$$\left|\partial_t M^{\text{int}}(T) - \partial_t M^{\text{int}}(0)\right| \leq 2\|u\|_{L^\infty_t L^2_x}^3 \|u\|_{L^\infty_t \dot{H}^1}.$$

Also by (2.30),

$$8\pi \int_I\!\!\int_{\mathbf{R}^d}|u(t,x)|^4 dx\,dt + 2\int_I\!\!\int\!\!\int |u(t,y)|^2 \frac{(x-y)}{|x-y|}\cdot\{\mathcal{N},u\}_p(t,x)\,dx\,dy\,dt$$

$$\leq 2\|u\|_{L^\infty_t L^2_x(I\times\mathbf{R}^3)}^3 \|u\|_{L^\infty_t \dot{H}^1(I\times\mathbf{R}^d)} \quad (3.114)$$

$$+4\int_I\!\!\int\!\!\int |\{\mathcal{N},u\}_m(t,y)||u(t,x)||\nabla u(t,x)|\,dx\,dy\,dt. \quad (3.115)$$

When $d \geq 4$, we have

$$-\Delta\Delta|x| = -\left(\partial_{rr} + \frac{d-1}{r}\partial_r\right)\left(\partial_{rr} + \frac{d-1}{r}\partial_r\right)r = \frac{(d-1)(d-3)}{r^3},$$

and so in this case,

$$(d-1)(d-3)\int_I\int\int\frac{|u(t,x)|^2|u(t,y)|^2}{|x-y|^3}dxdt$$
$$+2\int_I\int\int|u(t,y)|^2\left[\frac{(x-y)}{|x-y|}\right]\cdot\{\mathcal{N},u\}_p(t,x)dxdydt \quad (3.116)$$
$$\leq 2\|u\|^3_{L_t^\infty L_x^2(I\times\mathbf{R}^d)}\|u\|_{L_t^\infty \dot H^1(I\times\mathbf{R}^d)}$$
$$+4\int_I\int\int|\{\mathcal{N},u\}_m(t,y)||u(t,x)||\nabla u(t,x)|dxdydt. \quad (3.117)$$

When $\mathcal{N}=|u|^p u$, recalling (1.113), we get $\{\mathcal{N},u\}_m=0$. By (1.118) and integrating by parts,

$$\int\int|u(t,y)|^2\frac{(x-y)_j}{|x-y|}\{\mathcal{N},u\}_p^j dxdy$$
$$=\frac{2p}{p+2}\int\int|u(t,y)|^2\frac{1}{|x-y|}|u(t,x)|^{p+2}dxdy\geq 0.$$

Therefore, when $d=3$,

$$\frac{2p}{p+2}\int_I\int\int|u(t,y)|^2\frac{1}{|x-y|}|u(t,x)|^{p+2}dxdydt+\|u\|^4_{L_{t,x}^4(I\times\mathbf{R}^3)}$$
$$\lesssim \|u\|^3_{L_t^\infty L_x^2(I\times\mathbf{R}^3)}\|u\|_{L_t^\infty \dot H^1(I\times\mathbf{R}^3)},$$

and both terms in the left-hand side are positive. This proves Lemma 3.23 when $d=3$. When $d\geq 4$, by (3.116)–(3.117),

$$\frac{2p}{p+2}\int_I\int\int|u(t,y)|^2\frac{1}{|x-y|}|u(t,x)|^{p+2}dxdydt$$
$$+(d-1)(d-3)\int_I\int\int|u(t,x)|^2|u(t,y)|^2\frac{1}{|x-y|^3}dxdydt$$
$$\lesssim_d \|u\|^3_{L_t^\infty L_x^2(I\times\mathbf{R}^d)}\|u\|_{L_t^\infty \dot H^1(I\times\mathbf{R}^d)}.$$

By the Littlewood–Paley theorem, if $K(x)$ is the Littlewood–Paley kernel,

$$\||\nabla|^{\frac{3-d}{4}}f\|^4_{L^4(\mathbf{R}^d)}\sim_d \left\|\sum_j 2^{\frac{j(3-d)}{2}}|P_jf|^2\right\|^2_{L^2(\mathbf{R}^d)}$$
$$=\left\|\sum_j 2^{\frac{3j(d+1)}{2}}\left|\int K(2^j(x-y))f(y)dy\right|^2\right\|_{L^2(\mathbf{R}^d)}$$
$$\lesssim \left\|\sum_j\left(2^{jd}\int|K(2^j(x-y))|dy\right)\left(2^{\frac{j(d+3)}{2}}\int|K(2^j(x-y))||f(y)|^2 dy\right)\right\|_{L^2(\mathbf{R}^d)}$$
$$\lesssim \left\|\sum_j 2^{\frac{j(d+3)}{2}}\int|K(2^j(x-y))||f(y)|^2 dy\right\|_{L^2(\mathbf{R}^d)}. \quad (3.118)$$

3.4 Interaction Morawetz Estimate

By (1.33),
$$\sum_j 2^{\frac{j(d+3)}{2}} \left| K\left(2^j (x-y)\right) \right| \lesssim \frac{1}{|x-y|^{\frac{d+3}{2}}}.$$

Expanding the inner product,

$$\left\| \int \frac{1}{|x-y|^{\frac{d+3}{2}}} |f(y)|^2 \, dy \right\|^2_{L^2(\mathbf{R}^d)}$$

$$= \left\langle \int \frac{1}{|x-y|^{\frac{d+3}{2}}} |f(y)|^2 \, dy, \int \frac{1}{|x-z|^{\frac{d+3}{2}}} |f(z)|^2 \, dz \right\rangle_{L^2}$$

$$\sim \int\int \sum_{j \leq k} \frac{1}{2^{j(\frac{d+3}{2})}} \frac{1}{2^{k(\frac{d+3}{2})}} \left(\int_{\substack{|x-y| \sim 2^j, \\ |x-z| \sim 2^k}} dx \right) |f(y)|^2 |f(z)|^2 \, dy \, dz. \quad (3.119)$$

By the triangle inequality, $2^k \gtrsim |y-z|$, so by Hölder's inequality, when $d > 3$,

$$(3.119) \lesssim \int\int \sum_{j \leq k, 2^k \gtrsim |y-z|} 2^{(j-k)(\frac{d-3}{2})} 2^{-3k} |f(y)|^2 |f(z)|^2 \, dy \, dz$$

$$\lesssim \int\int \frac{1}{|z-y|^3} |f(y)|^2 |f(z)|^2 \, dy \, dz.$$

This completes the proof of Lemma 3.23. $\square$

Remark The identity

$$\int\int \frac{1}{|x-y|^3} |u(t,x)|^2 |u(t,y)|^2 \, dx \, dy \sim_d \left\| |\nabla|^{\frac{3-d}{2}} |u|^2 \right\|^2_{L^2_{t,x}(\mathbf{R}^3)} \quad (3.120)$$

implies that the same argument also proves

$$\left\| |\nabla|^{\frac{3-d}{2}} |u|^2 \right\|^2_{L^2_{t,x}} \lesssim_d \|u\|^3_{L^\infty_t L^2_x} \|u\|_{L^\infty_t \dot{H}^1} \quad (3.121)$$

when $d \geq 3$. See Tao et al. (2007a) for example. Colliander et al. (2009) and Planchon and Vega (2009) proved that (3.121) also holds in dimensions $d = 1, 2$; the latter's proof used bilinear estimates, and is useful in the study of the mass-critical problem. We consider it further in the next chapter.

4

Mass-Critical NLS Problem in Higher Dimensions

4.1 Bilinear Estimates

Interaction Morawetz estimates also yield bilinear estimates, an essential component to the profile decomposition in the mass-critical case. We begin with the one-dimensional bilinear result of Planchon and Vega (2009).

Theorem 4.1 *Suppose that $\mu \geq 0$ and that u and v solve the nonlinear problems*

$$iu_t + \Delta u = F(u) = \mu |u|^p u, \quad u : \mathbf{R} \times \mathbf{R} \to \mathbf{C} \tag{4.1}$$

and

$$iv_t + \Delta v = F(v) = \mu |v|^p v, \quad v : \mathbf{R} \times \mathbf{R} \to \mathbf{C}. \tag{4.2}$$

Then

$$4\|\partial_x(u\bar{v})\|_{L^2_{t,x}(\mathbf{R}\times\mathbf{R})}^2 + \frac{p\mu}{p+2}\int\int |u(t,x)|^{p+2}|v(t,x)|^2 dx\,dt$$
$$+ \frac{p\mu}{p+2}\int\int |u(t,x)|^2 |v(t,x)|^{p+2} dx\,dt$$
$$\lesssim \|u\|_{L^\infty_t \dot{H}^1(\mathbf{R}\times\mathbf{R})} \|u\|_{L^\infty_t L^2_x(\mathbf{R}\times\mathbf{R})} \|v\|_{L^\infty_t L^2_x(\mathbf{R}\times\mathbf{R})}^2$$
$$+ \|v\|_{L^\infty_t \dot{H}^1(\mathbf{R}\times\mathbf{R})} \|v\|_{L^\infty_t L^2_x(\mathbf{R}\times\mathbf{R})} \|u\|_{L^\infty_t L^2_x(\mathbf{R}\times\mathbf{R})}^2. \tag{4.3}$$

Proof Define the quantity

$$M(t) = \int\int |x-y| |v(t,x)|^2 |u(t,y)|^2 dx\,dy. \tag{4.4}$$

4.1 Bilinear Estimates

Taking a derivative in time, (1.112) yields

$$\partial_t M(t) = 2 \int\int \frac{(x-y)}{|x-y|} \text{Im}[\bar{u}(t,x) \partial_x u(t,x)] |v(t,y)|^2 dx dy$$
$$+ 2 \int\int \frac{(x-y)}{|x-y|} \text{Im}[\bar{v}(t,x) \partial_x v(t,x)] |u(t,y)|^2 dx dy. \quad (4.5)$$

Then (1.112), (1.119), and (1.120) imply

$$\partial_{tt} M(t)$$

$$\left. \begin{aligned} &= \int\int \frac{(x-y)}{|x-y|} |v(t,y)|^2 \partial_x^3 \left(|u(t,x)|^2\right) dx dy \\ &+ \int\int \frac{(x-y)}{|x-y|} |u(t,y)|^2 \partial_x^3 \left(|v(t,x)|^2\right) dx dy \end{aligned} \right\} \quad (4.6)$$

$$\left. \begin{aligned} &- 4 \int\int \frac{(x-y)}{|x-y|} |v(t,y)|^2 \partial_x \left(|\partial_x u(t,x)|^2\right) dx dy \\ &- 4 \int\int \frac{(x-y)}{|x-y|} |u(t,y)|^2 \partial_x \left(|\partial_x v(t,y)|^2\right) dx dy \end{aligned} \right\} \quad (4.7)$$

$$\left. \begin{aligned} &- \frac{2p\mu}{p+2} \int\int \frac{(x-y)}{|x-y|} |v(t,y)|^2 \partial_x \left(|u(t,x)|^{p+2}\right) dx dy \\ &- \frac{2p\mu}{p+2} \int\int \frac{(x-y)}{|x-y|} |u(t,y)|^2 \partial_x \left(|v(t,x)|^{p+2}\right) dx dy \end{aligned} \right\} \quad (4.8)$$

$$\left. \begin{aligned} &- 4 \int\int \frac{(x-y)}{|x-y|} \text{Im}[\bar{u}(t,x) \partial_x u(t,x)] \partial_y \text{Im}[\bar{v}(t,y) \partial_y v(t,y)] dx dy \\ &- 4 \int\int \frac{(x-y)}{|x-y|} \text{Im}[\bar{v}(t,x) \partial_x v(t,x)] \partial_y \text{Im}[\bar{u}(t,y) \partial_y u(t,y)] dx dy. \end{aligned} \right\} \quad (4.9)$$

Approximating u and v with Schwartz functions, $\frac{(x-y)}{|x-y|}$ may be treated as a distribution in $\mathbf{R}^2_{x,y}$. Distributionally, $\partial_x \frac{(x-y)}{|x-y|} = 2\delta(x-y)$, so integrating by parts, we obtain

$$(4.8) = \frac{4p\mu}{p+2} \int\int |v(t,x)|^2 |u(t,x)|^{p+2} + |u(t,x)|^2 |v(t,x)|^{p+2} dx dy \geq 0. \quad (4.10)$$

Also,

$$(4.9) = -16 \int\int \text{Im}[\bar{u}(t,x) \partial_x u(t,x)] \text{Im}[\bar{v}(t,x) \partial_x v(t,x)] dx dy \quad (4.11)$$

and

$$(4.7) = 8 \int\int |v(t,y)|^2 |\partial_x u(t,x)|^2 dx dy + 8 \int\int |u(t,y)|^2 |\partial_x v(t,x)|^2 dx dy. \quad (4.12)$$

Finally, integrating by parts, we get

$$(4.6) = -2\int |v(t,x)|^2 \partial_x^2 \left(|u(t,x)|^2\right) dx - 2\int |u(t,x)|^2 \partial_x^2 \left(|v(t,x)|^2\right) dx$$
$$= 4\int \left(\partial_x |v(t,x)|^2\right) \left(\partial_x |u(t,x)|^2\right) dx. \tag{4.13}$$

Adding up yields

$$(4.11) + (4.12) + (4.13) = 8\int |\partial_x (u\bar{v})(t,x)|^2 dx \geq 0. \tag{4.14}$$

Therefore, by the fundamental theorem of calculus, Theorem 4.1 follows. □

Corollary 4.2 *If u solves (4.1) then*

$$\left\|\partial_x \left(|u|^2\right)\right\|_{L_{t,x}^2(\mathbf{R}\times\mathbf{R})}^2 \lesssim \|u\|_{L_t^\infty \dot{H}^1(\mathbf{R}\times\mathbf{R})} \|u\|_{L_t^\infty L_x^2(\mathbf{R}\times\mathbf{R})}^3. \tag{4.15}$$

Proof Take $u = v$. □

The estimate (4.15) is exactly (3.121) when $d = 1$.

Also notice that Theorem 4.1 implies that if u and v solve (4.1) and (4.2) respectively with $\mu = 0$, and if $k \leq j - 5$,

$$(P_j u)\left(\overline{P_k v}\right) = \tilde{P}_j \left((P_j u)\left(\overline{P_k v}\right)\right), \tag{4.16}$$

so by Bernstein's inequality and (4.3),

$$\left\|(P_j u)\left(\overline{P_k v}\right)\right\|_{L_{t,x}^2(\mathbf{R}\times\mathbf{R})}^2 \lesssim 2^{-2j}(2^j + 2^k)\|u_0\|_{L_x^2(\mathbf{R})}^2 \|v_0\|_{L_x^2(\mathbf{R})}^2$$
$$\lesssim 2^{-j}\|u_0\|_{L_x^2(\mathbf{R})}^2 \|v_0\|_{L_x^2(\mathbf{R})}^2. \tag{4.17}$$

Then since

$$\left\|(P_j u)(P_k v)\right\|_{L_{t,x}^2}^2 = \left\|(P_j u)\left(\overline{P_k v}\right)\overline{(P_j u)}(P_k v)\right\|_{L_{t,x}^1},$$

by (4.17) and Hölder's inequality it is easy to see that

$$\left\|(P_j u)(P_k v)\right\|_{L_{t,x}^2(\mathbf{R}\times\mathbf{R})}^2 \lesssim 2^{-2j}(2^j + 2^k)\|u_0\|_{L_x^2(\mathbf{R})}^2 \|v_0\|_{L_x^2(\mathbf{R})}^2$$
$$\lesssim 2^{-j}\|u_0\|_{L_x^2(\mathbf{R})}^2 \|v_0\|_{L_x^2(\mathbf{R})}^2.$$

Such bilinear estimates are invariant under the Galilean transformation.

Theorem 4.3 *Let $\xi(t) : I \to \mathbf{R}$ be a function of time. If u and v solve (4.1) and (4.2) respectively, then*

$$\left\|\partial_x (u\bar{v})\right\|_{L_{t,x}^2(I\times\mathbf{R})}^2 \lesssim \|v\|_{L_t^\infty L_x^2(I\times\mathbf{R})}^2 \left\|e^{-ix\cdot\xi(t)}u\right\|_{L_t^\infty \dot{H}^1(I\times\mathbf{R}^d)} \|u\|_{L_t^\infty L_x^2(I\times\mathbf{R})}$$
$$+ \|u\|_{L_t^\infty L_x^2(I\times\mathbf{R})}^2 \left\|e^{-ix\cdot\xi(t)}v\right\|_{L_t^\infty \dot{H}^1(I\times\mathbf{R}^d)} \|v\|_{L_t^\infty L_x^2(I\times\mathbf{R})}.$$
$$\tag{4.18}$$

4.1 Bilinear Estimates

Proof Split $\operatorname{Im}[\bar{u}\partial_x u]$ as follows:
$$\operatorname{Im}[\bar{u}\partial_x u] = \operatorname{Im}\left[\bar{u}\partial_x\left(e^{-ix\cdot\xi(t)}u\right)\right] + \xi(t)|u|^2.$$

Since $\frac{(x-y)}{|x-y|}$ is an odd function, we get
$$\xi(t)\int\int \frac{(x-y)}{|x-y|}|u(t,x)|^2|v(t,y)|^2 dxdy$$
$$+\xi(t)\int\int \frac{(x-y)}{|x-y|}|v(t,x)|^2|u(t,y)|^2 dxdy = 0,$$

and therefore $|\partial_t M(t)|$ is bounded by the right-hand side of (4.18). Then (4.14) proves Theorem 4.3. □

For a solution to the linear Schrödinger equation, more may be said. Let $\chi \in C_0^\infty(\mathbf{R})$ be a partition of unity function, that is, a function supported on $[-2,2]$ such that for any $x \in \mathbf{R}$,
$$\sum_{m\in\mathbf{Z}} \chi(x-m) = 1.$$

Furthermore, let P_m be the Fourier multiplier such that for any f,
$$\mathscr{F}(P_m f) = \chi\left(2^{-j}(\xi - m)\right)\hat{f}(\xi).$$

Then by Theorem 4.3, we have the next result.

Theorem 4.4 *Suppose u and v solve (4.1) and (4.2) with initial data u_0 and v_0 respectively. Then for any $j \in \mathbf{Z}$,*
$$\left\|P_j(u\bar{v})\right\|_{L^2_{t,x}(\mathbf{R}\times\mathbf{R})} \lesssim 2^{-j/2}\|u_0\|_{L^2(\mathbf{R})}\|v_0\|_{L^2(\mathbf{R})}.$$

Proof By the triangle inequality and the Cauchy–Schwarz inequality,
$$\left\|P_j(u\bar{v})\right\|_{L^2_{t,x}(\mathbf{R}\times\mathbf{R})} \lesssim 2^{-j/2} \sum_{|m+n|\leq 3} \|P_m u_0\|_{L^2(\mathbf{R})}\|P_n v_0\|_{L^2(\mathbf{R})}$$
$$\lesssim 2^{-j/2}\|u_0\|_{L^2}\|v_0\|_{L^2}.$$
□

Theorem 4.4 may also be proved directly using Fourier analysis.

Proof using Fourier analysis Let $\|F(t,x)\|_{L^2_{t,x}(\mathbf{R}\times\mathbf{R})} = 1$, such that $\hat{F}(t,\xi)$ is supported on $|\xi| \sim 2^j$. By Hölder's inequality and the Plancherel identity it suffices to estimate
$$\int\int \hat{F}(t,\xi+\eta)\hat{u}(t,\eta)\hat{\bar{v}}(t,\xi)d\xi d\eta dt$$
$$= C\int\int \tilde{F}\left(-\xi^2+\eta^2,\eta+\xi\right)\hat{u}_0(\eta)\hat{\bar{v}}_0(\xi)d\eta d\xi. \qquad (4.19)$$

Making a change of variables,

$$\int |\tilde{F}(\eta^2 - \xi^2, \eta + \xi)|^2 d\eta\, d\xi \sim \int\int \frac{1}{|\xi + \eta|} |\tilde{F}(\tilde{\eta}, \tilde{\xi})|^2 d\tilde{\xi}\, d\tilde{\eta} \lesssim 2^{-j}.$$

Therefore,

$$(4.19) \lesssim 2^{-j/2} \|u_0\|_{L^2(\mathbf{R}^d)} \|v_0\|_{L^2(\mathbf{R}^d)}. \qquad \square$$

Corollary 4.5 *If the supports of $\hat{u}_0(\xi)$ and $\hat{v}_0(\xi)$ are separated by distance $\sim 2^j$, then*

$$\|uv\|_{L^2_{t,x}(\mathbf{R} \times \mathbf{R})} \lesssim 2^{-j/2} \|u_0\|_{L^2(\mathbf{R})} \|v_0\|_{L^2(\mathbf{R})}. \tag{4.20}$$

Proof Again write

$$\|uv\|^2_{L^2_{t,x}} = \|u\bar{v}\bar{u}v\|_{L^1_{t,x}} = \|u\bar{v}\|^2_{L^2_{t,x}}. \tag{4.21}$$

By the Fourier support properties of u and v we have

$$u\bar{v} = \widetilde{P}_j(u\bar{v}), \tag{4.22}$$

and so by Theorem 4.4 the proof is complete. $\qquad \square$

Theorem 4.4 and Corollary 4.5 may be extended to higher dimensions. Both the Fourier-analytic arguments of Bourgain (1998), Colliander *et al.* (2008), and Killip and Visan (2013), and the interaction Morawetz estimates of Planchon and Vega (2009) will be presented here, since both are useful in the study of the mass-critical problem.

Theorem 4.6 *Suppose u and v are solutions to (4.1) and (4.2) with $\mu = 0$ (linear problem) with initial data u_0 and v_0 in dimension $d \geq 2$. Suppose $\hat{u}_0(\xi)$ is supported on $|\xi| \sim 2^j$ and $\hat{v}_0(\xi)$ is supported on $|\xi| \sim 2^k$ with $k \leq j - 5$. Then*

$$\|(P_j u)(P_k v)\|_{L^2_{t,x}(\mathbf{R} \times \mathbf{R}^d)} + \|(P_j u)\overline{(P_k v)}\|_{L^2_{t,x}(\mathbf{R} \times \mathbf{R}^d)}$$

$$\lesssim_d \frac{2^{k(\frac{d-1}{2})}}{2^{j/2}} \|P_j u_0\|_{L^2(\mathbf{R}^d)} \|P_k v_0\|_{L^2(\mathbf{R}^d)}. \tag{4.23}$$

Proof Because $\hat{u}_0(\xi)$ is supported on $|\xi| \sim 2^j$, it is possible to decompose u_0 such that

$$u_0 = u_{0,1} + \cdots + u_{0,d},$$

where $\hat{u}_{0,i}(\xi)$ is supported on $|\xi_i| \gtrsim_d 2^j$. For example, in the case $d = 2$, let $\chi: \mathbf{R} \to \mathbf{R}$ be a cutoff function such that $\chi(\xi) = 1$ for $2^{j-5} \leq |\xi| \leq 2^{j+5}$ and χ is supported on $2^{j-6} \leq |\xi| \leq 2^{j+6}$. Then let

$$\hat{u}_{0,1}(\xi) = \chi(\xi_1)\hat{u}_0(\xi) \quad \text{and} \quad \hat{u}_{0,2}(\xi) = (1 - \chi(\xi_1))\hat{u}_0(\xi).$$

4.1 Bilinear Estimates

By construction, $\hat{u}_{0,1}$ is supported on $|\xi_1| \sim 2^j$. Also, since $\hat{u}_{0,2}(\xi)$ is supported on $|\xi_1| \ll 2^j$ and $\hat{u}_0(\xi)$ is supported on $|\xi| \sim 2^j$, it follows that $\hat{u}_{0,2}(\xi)$ is supported on $|\xi_2| \sim 2^j$. The decomposition for a general $d > 2$ is similar.

Relabeling $u_0 = u_{0,1}$ and integrating in ξ_1, η_1,

$$\int\int\int \hat{F}(t, \xi + \eta) u(t, \xi) v(t, \eta) d\xi\, d\eta\, dt$$

$$= C(d) \int\int \widetilde{F}(|\xi|^2 - |\eta|^2, \xi + \eta) \hat{u}_0(\xi) \hat{v}_0(\eta) d\xi\, d\eta$$

$$\lesssim \int \left\| \widetilde{F}(|\xi|^2 - |\eta|^2, \xi_1 + \eta_1, \xi_2 + \eta_2, \ldots, \xi_d + \eta_d) \right\|_{L^2_{\xi_1, \eta_1}(\mathbf{R} \times \mathbf{R})}$$

$$\times \left\| \hat{u}_0(\xi_1, \xi_2, \ldots, \xi_d) \right\|_{L^2_{\xi_1}(\mathbf{R})} \left\| \hat{v}_0(\eta_1, \eta_2, \ldots, \eta_d) \right\|_{L^2_{\eta_1}(\mathbf{R})} d\underline{\eta}\, d\underline{\xi}, \quad (4.24)$$

where for clarity we have written $d\underline{\eta} = d\eta_2 \cdots d\eta_d$ and $d\underline{\xi} = d\xi_2 \cdots d\xi_d$. Then by the proof of Theorem 4.4, the support of $\hat{v}_0(\xi)$, and Hölder's inequality,

$$(4.24) \lesssim_d \frac{2^{k\left(\frac{d-1}{2}\right)}}{2^{j/2}} \|u_0\|_{L^2(\mathbf{R}^d)} \|v_0\|_{L^2(\mathbf{R}^d)}.$$

A similar argument may be made for $u_{0,2}$. Then by (4.21) and (4.22), the proof is complete. $\square$

For the second proof we will need the following definition.

Definition 4.7 (Radon transform) For a general function f, the Radon transform is given by

$$\mathscr{R}(f)(s, \omega) = \int_{x \cdot \omega = s} f\, d\mu_{s,\omega},$$

where $d\mu_{s,\omega}$ is the Lebesgue measure on the hyperplane $\{x : x \cdot \omega = s\}$.

Second proof of Theorem 4.6 For any $\omega \in S^{d-1}$, let $x_\omega = x \cdot \omega$ and let $\partial_\omega = \omega \cdot \nabla$. Then there exists some $C(d)$ such that

$$\int |x \cdot \omega| d\omega = \frac{1}{C(d)} |x|, \quad \int \frac{x_\omega}{|x_\omega|} (\omega \cdot \nabla) d\omega = \frac{1}{C(d)} \nabla, \quad (4.25)$$

and

$$M(t) = \int\int |u(t,x)|^2 |u(t,y)|^2 |x - y| dx\, dy$$

$$= C(d) \int\int |u(t,x)|^2 |u(t,y)|^2 \int_{S^{d-1}} |x_\omega - y_\omega| d\omega\, dx\, dy.$$

Now fix $\omega \in S^{d-1}$, and let

$$M_\omega(t) = \int\int |x_\omega - y_\omega| |v(t,x)|^2 |u(t,y)|^2 dx\, dy.$$

Without loss of generality suppose $\omega = (1,0,0,\ldots,0)$. Taking a derivative in time and following (4.4)–(4.14),

$$\partial_t M_\omega(t) = 2\int\int \frac{(x_1-y_1)}{|x_1-y_1|} \operatorname{Im}[\bar{u}(t,x)\partial_{x_1}u(t,x)]|v(t,y)|^2 dx dy$$
$$+ 2\int\int \frac{(x_1-y_1)}{|x_1-y_1|} \operatorname{Im}[\bar{v}(t,x)\partial_{x_1}v(t,x)]|u(t,y)|^2 dx dy, \quad (4.26)$$

and

$$\partial_{tt} M_\omega(t)$$
$$= \int\int \frac{(x_1-y_1)}{|x_1-y_1|}|v(t,y)|^2 \partial_{x_1}\Delta\left(|u(t,x)|^2\right) dx dy$$
$$+ \int\int \frac{(x_1-y_1)}{|x_1-y_1|}|u(t,y)|^2 \partial_{x_1}\Delta\left(|v(t,x)|^2\right) dx dy$$
$$\left.\begin{array}{l}-4\int\int \frac{(x_1-y_1)}{|x_1-y_1|}|v(t,y)|^2 \partial_k \operatorname{Re}\left(\partial_{x_1}u \cdot \overline{\partial_k u}\right)(t,x) dx dy \\ -4\int\int \frac{(x_1-y_1)}{|x_1-y_1|}|u(t,y)|^2 \partial_k \operatorname{Re}\left(\partial_{x_1}v \cdot \overline{\partial_k v}\right)(t,x) dx dy\end{array}\right\} \quad (4.27)$$

$$\left.\begin{array}{l}-\frac{2p\mu}{p+2}\int\int \frac{(x_1-y_1)}{|x_1-y_1|}|v(t,y)|^2 \partial_{x_1}\left(|u(t,x)|^{p+2}\right) dx dy \\ -\frac{2p\mu}{p+2}\int\int \frac{(x_1-y_1)}{|x_1-y_1|}|u(t,y)|^2 \partial_{x_1}\left(|v(t,x)|^{p+2}\right) dx dy\end{array}\right\} \quad (4.28)$$

$$\left.\begin{array}{l}-4\int\int \frac{(x_1-y_1)}{|x_1-y_1|} \operatorname{Im}[\bar{u}(t,x)\partial_{x_1}u(t,x)]\partial_k \operatorname{Im}[\bar{v}(t,y)\partial_k v(t,y)] dx dy \\ -4\int\int \frac{(x_1-y_1)}{|x_1-y_1|} \operatorname{Im}[\bar{v}(t,x)\partial_{x_1}v(t,x)]\partial_k \operatorname{Im}[\bar{u}(t,y)\partial_k u(t,y)] dx dy.\end{array}\right\} \quad (4.29)$$

Integrating by parts, since $\mu \geq 0$,

$$(4.28) = \frac{2p\mu}{p+2}\int\int_{x_1=y_1} |v(t,y)|^2|u(t,x)|^{p+2} + |u(t,y)|^2|v(t,x)|^{p+2} dx dy \geq 0.$$

Remark When $\mu = 0$ this term does not appear, but this computation with $\mu > 0$ will prove to be useful in later study of the nonlinear Schrödinger equation.

Adding up as in (4.14),

$$(4.27) + (4.28) + (4.29)$$
$$= 8\int\int \left|\partial_{x_1}\left(u(t,x_1,x_2,\ldots,x_d)\overline{v(t,x_1,y_2,\ldots,y_d)}\right)\right|^2 dx dy. \quad (4.30)$$

Since the same computations may be performed for any $\omega \in S^{d-1}$, by the fundamental theorem of calculus, (4.30) gives an estimate for the Radon transform of u,

$$\int_0^T \int_{-\infty}^\infty \left(\partial_s \mathscr{R}\left(|u|^2\right)(s,\omega)\right)^2 ds\, dt \lesssim |\partial_t M_\omega(T) - \partial_t M_\omega(0)|. \tag{4.31}$$

Since

$$\int_0^T \int_{S^{d-1}} \int_{-\infty}^\infty \left(\partial_s \mathscr{R}\left(|u(t,x)|^2\right)(s,\omega)\right)^2 ds\, dt\, d\omega$$
$$= 2\left\| |\nabla|^{\frac{d-3}{2}} |u(t,x)|^2 \right\|_{L^2_{t,x}([0,T]\times \mathbf{R}^d)}^2, \tag{4.32}$$

it follows that (3.93) holds for all $d \geq 1$.

It is possible to wring even more information from (4.30). Suppose for example that $u = P_j u$ and $v = P_k v$ with $k \leq j - 5$. Then by Hölder's inequality, for any $\omega \in S^{d-1}$,

$$\int\int\int_{h\in \mathbf{R}^d: |h|\leq R} \left|\partial_\omega\left(u(t,x)\overline{v(t,x+h)}\right)\right|^2 dx\, dh\, dt \lesssim 2^j R \|u_0\|_{L^2}^2 \|v_0\|_{L^2}^2. \tag{4.33}$$

Averaging over $\omega \in S^{d-1}$, by (4.25) and the fact that $\mathscr{F}(u\bar{v})$ is supported on $|\xi| \sim 2^j$, we obtain

$$\int\int\int_{h\in \mathbf{R}^d: |h|\leq R} \left|\nabla\left(u(t,x)\overline{v(t,x+h)}\right)\right|^2 dx\, dh\, dt \lesssim 2^j R \|u_0\|_{L^2}^2 \|v_0\|_{L^2}^2,$$

and therefore by Bernstein's inequality, and the support of $\mathscr{F}(u\bar{v})$,

$$\int\int\int_{h\in \mathbf{R}^d: |h|\leq R} \left|\left(u(t,x)\overline{v(t,x+h)}\right)\right|^2 dx\, dh\, dt \lesssim 2^{-j} R \|u_0\|_{L^2}^2 \|v_0\|_{L^2}^2. \tag{4.34}$$

By the Littlewood–Paley decomposition,

$$u(t,x)\overline{v(t,x)} = u(t,x)\int \overline{K_k(x-y)v(t,y)}\, dy,$$

and then by (1.34) and (4.34),

$$\|u\bar{v}\|_{L^2_{t,x}(\mathbf{R}\times\mathbf{R}^d)}^2 \lesssim 2^{-j} 2^{k(d-1)} \|u_0\|_{L^2}^2 \|v_0\|_{L^2}^2. \tag{4.35}$$

This completes the second proof of Theorem 4.6. $\square$

4.2 Mass-Critical Profile Decomposition

The profile decomposition for the mass-critical nonlinear Schrödinger equation,

$$iu_t + \Delta u = F(u) = |u|^{\frac{4}{d}}u, \qquad u(0,x) = u_0 \in L^2(\mathbf{R}^d), \tag{4.36}$$

has an additional complication over the energy-critical profile decomposition, namely that there is no canonical choice for the origin in frequency. For the energy-critical problem,

$$\|u\|_{\dot{H}^1(\mathbf{R}^d)} = \big\| |\xi| \hat{u}(\xi) \big\|_{L^2(\mathbf{R}^d)}$$

is a weighted L^2-space in frequency, and therefore has a clear, canonical choice of origin. This is not so for the mass-critical problem. Instead, a profile decomposition for the mass-critical problem must take into account the Galilean transform. While the Galilean transform will certainly change the energy of a solution to the nonlinear Schrödinger problem, and thus a solution will no longer be a minimal energy blowup solution after applying a Galilean transform, the Galilean transform will not change the L^2-norm of a solution, and therefore, the symmetry group for solutions of (4.36) must include the Galilean transformation. The existence of a profile localized around some $\xi_n \in \mathbf{R}^d$ may be quantified by showing that the interaction of u with itself is large. This is accomplished using a bilinear Strichartz estimate.

Define the group action

$$g_{\theta,\xi_0,x_0,\lambda} u = e^{i\theta} e^{ix\cdot\xi_0} \frac{1}{\lambda^{d/2}} u\left(\frac{x-x_0}{\lambda}\right). \tag{4.37}$$

Remark The group action $f \mapsto e^{i\theta} f$ is a compact group action, since $S^1 \subset \mathbf{C}$ is compact. The other components ξ_0, x_0, and λ belong to the noncompact groups $\mathbf{R}^d$, $\mathbf{R}^d$, and $(0,\infty)$.

This group obeys the multiplication law

$$g_{\theta,\xi_0,x_0,\lambda} \cdot g_{\theta',\xi_0',x_0',\lambda'} = g_{\theta+\theta'-\frac{x_0\cdot\xi_0'}{\lambda},\xi_0+\frac{\xi_0'}{\lambda},x_0+\lambda x_0',\lambda\lambda'}, \tag{4.38}$$

and each element has the inverse

$$\left(g_{\theta,\xi_0,x_0,\lambda}\right)^{-1} = g_{-\theta-x_0\cdot\xi_0,-\lambda\xi_0,-\frac{x_0}{\lambda},\frac{1}{\lambda}}. \tag{4.39}$$

By (1.39) and (1.42),

$$e^{it\Delta} g_{\theta,\xi_0,x_0,\lambda} u_0 = g_{\theta-t|\xi_0|^2,\xi_0,x_0+2t\xi_0,\lambda} \left[e^{i\frac{t}{\lambda^2}\Delta} u_0\right]. \tag{4.40}$$

A profile may be extracted modulo these symmetries.

Theorem 4.8 *Suppose f_n is a bounded sequence of functions such that*

$$\lim_{n\to\infty} \|f_n\|_{L^2(\mathbf{R}^d)} = A \tag{4.41}$$

4.2 Mass-Critical Profile Decomposition

and
$$\lim_{n\to\infty} \left\| e^{it\Delta} f_n \right\|_{L_{t,x}^{\frac{2(d+2)}{d}}(\mathbf{R}\times\mathbf{R}^d)} = \varepsilon > 0. \tag{4.42}$$

Then passing to a subsequence, there exists $\phi \in L^2(\mathbf{R}^d)$, $\lambda_n \in (0,\infty)$, $\xi_n, x_n \in \mathbf{R}^d$, $t_n \in \mathbf{R}$, and $\alpha(d) > 0$, such that

$$\lambda_n^{d/2} e^{-i\xi_n \cdot (\lambda_n x + x_n)} \left[e^{it_n\Delta} f_n \right](\lambda_n x + x_n) \rightharpoonup \phi(x), \text{ weakly}, \tag{4.43}$$

and
$$\lim_{n\to\infty} \|f_n\|_{L^2(\mathbf{R}^d)}^2 - \|f_n - \phi_n\|_{L^2(\mathbf{R}^d)}^2 = \|\phi\|_{L^2(\mathbf{R}^d)}^2 \gtrsim \varepsilon^2 \left(\frac{\varepsilon}{A}\right)^\alpha, \tag{4.44}$$

where
$$\phi_n(x) = e^{-it_n\Delta} g_{0,\xi_n,x_n,\lambda_n}\phi = \frac{1}{\lambda_n^{d/2}} e^{-it_n\Delta}\left[e^{i\xi_n \cdot x} \phi\left(\frac{x - x_n}{\lambda_n}\right)\right].$$

Proof First, generalize dyadic intervals to any dimension d.

Definition 4.9 (Dyadic cubes) Let D_j be the set of all dyadic cubes of side length 2^j,
$$D_j = \left\{ \prod_{l=1}^d [2^j k_l, 2^j (k_l + 1)) \subset \mathbf{R}^d : k \in \mathbf{Z}^d \right\}.$$

Then let
$$\mathcal{D} = \bigcup_{j\in\mathbf{Z}} D_j,$$

and for any $Q \in \mathcal{D}$, let
$$\hat{f}_Q(\xi) = \chi_Q(\xi) \hat{f}(\xi),$$

where $\chi_Q(\xi)$ is the characteristic function of Q.

Then, make a Whitney decomposition in $\mathbf{R}^d$ in order to effectively utilize bilinear Strichartz estimates for estimating $\left\| [e^{it\Delta} u_0][\overline{e^{it\Delta} u_0}] \right\|_{L_{t,x}^p}$, for some $1 \le p \le \infty$.

By Lemma 1.15 it is possible to partition $\mathbf{R}^d \times \mathbf{R}^d$ into a union of pairs of dyadic cubes:
$$\mathbf{R}^d \times \mathbf{R}^d = \bigcup_{Q\in\mathcal{D}: Q \sim Q'} Q \times Q', \tag{4.45}$$

where $Q \sim Q'$ if $\text{dist}(Q,Q') \sim \text{diam}(Q) = \text{diam}(Q')$. First consider the case when $d = 1$.

Theorem 4.10 *Suppose* $\text{dist}(Q,Q') \gtrsim \text{diam}(Q) = \text{diam}(Q')$. *Then,*
$$\left\| [e^{it\Delta} f_Q][e^{it\Delta} f_{Q'}] \right\|_{L_{t,x}^{13/5}(\mathbf{R}\times\mathbf{R})} \lesssim |Q|^{-5/13} \|\hat{f}\|_{L_\xi^{13/8}(Q)} \|\hat{f}\|_{L_\xi^{13/8}(Q')}. \tag{4.46}$$

Proof By (1.11),
$$\| [e^{it\Delta} f_Q] [e^{it\Delta} f_{Q'}] \|_{L^\infty_{t,x}(\mathbf{R}\times\mathbf{R})} \lesssim \|\hat{f}\|_{L^1_\xi(Q)} \|\hat{f}\|_{L^1_\xi(Q')}. \tag{4.47}$$

Also, by (4.20),
$$\| [e^{it\Delta} f_Q] [e^{it\Delta} f_{Q'}] \|_{L^2_{t,x}(\mathbf{R}\times\mathbf{R})} \lesssim |Q|^{-1/2} \|\hat{f}\|_{L^2_\xi(Q)} \|\hat{f}\|_{L^2_\xi(Q')}.$$

Then (4.46) holds by interpolation. □

The importance of the exponents $\frac{13}{5}$ and $\frac{13}{8}$ in (4.46) is that $\frac{13}{5} < 3$ and $\frac{13}{8} < 2$. Inequality (4.46) with $\frac{13}{5}$ replaced by 3 and $\frac{13}{8}$ replaced by 2 follows directly from Hölder's inequality and Strichartz estimates. The fact that $\frac{13}{5} < 3$ means that the left-hand side of (4.46) is mass subcritical, and therefore may be interpolated with an $L^\infty_{t,x}$ estimate, which can then be used to extract a profile from a subsequence, as in the energy-critical profile decomposition.

If $d \geq 2$, then $2 \geq \frac{d+2}{d}$, so any interpolation of (4.23) with (4.47) will necessarily be mass supercritical. Therefore, the profile decomposition will instead utilize the following bilinear estimate of Tao (2003), given without proof.

Theorem 4.11 *Suppose u_0 and u_1 are L^2 functions supported on sets $K_0, K_1 \subset \mathbf{R}^d$ with diameter $\leq R$ and separated by distance $\sim R$. Then for any $q > \frac{d+3}{d+1}$,*
$$\| [e^{it\Delta} u_0] [e^{it\Delta} u_1] \|_{L^q_{t,x}(\mathbf{R}\times\mathbf{R}^d)} \lesssim_{q,d} R^{d-\frac{d+2}{q}} \|u_0\|_{L^2(\mathbf{R}^d)} \|u_1\|_{L^2(\mathbf{R}^d)}.$$

Proof See Tao (2003). □

This yields the analogue of Theorem 4.10 in dimensions $d \geq 2$.

Theorem 4.12 *Suppose $\mathrm{dist}(Q, Q') \gtrsim \mathrm{diam}(Q) = \mathrm{diam}(Q')$. Then for some $p < 2$,*
$$\| [e^{it\Delta} f_Q] [e^{it\Delta} f_{Q'}] \|_{L^{\frac{d^2+3d+1}{d(d+1)}}_{t,x}(\mathbf{R}\times\mathbf{R}^d)} \lesssim_{d,p} |Q|^{1-\frac{2}{p}-\frac{1}{d^2+3d+1}} \|\hat{f}\|_{L^p_\xi(Q)} \|\hat{f}\|_{L^p_\xi(Q')}. \tag{4.48}$$

Remark As in Theorem 4.10 for $d = 1$, (4.48) is mass subcritical since $\frac{d^2+3d+1}{d(d+1)} < \frac{d+2}{d}$ and $p < 2$.

Proof Again by (1.11),
$$\| [e^{it\Delta} f_Q] [e^{it\Delta} f_{Q'}] \|_{L^\infty_{t,x}(\mathbf{R}\times\mathbf{R}^d)} \lesssim_d \|\hat{f}\|_{L^1_\xi(Q)} \|\hat{f}\|_{L^1_\xi(Q')}. \tag{4.49}$$

By Theorem 4.11, for some $q > \frac{d+3}{d+1}$,
$$\| [e^{it\Delta} f_Q] [e^{it\Delta} f_{Q'}] \|_{L^q_{t,x}(\mathbf{R}\times\mathbf{R}^d)} \lesssim_q |Q|^{1-\frac{d+2}{dq}} \|\hat{f}\|_{L^2_\xi(Q)} \|\hat{f}\|_{L^2_\xi(Q')}. \tag{4.50}$$

4.2 Mass-Critical Profile Decomposition

Therefore, (4.48) holds for $\frac{2(d^2+3d+1)}{d^2+3d+2} < p \leq 2$. $\square$

Remark If the endpoint case ($q = \frac{d+3}{d+1}$) of Theorem 4.11 were proved then (4.48) would hold for $\frac{2(d^2+3d+1)}{d^2+3d+2} \leq p \leq 2$. When $p = 2$, (4.48) follows directly from (4.50). If (4.50) also held for $q = \frac{d+3}{d+1}$, then interpolating (4.50) (for $q = \frac{d+3}{d+1}$) with (4.49) would yield (4.48) with $p = \frac{2(d^2+3d+1)}{d^2+3d+2}$.

Theorems 4.10 and 4.12 combined with the Whitney decomposition imply the following proposition.

Proposition 4.13 When $d = 1$,

$$\|e^{it\Delta}f\|_{L^6_{t,x}(\mathbf{R}\times\mathbf{R})} \lesssim \|f\|_{L^2_x(\mathbf{R})}^{13/15} \left(\sup_{Q\in\mathscr{D}} |Q|^{-\frac{1}{2}} \|e^{it\Delta}f_Q\|_{L^\infty_{t,x}(\mathbf{R}\times\mathbf{R}^d)}\right)^{2/15}. \quad (4.51)$$

For $d \geq 2$, let $q = \frac{2(d^2+3d+1)}{d^2}$. Then

$$\|e^{it\Delta}f\|_{L^{\frac{2(d+2)}{d}}_{t,x}(\mathbf{R}\times\mathbf{R}^d)} \lesssim_d \|f\|_{L^2_x(\mathbf{R}^d)}^{\frac{d+1}{d+2}} \left(\sup_{Q\in\mathscr{D}} |Q|^{\frac{d+2}{dq}-\frac{1}{2}} \|e^{it\Delta}f_Q\|_{L^q_{t,x}(\mathbf{R}\times\mathbf{R}^d)}\right)^{\frac{1}{d+2}}. \quad (4.52)$$

Proposition 4.13 directly implies the analogue (3.27) for the mass-critical problem. As in the energy-critical case, $\alpha(d)$ may change from line to line.

By Alaoglu's theorem, a ball in $L^2_x(\mathbf{R}^d)$ is weakly compact. When $d = 1$, (4.41), (4.42), and (4.51) directly imply that after passing to a subsequence there exists a sequence of cubes $\{Q_n\} \subset \mathscr{D}$ such that

$$\varepsilon\left(\frac{\varepsilon}{A}\right)^{\frac{13}{2}} \lesssim |Q_n|^{-1/2} \|e^{it\Delta}(f_n)_{Q_n}\|_{L^\infty_{t,x}(\mathbf{R}\times\mathbf{R})}.$$

When $d \geq 2$, (4.41), (4.42), and (4.52) imply

$$\varepsilon^{d+2}A^{-(d+1)} \lesssim \lim_{n\to\infty} |Q_n|^{\frac{d+2}{dq}-\frac{1}{2}} \|e^{it\Delta}(f_n)_{Q_n}\|_{L^q_{t,x}(\mathbf{R}\times\mathbf{R}^d)}. \quad (4.53)$$

Let λ_n^{-1} be the side length of Q_n and let ξ_n be the center of Q_n. By Hölder's inequality, (4.53), and Strichartz estimates,

$$\varepsilon^{d+2}A^{-(d+1)} \lesssim \liminf_{n\to\infty} \lambda_n^{\frac{d}{2}-\frac{d+2}{q}} \|e^{it\Delta}(f_n)_{Q_n}\|_{L^{\frac{2(d+2)}{d}}_{t,x}}^{\frac{d(d+2)}{d^2+3d+1}} \|e^{it\Delta}(f_n)_{Q_n}\|_{L^\infty_{t,x}}^{\frac{d+1}{d^2+3d+1}}$$

$$\lesssim \liminf_{n\to\infty} \lambda_n^{\frac{d}{2}-\frac{d+2}{q}} A^{\frac{d(d+2)}{d^2+3d+1}} \|e^{it\Delta}(f_n)_{Q_n}\|_{L^\infty_{t,x}}^{\frac{d+1}{d^2+3d+1}}.$$

Therefore, there exist sequences $t_n \in \mathbf{R}$, $x_n \in \mathbf{R}^d$ such that

$$\varepsilon\left(\frac{\varepsilon}{A}\right)^\alpha \lesssim \liminf_{n\to\infty} \lambda_n^{-\frac{d}{2}} \left|e^{it_n\Delta}(f_n)_{Q_n}\right|(x_n).$$

By the Sobolev embedding theorem,

$$\left\|\nabla\left(e^{-ix\cdot\xi_n}e^{it_n\Delta}(f_n)_{Q_n}\right)\right\|_{L_x^\infty(\mathbf{R}^d)} \lesssim \lambda_n^{-\frac{d}{2}-1} A,$$

and therefore there exists some $\psi \in L_x^2(\mathbf{R}^d)$, $\|\psi\|_{L^2} = 1$, such that

$$\liminf_{n\to\infty}\left|\left\langle e^{-i(x+x_n)\cdot\xi_n}e^{it_n\Delta}(f_n)_{Q_n}(x+x_n), \frac{1}{\lambda_n^{d/2}}\psi\left(\frac{x}{\lambda_n}\right)\right\rangle\right|$$

$$= \liminf_{n\to\infty}\left|\left\langle \lambda_n^{d/2} e^{-i(\lambda_n x+x_n)\cdot\xi_n}\left[e^{it_n\Delta}f_n\right](\lambda_n x+x_n), \psi\right\rangle\right| \gtrsim \varepsilon\left(\frac{\varepsilon}{A}\right)^\alpha.$$

Therefore,

$$g_{-x_n\cdot\xi_n,-\lambda_n\xi_n,\frac{-x_n}{\lambda_n},\frac{1}{\lambda_n}}\left[e^{it_n\Delta}f_n\right] = g_{0,\xi_n,x_n,\lambda_n}^{-1}\left[e^{it_n\Delta}f_n\right] \quad (4.54)$$

has a subsequence which converges weakly to some $\phi(x) \in L_x^2(\mathbf{R}^d)$ with $\|\phi\|_{L_x^2(\mathbf{R}^d)} \gtrsim \varepsilon\left(\frac{\varepsilon}{A}\right)^\alpha$. This proves (4.43).

Weak convergence of (4.54) implies

$$\left\langle g_{0,\xi_n,x_n,\lambda_n}^{-1}\left[e^{it_n\Delta}f_n\right], g_{0,\xi_n,x_n,\lambda_n}^{-1}\left[e^{it_n\Delta}f_n\right]\right\rangle_{L^2}$$

$$- \left\langle g_{0,\xi_n,x_n,\lambda_n}^{-1}\left[e^{it_n\Delta}f_n\right] - \phi, g_{0,\xi_n,x_n,\lambda_n}^{-1}\left[e^{it_n\Delta}f_n\right] - \phi\right\rangle_{L^2}$$

$$= \left\langle 2g_{0,\xi_n,x_n,\lambda_n}^{-1}\left[e^{it_n\Delta}f_n\right] - \phi, \phi\right\rangle_{L^2} \to \|\phi\|_{L^2}^2.$$

Since the $L^2(\mathbf{R}^d)$ inner product is invariant under the Galilean transform, translation in space, scaling, and the action of $e^{it\Delta}$, we have

$$\langle f_n, f_n\rangle_{L^2} - \langle f_n - \phi_n, f_n - \phi_n\rangle_{L^2} \to \|\phi\|_{L^2(\mathbf{R}^d)}^2,$$

which proves (4.44). $\square$

To complete the proof of Theorem 4.8 it remains only to prove Proposition 4.13.

Proof of Proposition 4.13 Without loss of generality suppose $\|f\|_{L^2(\mathbf{R}^d)} = 1$. By (4.45), for almost every $\xi, \xi' \in \mathbf{R}^d \times \mathbf{R}^d$,

$$\sum_{Q \sim Q'} \chi_Q(\xi)\chi_{Q'}(\xi') = 1.$$

4.2 Mass-Critical Profile Decomposition

Since $\text{dist}(Q, Q') \leq 10\,\text{diam}(Q)$, let

$$Q'' = Q + Q' = \{\xi'' : \xi'' = \xi + \xi', \xi \in Q, \xi' \in Q'\}.$$

Clearly, $\text{diam}(Q'') = \text{diam}(Q) + \text{diam}(Q')$. Let $c(Q'')$ be the center of Q''. Adding two paraboloids together gives

$$|\xi|^2 + |\xi'|^2 = \frac{1}{2}|\xi + \xi'|^2 + \frac{1}{2}|\xi - \xi'|^2$$
$$= \frac{1}{2}|c(Q'')|^2 + c(Q'') \cdot (\xi + \xi' - c(Q'')) + \frac{1}{2}|\xi + \xi' - c(Q'')|^2$$
$$+ \frac{1}{2}|\xi - \xi'|^2.$$

Then let

$$R(Q + Q') = \left\{(\tau, \eta) : \frac{1}{8} \leq \frac{\tau - \frac{1}{2}|c(Q'')|^2 - c(Q'') \cdot [\eta - c(Q'')]}{\text{diam}(Q'')^2} \leq 1\right\}.$$

If $\widetilde{F}(\tau, \xi)$ is the Fourier transform of $F(t, x)$ in space and in time, then

$$\text{supp}\left(\widetilde{[e^{it\Delta}f_Q][e^{it\Delta}f_{Q'}]}(\tau, \eta)\right) \subset R(Q + Q').$$

The supports of $R(Q + Q')$ are approximately disjoint.

Lemma 4.14 *For $\alpha \leq 1.01$,*

$$\sup_{\tau, \eta} \sum_{(Q, Q') \in \mathscr{F}} \chi_{\alpha R(Q + Q')}(\tau, \eta) \lesssim_d 1.$$

Proof For $(\tau_1, \xi_1), (\tau_2, \xi_2)$ on the paraboloid $\tau = |\xi|^2$, $\xi_1 \in Q$, $\xi_2 \in Q'$, $Q \sim Q'$,

$$\tau_1 + \tau_2 - \frac{1}{2}|\xi_1 + \xi_2|^2 = \frac{1}{2}|\xi_1 - \xi_2|^2 \sim \text{diam}(Q)^2.$$

So for a given (τ, η) it suffices to consider Q such that $\text{diam}(Q)^2 \sim \tau - \frac{1}{2}|\eta|^2$ and $\text{dist}(Q, Q') \sim \text{diam}(Q)$. Since $|Q| = |Q'|$,

$$\text{dist}\left(\frac{1}{2}\eta, Q\right), \text{dist}\left(\frac{1}{2}\eta, Q'\right) \lesssim \text{diam}(Q).$$

Therefore, for a fixed τ, η, there are only finitely many $(Q, Q') \in \mathscr{F}$ such that $(\tau, \eta) \in R(Q + Q')$. $\square$

Therefore, by the Plancherel identity,

$$\|e^{it\Delta}f\|_{L^4_{t,x}}^4 = \left\|\sum_{Q\sim Q'} [e^{it\Delta}f_Q][e^{it\Delta}f_{Q'}]\right\|_{L^2_{t,x}}^2$$
$$\lesssim \sum_{Q\sim Q'} \|[e^{it\Delta}f_Q][e^{it\Delta}f_{Q'}]\|_{L^2_{t,x}}^2. \qquad (4.55)$$

Meanwhile, by the triangle inequality,

$$\|e^{it\Delta}f\|_{L^2_{t,x}}^2 \leq \sum_{Q\sim Q'} \|[e^{it\Delta}f_Q][e^{it\Delta}f_{Q'}]\|_{L^1_{t,x}} \qquad (4.56)$$

and

$$\|[e^{it\Delta}f]^2\|_{L^\infty_{t,x}} \lesssim \sum_{Q\sim Q'} \|[e^{it\Delta}f_Q][e^{it\Delta}f_{Q'}]\|_{L^\infty_{t,x}} \qquad (4.57)$$

also hold. Interpolating (4.55) and (4.57) gives, when $d = 1$,

$$\|[e^{it\Delta}f]^2\|_{L^3_{t,x}(\mathbf{R}\times\mathbf{R})}^{3/2} \lesssim \sum_{Q\sim Q'} \|[e^{it\Delta}f_Q][e^{it\Delta}f_{Q'}]\|_{L^3_{t,x}(\mathbf{R}\times\mathbf{R})}^{3/2}. \qquad (4.58)$$

Therefore, by (4.46),

(4.58)
$$\lesssim \sum_{(Q,Q')\in\mathscr{F}} \|e^{it\Delta}f_Q\|_{L^\infty_{t,x}(\mathbf{R}\times\mathbf{R})}^{1/5} \|e^{it\Delta}f_{Q'}\|_{L^\infty_{t,x}(\mathbf{R}\times\mathbf{R})}^{1/5} |Q|^{-1/2} \|\hat{f}\|_{L^{13/8}_\xi(Q)}^{13/10} \|\hat{f}\|_{L^{13/8}_\xi(Q')}^{13/10}$$
$$\lesssim \left(\sup_{Q\in\mathscr{D}} |Q|^{-1/2}\|e^{it\Delta}f_Q\|_{L^\infty_{t,x}(\mathbf{R}\times\mathbf{R})}\right)^{2/5} \left(\sum_{Q\in\mathscr{D}} |Q|^{-3/10}\|\hat{f}\|_{L^{13/8}_\xi(Q)}^{13/5}\right).$$

Now for $|Q| = 2^j$ for some $j \in \mathbf{Z}$, make the following split:

$$\hat{f}(\xi) = \hat{f}^j(\xi) + \hat{f}_j(\xi),$$

where $\hat{f}^j(\xi) = \hat{f}(\xi)$ when $|\hat{f}(\xi)| \geq 2^{-j/2}$ and $\hat{f}_j(\xi) = \hat{f}(\xi)$ when $|\hat{f}(\xi)| < 2^{-j/2}$. By the Plancherel identity, the fact that $\|f\|_{L^2} = 1$, and Young's inequality, we have

$$\sum_j \sum_{Q \in \mathcal{D}_j} 2^{-\frac{3j}{10}} \|\hat{f}^j\|_{L_\xi^{13/8}(Q)}^{13/5}$$

$$= \sum_j \sum_{Q \in \mathcal{D}_j} 2^{-\frac{3j}{10}} \left(\int_Q |\hat{f}^j(\xi)|^{13/8} d\xi \right)^{8/5}$$

$$\lesssim \sum_j \sum_{Q \in \mathcal{D}_j} 2^{-\frac{3j}{10}} \left(\sum_{k \geq -\frac{j}{2}} \int_{\{\xi \in Q : 2^k \leq |\hat{f}(\xi)| < 2^{k+1}\}} |\hat{f}(\xi)|^{13/8} d\xi \right)^{8/5}$$

$$\lesssim \sum_j \sum_{Q \in \mathcal{D}_j} 2^{-\frac{3j}{10}} \left(\sum_{k \geq -\frac{j}{2}} 2^{-\frac{3k}{8}} \int_{\{\xi \in Q : 2^k \leq |\hat{f}(\xi)| < 2^{k+1}\}} |\hat{f}(\xi)|^2 d\xi \right)^{8/5}$$

$$\lesssim \sum_j \sum_{Q \in \mathcal{D}_j} \sum_{k \geq -\frac{j}{2}} 2^{-\frac{15j}{80}} 2^{-\frac{15k}{40}} \left(\int_{\{\xi \in Q : 2^k \leq |\hat{f}(\xi)| < 2^{k+1}\}} |\hat{f}(\xi)|^2 d\xi \right)$$

$$\lesssim 1. \tag{4.59}$$

Next, by Hölder's inequality and Young's inequality,

$$\sum_j \sum_{Q \in \mathcal{D}_j} 2^{-\frac{3j}{10}} \|\hat{f}_j\|_{L_\xi^{13/8}(Q)}^{13/5}$$

$$= \sum_j \sum_{Q \in \mathcal{D}_j} 2^{-\frac{3j}{10}} \left(\int_Q |\hat{f}_j(\xi)|^{13/8} d\xi \right)^{8/5}$$

$$\lesssim \sum_j \sum_{Q \in \mathcal{D}_j} 2^{-\frac{3j}{10}} \left(\sum_{k < -\frac{j}{2}} \int_{\{\xi \in Q : 2^k \leq |\hat{f}(\xi)| < 2^{k+1}\}} |\hat{f}(\xi)|^{13/8} d\xi \right)^{8/5}$$

$$\lesssim \sum_j \sum_{Q \in \mathcal{D}_j} 2^{-\frac{3j}{10}} \left(\sum_{k < -\frac{j}{2}} 2^{\frac{3j}{16}} \left(\int_{\{\xi \in Q : 2^k \leq |\hat{f}(\xi)| < 2^{k+1}\}} |\hat{f}(\xi)|^2 d\xi \right)^{13/16} \right)^{8/5}$$

$$\lesssim \sum_j \sum_{Q \in \mathcal{D}_j} \left(\sum_{k < -\frac{j}{2}} \left(\int_{\{\xi \in Q : 2^k \leq |\hat{f}(\xi)| < 2^{k+1}\}} |\hat{f}(\xi)|^2 d\xi \right)^{5/8} \left(2^{2k} 2^j \right)^{3/16} \right)^{8/5}$$

$$\lesssim 1. \tag{4.60}$$

This completes the proof of (4.51).

When $d \geq 2$, interpolating (4.55) and (4.56),

$$\left\| e^{it\Delta} f \right\|_{L_{t,x}^{\frac{2(d+2)}{d}}(\mathbf{R} \times \mathbf{R}^d)}^{\frac{2(d+2)}{d}} \lesssim_d \sum_{Q \sim Q'} \left\| [e^{it\Delta} f_Q] [e^{it\Delta} f_{Q'}] \right\|_{L_{t,x}^{\frac{d+2}{d}}(\mathbf{R} \times \mathbf{R}^d)}^{\frac{d+2}{2}}. \tag{4.61}$$

Then by (4.48), for $q = \frac{2(d^2+3d+1)}{d^2}$ and $\frac{2(d^2+3d+1)}{d^2+3d+2} < p \leq 2$,

$$(4.61) \lesssim_d \sum_{Q \sim Q'} \left\| e^{it\Delta} f_Q \right\|_{L_{t,x}^q}^{\frac{1}{d}} \left\| e^{it\Delta} f_{Q'} \right\|_{L_{t,x}^q}^{\frac{1}{d}} \left\| [e^{it\Delta} f_Q][e^{it\Delta} f_{Q'}] \right\|_{L_{t,x}^{\frac{d^2+3d+1}{d(d+1)}}}^{\frac{d+1}{d}}$$

$$\lesssim_d \sum_{Q \sim Q'} \left\| e^{it\Delta} f_Q \right\|_{L_{t,x}^q}^{\frac{1}{d}} \left\| e^{it\Delta} f_{Q'} \right\|_{L_{t,x}^q}^{\frac{1}{d}} |Q|^{\left(1 - \frac{2}{p} - \frac{1}{d^2+3d+1}\right)\frac{d+1}{d}} \|\hat{f}\|_{L_\xi^p(Q)}^{\frac{d+1}{d}} \|\hat{f}\|_{L_\xi^p(Q')}^{\frac{d+1}{d}}$$

$$\lesssim \left(\sup_{Q \in \mathscr{D}} |Q|^{\frac{d+2}{dq} - \frac{1}{2}} \left\| e^{it\Delta} f_Q \right\|_{L_{t,x}^q} \right)^{\frac{2}{d}} \sum_{Q \in \mathscr{D}} \left(|Q|^{-\frac{2-p}{p}} \|\hat{f}\|_{L_\xi^p(Q)}^2 \right)^{\frac{d+1}{d}}.$$

Then if $Q \in \mathscr{D}_j$, and $|Q| = 2^{jd}$, split $\hat{f}(\xi)$ as $\hat{f}^j(\xi) + \hat{f}_j(\xi)$, where

$$\hat{f}^j(\xi) = \hat{f}(\xi) \quad \text{when } |\hat{f}(\xi)| \geq 2^{-jd/2},$$
$$\hat{f}_j(\xi) = \hat{f}(\xi) \quad \text{when } |\hat{f}(\xi)| < 2^{-jd/2}.$$

Because $p < 2$, we have $l^p \subset l^2$ and then, following (4.59), and since $\|\hat{f}\|_{L_\xi^2} = 1$, we get

$$\sum_j \sum_{Q \in \mathscr{D}_j} \left(|Q|^{-\frac{2-p}{p}} \|\hat{f}^j\|_{L_\xi^p(Q)}^2 \right)^{\frac{d+1}{d}} \lesssim_d \sum_j \left(\sum_{Q \in \mathscr{D}_j} |Q|^{-\frac{2-p}{2}} \|\hat{f}^j\|_{L_\xi^p(Q)}^p \right)^{\frac{2(d+1)}{pd}}$$

$$\lesssim \sum_j \left(2^{-jd\frac{2-p}{2}} \int_{|\hat{f}| \geq 2^{-jd/2}} |\hat{f}(\xi)|^p d\xi \right)^{\frac{2(d+1)}{pd}}$$

$$\lesssim 1.$$

Also by Hölder's inequality, following (4.60), we have

$$\sum_j \sum_{Q \in \mathscr{D}_j} \left(|Q|^{-\frac{2-p}{p}} \|\hat{f}_j\|_{L_\xi^p(Q)}^2 \right)^{\frac{d+1}{d}}$$

$$\lesssim_d \left(\sum_j \sum_{Q \in \mathscr{D}_j} \left[|Q|^{-\frac{2-p}{p}} |Q|^{\frac{2}{p} - \frac{d}{d+1}} \right]^{\frac{d+1}{d}} \|\hat{f}_j\|_{L_\xi^{\frac{2(d+1)}{d}}(Q)}^{\frac{2(d+1)}{d}} \right)$$

$$= \left(\sum_j \sum_{Q \in \mathscr{D}_j} |Q|^{1/d} \|\hat{f}_j\|_{L_\xi^{\frac{2(d+1)}{d}}(Q)}^{\frac{2(d+1)}{d}} \right)$$

$$\lesssim \sum_j 2^j \int_{\{\xi : |\hat{f}(\xi)| \leq 2^{-jd/2}\}} |\hat{f}(\xi)|^{\frac{2(d+1)}{d}} d\xi \lesssim \int |\hat{f}(\xi)|^2 d\xi = 1.$$

This finally proves (4.52), proving Proposition 4.13. □

Theorem 4.15 (Profile decomposition for the L^2-critical problem) *Let u_n be*

4.2 Mass-Critical Profile Decomposition

a bounded sequence in $L^2(\mathbf{R}^d)$, $\|u_n\|_{L^2} \leq A$. Then for any $1 \leq j < \infty$, there exist $\phi^j \in L^2(\mathbf{R}^d)$, with ϕ^j possibly being zero, $(t_n^j, x_n^j) \in \mathbf{R} \times \mathbf{R}^d$, $\xi_n^j \in \mathbf{R}^d$, and $\lambda_n^j \in (0,\infty)$, such that

$$u_n = \sum_{j=1}^{J} e^{-it_n^j \Delta} g_n^j \phi^j + w_n^J,$$

where

$$g_n^j = g_{0,\xi_n^j,x_n^j,\lambda_n^j}$$

and

$$\lim_{J\to\infty} \limsup_{n\to\infty} \|e^{it\Delta} w_n^J\|_{L_{t,x}^{\frac{2(d+2)}{d}}} = 0. \quad (4.62)$$

Moreover, for any J, $1 \leq j \leq J$,

$$(g_n^j)^{-1} e^{it_n^j \Delta} w_n^J \rightharpoonup 0, \text{ weakly}, \quad (4.63)$$

$$\lim_{n\to\infty} \left[\|u_n\|_{L^2(\mathbf{R}^d)}^2 - \sum_{j=1}^{J} \|\phi^j\|_{L^2(\mathbf{R}^d)}^2 - \|w_n^J\|_{L^2(\mathbf{R}^d)}^2 \right] = 0, \quad (4.64)$$

and for $j \neq k$,

$$\frac{\lambda_n^j}{\lambda_n^k} + \frac{\lambda_n^k}{\lambda_n^j} + \lambda_n^k |\xi_n^j - \xi_n^k| + \frac{|t_n^j - t_n^k|}{(\lambda_n^j)^2} + \frac{|x_n^j - x_n^k + 2(t_n^k - t_n^j)\xi_n^j|}{\lambda_n^k} \to \infty. \quad (4.65)$$

Proof We use Theorem 4.8 and induction. If $\liminf_{n\to\infty} \|e^{it\Delta} u_n\|_{L_{t,x}^{\frac{2(d+2)}{d}}} = 0$, then pass to a subsequence and let $w_n^J = u_n$.

If

$$\liminf_{n\to\infty} \|e^{it\Delta} u_n\|_{L_{t,x}^{\frac{2(d+2)}{d}}(\mathbf{R}\times\mathbf{R}^d)} = \varepsilon > 0,$$

then by (4.43) and (4.44), possibly after passing to a subsequence, there exists $\phi^1 \in L^2$, $\|\phi^1\|_{L^2} \gtrsim \varepsilon \left(\frac{\varepsilon}{A}\right)^\alpha$, and $\lambda_n^1 \in (0,\infty)$, $t_n^1 \in \mathbf{R}$, $x_n^1, \xi_n^1 \in \mathbf{R}^d$, such that

$$(\lambda_n^1)^{d/2} e^{-i\xi_n^1 \cdot (\lambda_n^1 x + x_n^1)} \left[e^{it_n^1 \Delta} u_n \right](\lambda_n^1 x + x_n^1) \rightharpoonup \phi^1(x).$$

Letting

$$w_n^1 = u_n - \frac{1}{(\lambda_n^1)^{d/2}} e^{-it_n^1 \Delta} \left[e^{ix\cdot \xi_n^1} \phi^1 \left(\frac{x - x_n^1}{\lambda_n^1} \right) \right], \quad (4.66)$$

then (4.44) implies

$$\lim_{n\to\infty} \left[\|u_n\|_{L^2}^2 - \|\phi^1\|_{L^2}^2 - \|w_n^1\|_{L^2}^2 \right] = 0,$$

and (4.43) implies

$$(\lambda_n^1)^{d/2} e^{-i\xi_n^1 \cdot (\lambda_n^1 x + x_n^1)} \left[e^{it_n^1 \Delta} w_n^1 \right] (\lambda_n^1 x + x_n^1) \rightharpoonup 0.$$

Now suppose (4.63), (4.64), and (4.65) hold for $1 \le j < J_0$, and

$$\liminf_{n \to \infty} \left\| e^{it\Delta} w_n^{J_0-1} \right\|_{L_{t,x}^{\frac{2(d+2)}{d}} (\mathbf{R} \times \mathbf{R}^d)} = 0.$$

In that case let $\phi^j = 0$ for all $j \ge J_0$ and the proof is complete.

If

$$\liminf_{n \to \infty} \left\| e^{it\Delta} w_n^{J_0-1} \right\|_{L_{t,x}^{\frac{2(d+2)}{d}}} = \varepsilon(J_0) > 0,$$

then by Theorem 4.8, there exists a nonzero ϕ^{J_0} such that after passing to a subsequence,

$$(\lambda_n^{J_0})^{d/2} e^{-i\xi_n^{J_0} \cdot (\lambda_n^{J_0} x + x_n^{J_0})} \left[e^{it_n^{J_0} \Delta} w_n^{J_0-1} \right] (\lambda_n^{J_0} x + x_n^{J_0}) \rightharpoonup \phi^{J_0}. \tag{4.67}$$

Let

$$w_n^{J_0} = w_n^{J_0-1} - \frac{1}{(\lambda_n^{J_0})^{d/2}} e^{-it_n^{J_0} \Delta} \left[e^{ix \cdot \xi_n^{J_0}} \phi^{J_0} \left(\frac{x - x_n^{J_0}}{\lambda_n^{J_0}} \right) \right]. \tag{4.68}$$

Again by (4.44),

$$\lim_{n \to \infty} \left[\left\| w_n^{J_0-1} \right\|_{L^2}^2 - \left\| \phi^{J_0} \right\|_{L^2}^2 - \left\| w_n^{J_0} \right\|_{L^2}^2 \right] = 0,$$

and so (4.64) continues to hold.

Now, by induction, (4.65) holds for any pair $1 \le j < k < J_0$. Suppose (4.65) does not hold for some $1 \le j < J_0$ and J_0. Then there exists some $\lambda \in (0, \infty)$, $x_0, \xi_0 \in \mathbf{R}^d$, and $t_0 \in \mathbf{R}$, such that after passing to a subsequence

$$\frac{\lambda_n^j}{\lambda_n^{J_0}} \to \lambda_0, \quad (\lambda_n^{J_0})(\xi_n^j - \xi_n^{J_0}) \to \xi_0,$$

$$\frac{x_n^j - x_n^{J_0} + 2(t_n^{J_0} - t_n^j)\xi_n^j}{\lambda_n^{J_0}} \to x_0, \quad \frac{t_n^{J_0} - t_n^j}{(\lambda_n^{J_0})^2} \to t_0. \tag{4.69}$$

By (4.67),

$$(g_n^{J_0})^{-1} e^{it_n^{J_0} \Delta} w_n^{J_0-1} \rightharpoonup \phi^{J_0}(x), \tag{4.70}$$

and by induction, since (4.63) holds for $1 \le j < J_0$,

$$(g_n^j)^{-1} e^{it_n^j \Delta} w_n^{J_0-1} \rightharpoonup 0,$$

4.2 Mass-Critical Profile Decomposition

which is equivalent to

$$\left\langle \left(g_n^j\right)^{-1} e^{it_n^j \Delta} w_n^{J_0-1}, \psi \right\rangle \to 0, \tag{4.71}$$

for any $\psi \in L_x^2(\mathbf{R}^d)$. Now, by (4.40),

$$\left\langle \left(g_n^j\right)^{-1} e^{it_n^{J_0} \Delta} e^{i(t_n^j - t_n^{J_0})\Delta} w_n^{J_0-1}, \psi \right\rangle$$

$$= \left\langle e^{it_n^{J_0} \Delta} w_n^{J_0-1}, e^{i(t_n^{J_0} - t_n^j)\Delta} \left(g_n^j\right) \psi \right\rangle$$

$$= \left\langle \left(g_n^{J_0}\right)^{-1} e^{it_n^{J_0} \Delta} w_n^{J_0-1}, \left(g_n^{J_0}\right)^{-1} \left(g_{-(t_n^{J_0}-t_n^j)|\xi_n^j|^2, \xi_n^j, x_n^j + 2(t_n^{J_0}-t_n^j)\xi_n^j, \lambda_n^j}\right)\right.$$

$$\left. \times \exp\left(i\frac{(t_n^{J_0}-t_n^j)}{(\lambda_n^j)^2}\Delta\right)\psi\right\rangle.$$

By (4.69),

$$\exp\left(i\frac{(t_n^{J_0}-t_n^j)}{(\lambda_n^j)^2}\right)\psi \to e^{it_0 \Delta}\psi,$$

in $L_x^2(\mathbf{R}^d)$. Additionally, (4.38), (4.39), (4.69), and the fact that S^1 is compact in $\mathbf{C}$ imply that, possibly after passing to a subsequence, for any test function $\psi \in L_x^2(\mathbf{R}^d)$,

$$\left(g_n^{J_0}\right)^{-1} \left(g_{-(t_n^{J_0}-t_n^j)|\xi_n^j|^2, \xi_n^j, x_n^j + 2(t_n^{J_0}-t_n^j)\xi_n^j, \lambda_n^j}\right)\psi \to g_{\theta, \xi_0, x_0, \lambda_0}\psi,$$

for some $\theta \in [0, 2\pi)$. However, in this case (4.70) contradicts (4.71). Therefore, (4.65) must continue to hold.

To prove (4.62), observe that (4.64) implies that for any $J \geq 1$,

$$\lim_{n \to \infty} \|w_n^J\|_{L^2} \leq A. \tag{4.72}$$

By (4.43) and (4.44),

$$\liminf_{n \to \infty} \|e^{it\Delta} w_n^J\|_{L_{t,x}^{2(d+2)/d}(\mathbf{R} \times \mathbf{R}^d)} = \varepsilon(J) > 0 \tag{4.73}$$

implies $\|\phi^{J+1}\|_{L^2} \gtrsim \varepsilon(J) \left(\frac{\varepsilon(J)}{A}\right)^\alpha$. Then (4.64) and $\|u_n\|_{L^2} \leq A$ directly imply

$$\lim_{J \nearrow \infty} \varepsilon(J) = 0, \tag{4.74}$$

proving (4.62). □

As in the energy-critical problem, define the scattering size function as follows.

Definition 4.16 (Scattering size function) For any $0 < m < \infty$, let

$$A(m) = \sup \left\{ \|u\|_{L^{2(d+2)/d}_{t,x}(\mathbf{R}\times\mathbf{R}^d)} : u \text{ solves (4.36))}, \|u_0\|_{L^2(\mathbf{R}^d)} = m \right\},$$

and for any initial data $u_0 \in L^2_x(\mathbf{R}^d)$, let

$$A(u_0) = \|u\|_{L^{2(d+2)/d}_{t,x}(\mathbf{R}\times\mathbf{R}^d)},$$

where u is the solution to (4.36) with initial data $u(0) = u_0$.

Definition 4.17 (Almost-periodic) A solution to (4.36) is called almost-periodic if there exists a compact $K \subset L^2_x(\mathbf{R}^d)$ such that for all $t \in I$, there exists $g(t) \in G$ such that $g(t)u(t) \in K$, where G is the group of symmetries given by (4.37)–(4.40).

As in the energy-critical case, if scattering fails, there must exist a nonzero, almost-periodic solution to (4.36).

Theorem 4.18 *If global well-posedness and scattering of (4.36) do not hold for all $u_0 \in L^2_x(\mathbf{R}^d)$ then there exists a nonzero solution to (4.36) that is almost-periodic for all $t \in I$, where I is the maximal interval of existence of u.*

By Theorem 1.17, $A(m) \lesssim m$ for $m \leq \varepsilon_0(d)$. Theorem 1.24 implies that $A(m)$ is a continuous, increasing function of m. Hence, the set $\{m : A(m) = \infty\}$ is closed and disjoint from zero and there exists $0 < m_0 \leq \infty$ such that

$$\{m : A(m) = \infty\} = [m_0, \infty).$$

Let $u_n(0)$ be a sequence of initial data satisfying $A(u_n(0)) \nearrow +\infty$ and $\|u_n(0)\|_{L^2_x(\mathbf{R}^d)} \nearrow m_0$. Let u_n denote the solution to (4.36) with initial data $u_n(0)$, and suppose

$$\|u_n\|_{L^{2(d+2)/d}_{t,x}([0,\infty)\times\mathbf{R}^d)} \nearrow +\infty, \quad \|u_n\|_{L^{2(d+2)/d}_{t,x}((-\infty,0]\times\mathbf{R}^d)} \nearrow +\infty. \quad (4.75)$$

If $m_0 < \infty$ then apply Theorem 4.15, obtaining a profile decomposition of $u_n(0)$, possibly after passing to a subsequence.

First, suppose that there exists a sequence t_n^j such that, after passing to a subsequence, $|t_n^j| \nearrow \infty$, say $t_n^j \nearrow +\infty$. After relabeling, suppose that $j = 1$, and split:

$$u_n = g_n^1 e^{it_n^1 \Delta} \phi^1 + \left(u_n - g_n^1 e^{it_n^1 \Delta} \phi^1\right).$$

Now, $\|\phi^1\|_{L^2} > 0$, so by (4.64),

$$\lim_{n\to\infty} \left\|u_n - g_n^1 e^{it_n^1 \Delta} \phi^1\right\|_{L^2_x(\mathbf{R}^d)} \leq m_0 - \varepsilon,$$

4.2 Mass-Critical Profile Decomposition

for some $\varepsilon > 0$. But then the scattering size is bounded:

$$\lim_{n\to\infty} A\left(u_n - g_n^1 e^{it_n^1 \Delta} \phi^1\right) < \infty.$$

By the dominated convergence theorem and Strichartz estimates,

$$\lim_{n\to\infty} \left\| e^{it\Delta} g_n^1 e^{it_n^1 \Delta} \phi^1 \right\|_{L_{t,x}^{2(d+2)/d}([0,\infty)\times\mathbf{R}^d)} = 0,$$

which by Theorem 1.24 contradicts (4.75). Therefore, after passing to a subsequence and possibly making a diagonal argument, assume that for each j, t_n^j converges to some $t^j \in \mathbf{R}$, and then set $e^{it^j\Delta}\phi^j$ to be the new ϕ^j, setting $t_n^j \equiv 0$ and absorbing the error into w_n^J.

Now for each j let v^j be the solution to (4.36) with initial data ϕ^j. Let v_n^j be the solution to (4.36) with initial data $g_n^j \phi^j$.

Lemma 4.19 *If $v_n^j(t)$ and $v_n^k(t)$ are orthogonal profiles for $j \neq k$, that is, (4.65) holds for g_n^j and g_n^k, and*

$$\limsup_{n\to\infty} \left[\|v_n^j\|_{L_{t,x}^{2(d+2)/d}(\mathbf{R}\times\mathbf{R}^d)} + \|v_n^k\|_{L_{t,x}^{2(d+2)/d}(\mathbf{R}\times\mathbf{R}^d)} \right] < \infty,$$

then

$$\lim_{n\to\infty} \left\| F\left(v_n^j + v_n^k\right) - F\left(v_n^j\right) - F\left(v_n^k\right) \right\|_{L_{t,x}^{2(d+2)/(d+4)}(\mathbf{R}\times\mathbf{R}^d)} = 0. \quad (4.76)$$

Proof To prove (4.76) it suffices to show

$$\lim_{n\to\infty} \left\| (\lambda_n^j)^{-d/2} e^{ix\cdot\xi_n^j} v^j\left(\frac{t}{(\lambda_n^j)^2}, \frac{x - x_n^j - 2t\xi_n^j}{\lambda_n^j}\right) \right. $$
$$\left. \times (\lambda_n^k)^{-d/2} e^{ix\cdot\xi_n^k} v^k\left(\frac{t}{(\lambda_n^k)^2}, \frac{x - x_n^k - 2t\xi_n^k}{\lambda_n^k}\right) \right\|_{L_{t,x}^{\frac{d+2}{d}}(\mathbf{R}\times\mathbf{R}^d)} = 0.$$

This follows directly from (4.65) with $t_n^j = t_n^k \equiv 0$.

To make such calculations completely rigorous, approximate v^j and v^k by $C_0^\infty(\mathbf{R}^{d+1})$ functions. □

Now suppose that

$$\sup_j \|\phi^j\|_{L_x^2(\mathbf{R}^d)} \leq m_0 - \varepsilon.$$

By (4.64), there exist only finitely many j such that $\|\phi^j\|_{L^2} > \varepsilon_0(d)$. For each such j,

$$\|v^j\|_{L_{t,x}^{2(d+2)/d}(\mathbf{R}\times\mathbf{R}^d)} \leq A(m_0 - \varepsilon). \quad (4.77)$$

Also by Theorem 1.17, if $\|\phi^j\|_{L^2} \leq \varepsilon_0(d)$,

$$\|v^j\|_{L_{t,x}^{2(d+2)/d}(\mathbf{R}\times\mathbf{R}^d)} \lesssim \|\phi^j\|_{L^2}. \tag{4.78}$$

Combining Lemma 4.19, (4.77), (4.78), and (4.65),

$$\sup_J \lim_{n\to\infty} A\left(\sum_{j=1}^J g_n^j \phi^j\right) < \infty.$$

This fact, together with Theorem 1.24 and (4.62), contradicts (4.75).

Therefore, for such a sequence $\|\phi^1\|_{L^2} = m_0$, and by (4.64), $\phi^j \equiv 0$ for $j \geq 2$, and $w_n^1 \to 0$ in $L_x^2(\mathbf{R}^d)$. Thus, $u_n(0)$, modulo action from the group (4.37)–(4.40), should converge in $L_x^2(\mathbf{R}^d)$ to some $u_0 \in L_x^2(\mathbf{R}^d)$ with $A(u_0) = \infty$.

Furthermore, if u is the solution to (4.36) with initial data u_0 on the maximal interval I, then

$$\|u\|_{L_{t,x}^{2(d+2)/d}(I\cap[0,\infty)\times\mathbf{R}^d)} = \|u\|_{L_{t,x}^{2(d+2)/d}(I\cap(-\infty,0]\times\mathbf{R}^d)} = \infty. \tag{4.79}$$

Now, by Theorem 1.21, u must be the solution on some maximal open time interval I. Taking a sequence of times $t_n \in I$, and setting $u_n(0) = u(t_n)$, the previous argument shows that after passing to a subsequence and acting by some $g_n \in G$, $u(t_n)$ must have a convergent subsequence. Thus $u(t)$ must lie in GK for all $t \in I$, where $K \subset L_x^2(\mathbf{R}^d)$ is a compact set. Then by the Arzelà–Ascoli theorem, we get the following.

Theorem 4.20 (Mass-critical almost-periodicity) *If global well-posedness and scattering fails for the defocusing, mass-critical nonlinear Schrödinger equation, then there exists a solution $u(t)$ to (4.36) on a maximal interval I and*

$$x(t): \quad I \to \mathbf{R}^d,$$
$$\xi(t): \quad I \to \mathbf{R}^d,$$
$$N(t): \quad I \to (0,\infty),$$

such that for all $t \in I$ and for any $\eta > 0$,

$$\int_{|x-x(t)|\geq \frac{C(\eta)}{N(t)}} |u(t,x)|^2 dx + \int_{|\xi-\xi(t)|\geq C(\eta)N(t)} |\hat{u}(t,\xi)|^2 d\xi < \eta. \tag{4.80}$$

The observation on lengths of intervals in Lemma 2.13 also holds for almost-periodic solutions to the mass-critical and energy-critical problems. See also Theorem 3.13.

Theorem 4.21 *If u is a nonzero, almost-periodic solution to either the energy-critical equation or mass-critical equation on a maximal interval I, then for*

4.2 Mass-Critical Profile Decomposition

$J \subset I$ compact,

$$\int_J N(t)^2 \, dt \lesssim \int_J \int |u(t,x)|^p \, dx \, dt \lesssim 1 + \int_J N(t)^2 \, dt, \qquad (4.81)$$

where $p = \frac{2(d+2)}{d-2}$ for the energy-critical problem and $p = \frac{2(d+2)}{d}$ for the mass-critical problem.

Proof First consider the mass-critical problem. For any t_0, there exists $\delta(t_0) > 0$ such that (4.36) is locally well posed on

$$I_{t_0} = \left[t_0 - \frac{\delta(t_0)}{N(t_0)^2}, t_0 + \frac{\delta(t_0)}{N(t_0)^2} \right].$$

Moreover, because $u(t)$ lies in GK, where K is a compact set, there exists some $T_2(\varepsilon_0) > 0$ such that

$$\left\| e^{it\Delta} u(t_0) \right\|_{L_{t,x}^{2(d+2)/d}\left(\left[t_0 - \frac{T_2(\varepsilon_0)}{N(t_0)^2}, t_0 + \frac{T_2(\varepsilon_0)}{N(t_0)^2} \right] \times \mathbf{R}^d \right)} \leq \varepsilon_0. \qquad (4.82)$$

Also, for any nonzero $u(t_0)$,

$$\left\| e^{it\Delta} u(t_0) \right\|_{L_{t,x}^{2(d+2)/d}\left(\left[t_0 - \frac{T_2(\varepsilon_0)}{N(t_0)^2}, t_0 + \frac{T_2(\varepsilon_0)}{N(t_0)^2} \right] \times \mathbf{R}^d \right)} > 0,$$

so since $u(t_0)$ lies in a precompact set modulo symmetries,

$$\left\| e^{it\Delta} u(t_0) \right\|_{L_{t,x}^{2(d+2)/d}\left(\left[t_0 - \frac{T_2(\varepsilon_0)}{N(t_0)^2}, t_0 + \frac{T_2(\varepsilon_0)}{N(t_0)^2} \right] \times \mathbf{R}^d \right)} \sim \varepsilon_0. \qquad (4.83)$$

Inequality (4.82) and Strichartz estimates imply that for all $t \in \left[t_0 - \frac{T_2}{N(t_0)^2}, t_0 + \frac{T_2}{N(t_0)^2} \right]$,

$$\left\| u(t) - e^{i(t-t_0)\Delta} u(t_0) \right\|_{L_x^2(\mathbf{R}^d)} \lesssim \varepsilon_0^{1 + \frac{4}{d}}. \qquad (4.84)$$

By Hölder's inequality, the Sobolev embedding theorem, and the Arzelà–Ascoli theorem, there exists a constant $C(u)$ such that

$$\frac{999 m_0}{1000} \leq \int_{|x - x(t)| \leq \frac{C(m_0^2/1000)}{N(t)}} |u(t,x)|^2 dx$$

$$\leq C\left(\frac{m_0^2}{1000} \right) \frac{N(t_0)}{N(t)} + \frac{m_0^2}{1000} + C(u) \varepsilon_0^{1 + \frac{4}{d}}. \qquad (4.85)$$

This implies that for ε_0 sufficiently small, $N(t) \lesssim N(t_0)$ for all

$$t \in \left[t_0 - \frac{T_2(\varepsilon_0)}{N(t_0)^2}, t_0 + \frac{T_2(\varepsilon_0)}{N(t_0)^2} \right].$$

Time reversal symmetry proves that $N(t) \gtrsim N(t_0)$ as well. Relations (4.83)–(4.85) imply (4.81) in the mass-critical case.

The proof in the energy-critical case is similar, although Lemma 3.2 is used in place of the perturbative argument in (4.84). □

Additionally, for the mass-critical problem (4.84),

$$\left\{\xi : |\xi - \xi(t_0)| \leq C\left(\frac{m_0^2}{1000}\right) N(t_0)\right\} \cap \left\{\xi : |\xi - \xi(t)| \leq C\left(\frac{m_0^2}{1000}\right) N(t)\right\}$$
$$\neq \emptyset.$$

Therefore, for all $t \in \left[t_0 - \frac{T_2}{N(t_0)^2}, t_0 + \frac{T_2}{N(t_0)^2}\right]$, we have $|\xi(t) - \xi(t_0)| \lesssim N(t)$. This implies the following.

Theorem 4.22 *Possibly after modifying $C(\eta)$ in (4.80) by a fixed constant,*

$$\left|\frac{d}{dt} N(t)\right| \lesssim N(t)^3 \tag{4.86}$$

and

$$\left|\frac{d}{dt} \xi(t)\right| \lesssim N(t)^3. \tag{4.87}$$

Theorem 4.21 also implies that if $\sup(I) < \infty$, then

$$\lim_{t \nearrow \sup(I)} N(t) = +\infty.$$

In fact $N(t) \gtrsim (\sup(I) - t)^{-1/2}$. The same holds if $\inf(I) > -\infty$.

Now let us make a reduction similar to Theorem 3.14 to yield the next result.

Theorem 4.23 (Mass-critical scenarios) *If there is a nonzero almost-periodic solution to (4.36), then for at least one such solution, one of the following two scenarios holds:*

$$I = \mathbf{R}, \quad N(t) \leq 1 \quad \text{for all } t \in \mathbf{R} \tag{4.88}$$

or

$$I = (0, \infty), \quad N(t) = t^{-1/2}. \tag{4.89}$$

We call (4.89) a self-similar solution.

Proof The proof is nearly identical to Theorem 3.14. Once again define the quantity

$$\text{osc}(T) = \inf_{t_0 \in I} \frac{\sup\{N(t) : t \in I \text{ and } |t - t_0| \leq TN(t_0)^{-2}\}}{\inf\{N(t) : t \in I \text{ and } |t - t_0| \leq TN(t_0)^{-2}\}}.$$

4.2 Mass-Critical Profile Decomposition

If $\sup_T \operatorname{osc}(T) < \infty$ it is possible to extract a solution satisfying $N(t) \sim 1$. In the case that

$$\lim_{T \to \infty} \operatorname{osc}(T) = \infty,$$

modify $a(t_0)$ slightly, defining

$$a(t_0) = \frac{\inf\{N(t) : t \in I \text{ and } t \geq t_0\}}{N(t_0)} + \frac{\inf\{N(t) : t \in I \text{ and } t \leq t_0\}}{N(t_0)}.$$

This time, if

$$\inf_{t_0 \in I} a(t_0) = 0,$$

it is possible to extract an almost-periodic solution satisfying $N(t) \leq 1$ for all $t \in I$, rather than $N(t) \geq 1$ for all $t \in I$, as was the case in the energy-critical problem. The condition $N(t) \leq 1$ on I automatically implies $I = \mathbf{R}$.

Finally, if

$$\inf_{t_0 \in I} a(t_0) = 2\varepsilon > 0,$$

it is possible to extract an almost-periodic solution satisfying $N(t) \sim t^{-1/2}$ for all $t \in (0, \infty)$, as in Theorem 3.14. Because this is the only finite-time reduction for the mass-critical problem, (3.69) may be replaced by (4.89). $\square$

The quantity $N(t)^3$ occurs quite naturally in the study of the mass-critical nonlinear Schrödinger equation. We have already seen that

$$\left|\frac{d}{dt}\xi(t)\right|, \left|\frac{d}{dt}N(t)\right| \lesssim N(t)^3. \tag{4.90}$$

It is convenient to split (4.88) into two scenarios.

Definition 4.24 (Soliton-like solution)

$$\int_0^\infty N(t)^3 \, dt = \infty.$$

Definition 4.25 (Rapid cascade solution)

$$\int_0^\infty N(t)^3 \, dt < \infty. \tag{4.91}$$

As we shall see in the next section, the $L^2_{t,x}$ bilinear Strichartz estimates of Theorem 2.12 will also introduce scaling like $N(t)^3$. These bilinear Strichartz estimates are fundamentally tied to the fact that the Schrödinger equation has two derivatives.

Remark Notice that in case (4.91), $\xi(t)$ can travel only a finite total distance. Additionally,

$$\int_0^\infty \left| \frac{d}{dt} N(t) \right| dt < \infty,$$

which implies $\lim_{t \to \infty} N(t) = 0$. For a more general function that satisfies $\int_0^\infty f(t) \, dt < \infty$, all that can be said is $\liminf_{t \to \infty} f(t) = 0$.

Remark The reduction to three scenarios in Killip et al. (2009) divided (4.88) in a slightly different manner, separating it into the case when $N(t) = 1$, and the case when $N(t) \leq 1$ and $\liminf_{t \to \pm\infty} N(t) = 0$.

4.3 Radial Mass-Critical Problem in Dimensions $d \geq 2$

To show that there does not exist a self-similar solution, it is enough to prove that a self-similar solution has $\dot{H}^1$ regularity. Then by Theorem 1.30 this contradicts that the self-similar solution blows up in finite time.

Theorem 4.26 (Additional regularity for a self-similar solution) *Suppose u is a self-similar solution to (4.36), that is, u is of the form of (4.89). Then there exists some $\xi_\infty \in \mathbf{R}^d$ such that for any $0 \leq s < 1 + \frac{4}{d}$, $t \in \mathbf{R}$, we have*

$$\left\| e^{-ix \cdot \xi_\infty} u(t) \right\|_{\dot{H}_x^s(\mathbf{R}^d)} \lesssim_{d,s,u} t^{-s/2}.$$

Proof Killip et al. (2008, 2009) and Tao et al. (2007b) proved Theorem 4.26 for radial data in dimensions $d \geq 2$. The argument used the restriction theory of Shao (2009) for a radial solution to prove the base case of the inductive argument. The proof here will instead use the double Duhamel argument for the base case, and will hold for both radial and nonradial solutions. The inductive argument is the same as in Killip et al. (2008, 2009) and Tao et al. (2007b).

By (4.90), $\xi(t)$ converges to some $\xi_\infty \in \mathbf{R}^d$ as $t \to +\infty$. Moreover, (4.90) implies

$$|\xi(t) - \xi_\infty| \lesssim t^{-1/2}, \quad \text{for all } t \in (0, \infty).$$

Therefore, after modifying $C(\eta)$ by a constant, we may choose $\xi(t) = \xi_\infty$, and then without loss of generality make a Galilean transform so that $\xi_\infty = 0$.

4.3 Radial Mass-Critical Problem in Dimensions $d \geq 2$

Following Killip et al. (2009), define

$$\left.\begin{aligned}\mathscr{M}(k) &= \sup_{j \in \mathbf{Z}} \left\| u_{>k-j/2} \right\|_{L_t^\infty L_x^2\left([2^j, 2^{j+1}] \times \mathbf{R}^d\right)}, \\ \mathscr{S}(k) &= \sup_{j \in \mathbf{Z}} \left\| u_{>k-j/2} \right\|_{L_{t,x}^{\frac{2(d+2)}{d}}\left([2^j, 2^{j+1}] \times \mathbf{R}^d\right)}, \\ \mathscr{N}(k) &= \sup_{j \in \mathbf{Z}} \left\| P_{>k-j/2} F(u(t)) \right\|_{L_{t,x}^{\frac{2(d+2)}{(d+4)}}\left([2^j, 2^{j+1}] \times \mathbf{R}^d\right)}.\end{aligned}\right\} \quad (4.92)$$

By (4.81) we know

$$\mathscr{M}(k) + \mathscr{S}(k) + \mathscr{N}(k) \lesssim 1, \tag{4.93}$$

and for any $j \in \mathbf{Z}$, Strichartz estimates imply

$$\left.\begin{aligned}\|u\|_{L_t^2 L_x^{\frac{2d}{d-2}}\left([2^j, 2^{j+1}] \times \mathbf{R}^d\right)} &\lesssim 1, \quad \text{for } d \geq 3, \\ \|u\|_{L_t^{8/3} L_x^8\left([2^j, 2^{j+1}] \times \mathbf{R}^2\right)} &\lesssim 1, \quad \|u\|_{L_t^4 L_x^\infty\left([2^j, 2^{j+1}] \times \mathbf{R}\right)} \lesssim 1.\end{aligned}\right\} \quad (4.94)$$

Next, by (4.80),

$$\lim_{k \nearrow \infty} \mathscr{M}(k) = 0. \tag{4.95}$$

Also, by Strichartz estimates,

$$\mathscr{S}(k) \lesssim \mathscr{M}(k) + \mathscr{N}(k). \tag{4.96}$$

Make the following splitting:

$$\left\| P_{>k-\frac{j}{2}} F(u) \right\|_{L_{t,x}^{\frac{2(d+2)}{d+4}}\left([2^j, 2^{j+1}] \times \mathbf{R}^d\right)} \lesssim \left\| P_{>k-\frac{j}{2}} F\left(u_{\leq k-\frac{j}{2}}\right) \right\|_{L_{t,x}^{\frac{2(d+2)}{d+4}}\left([2^j, 2^{j+1}] \times \mathbf{R}^d\right)}$$
$$+ \left\| |u_{>k-\frac{j}{2}}| |u|^{\frac{4}{d}} \right\|_{L_{t,x}^{\frac{2(d+2)}{d+4}}\left([2^j, 2^{j+1}] \times \mathbf{R}^d\right)}.$$

By Bernstein's inequality, (4.94), and interpolation,

$$\left\| P_{>k-\frac{j}{2}} F\left(u_{\leq k-\frac{j}{2}}\right) \right\|_{L_{t,x}^{\frac{2(d+2)}{d+4}}\left([2^j, 2^{j+1}] \times \mathbf{R}^d\right)}$$
$$\lesssim \frac{1}{2^{k-\frac{j}{2}}} \left\| \nabla u_{\leq k-\frac{j}{2}} \right\|_{L_{t,x}^{\frac{2(d+2)}{d}}\left([2^j, 2^{j+1}] \times \mathbf{R}^d\right)} \|u\|_{L_{t,x}^{\frac{2(d+2)}{d}}\left([2^j, 2^{j+1}] \times \mathbf{R}^d\right)}^{\frac{4}{d}}$$
$$\lesssim 2^{-k+\frac{j}{2}} \sum_{l \leq k-\frac{j}{2}} 2^l \|u_{>l}\|_{L_t^{\frac{2(d+1)}{d}} L_x^{\frac{2(d+1)}{d-1}}\left([2^j, 2^{j+1}] \times \mathbf{R}^d\right)}^{\frac{d+1}{d+2}} \|u_{>l}\|_{L_t^\infty L_x^2\left([2^j, 2^{j+1}] \times \mathbf{R}^d\right)}^{\frac{1}{d+2}}$$
$$\lesssim 2^{-k+\frac{j}{2}} \sum_{l \leq k-\frac{j}{2}} 2^l \|P_{>l} u\|_{L_t^{\frac{2(d+1)}{d}} L_x^{\frac{2(d+1)}{d-1}}\left([2^j, 2^{j+1}] \times \mathbf{R}^d\right)}^{\frac{d+1}{d+2}} \|P_{>l} u\|_{L_t^\infty L_x^2\left([2^j, 2^{j+1}] \times \mathbf{R}^d\right)}^{\frac{1}{d+2}}$$
$$\lesssim \sum_{l \leq k} 2^{l-k} \mathscr{M}(l)^{\frac{1}{d+2}}. \tag{4.97}$$

Expressions (4.95) and (4.97) imply $\lim_{k\nearrow\infty} \mathcal{N}(k) = 0$, and therefore, by (4.96),

$$\lim_{k\nearrow\infty} \mathcal{M}(k) + \mathcal{S}(k) + \mathcal{N}(k) = 0. \tag{4.98}$$

This decay can be upgraded to a quantitative decay estimate.

Theorem 4.27 *When $d \geq 2$, equation (4.98) has the quantitative decay*

$$\mathcal{M}(k) + \mathcal{S}(k) + \mathcal{N}(k) \lesssim 2^{-\frac{k}{2d}}; \tag{4.99}$$

and when $d = 1$, the decay is

$$\mathcal{M}(k) + \mathcal{S}(k) + \mathcal{N}(k) \lesssim 2^{-\frac{k}{4}}.$$

The proof of this theorem will be postponed.

Theorem 4.27 may then be used as the base case in an inductive argument.

Lemma 4.28 (Inductive lemma) *Let $N > 0$, $0 < s < 1 + \frac{4}{d}$, and $d \geq 3$. For all $k > 100$, $0 < \beta \leq 1$,*

$$\mathcal{N}(k) \lesssim_s \sum_{k_1 \leq k\beta - N} 2^{(k_1-k)s} \mathcal{S}(k_1)$$

$$+ \left[\mathcal{S}\left(\frac{k\beta}{2(d-1)} - N \right) + \mathcal{S}(k\beta - N) \right]^{4/d} \mathcal{S}(k\beta - N)$$

$$+ 2^{-2k\beta/d^2} [\mathcal{M}(k\beta - N) + \mathcal{N}(k\beta - N)]. \tag{4.100}$$

Proof This lemma was proved in Killip *et al.* (2008) for $d \geq 3$: their argument will be the one presented here. First, split the nonlinearity:

$$F(u)$$
$$= F(u_{\leq k\beta - j/2 - N}) + O\left(|u_{\leq k\alpha - j/2 - N}|^{\frac{4}{d}} |u_{> k\beta - j/2 - N}| \right)$$
$$+ O\left(|u_{k\alpha - j/2 - N \leq \cdot \leq k\beta - j/2 - N}|^{\frac{4}{d}} |u_{> k\beta - j/2 - N}| \right) + O\left(|u_{> k\beta - j/2 - N}|^{1 + \frac{4}{d}} \right),$$

where α will be defined later. By Bernstein's inequality, the fractional product rule, and the fractional chain rule of Visan (2007), for any $0 \leq s < 1 + \frac{4}{d}$,

$$\left\| P_{> k - j/2} F(u_{\leq k\beta - j/2 - N}) \right\|_{L^{\frac{2(d+2)}{d+4}}_{t,x}([2^j, 2^{j+1}] \times \mathbf{R}^d)}$$

$$\lesssim 2^{\left(\frac{j}{2} - k\right)s} \left\| |\nabla|^s F(u_{\leq k\beta - j/2 - N}) \right\|_{L^{\frac{2(d+2)}{d+4}}_{t,x}([2^j, 2^{j+1}] \times \mathbf{R}^d)}$$

$$\lesssim_s \sum_{k_1 \leq k\beta - N} 2^{(k_1 - k)s} \mathcal{S}(k_1).$$

4.3 Radial Mass-Critical Problem in Dimensions $d \geq 2$

By Hölder's inequality,

$$\left\| |u_{k\alpha - j/2 - N \leq \cdot \leq k\beta - j/2 - N}|^{\frac{4}{d}} |u_{>k\beta - j/2 - N}| \right\|_{L_{t,x}^{\frac{2(d+2)}{d+4}}([2^j, 2^{j+1}] \times \mathbf{R}^d)}$$

$$\lesssim \mathscr{S}(k\alpha - N)^{\frac{4}{d}} \mathscr{S}(k\beta - N)$$

and

$$\left\| |u_{>k\beta - j/2 - N}|^{1 + \frac{4}{d}} \right\|_{L_{t,x}^{\frac{2(d+2)}{d+4}}([2^j, 2^{j+1}] \times \mathbf{R}^d)} \lesssim \mathscr{S}(k\beta - N)^{1 + \frac{4}{d}}.$$

Finally by the bilinear Strichartz estimates of (4.23) with $\alpha = \frac{\beta}{2(d-1)}$, (4.96), Hölder's inequality in time, and the Christ–Kiselev lemma (see Christ and Kiselev (2001), and Smith and Sogge (2000)),

$$\left\| O\left(|u_{\leq k\alpha - j/2 - N}|^{\frac{4}{d}} |u_{>k\beta - j/2 - N}| \right) \right\|_{L_{t,x}^{\frac{2(d+2)}{d+4}}([2^j, 2^{j+1}] \times \mathbf{R}^d)}$$

$$\lesssim \left\| (u_{\leq k\alpha - j/2 - N})(u_{>k\beta - j/2 - N}) \right\|_{L_{t,x}^2([2^j, 2^{j+1}] \times \mathbf{R}^d)}^{\frac{8}{d^2}}$$

$$\times \left\| u_{>k\beta - j/2 - N} \right\|_{L_{t,x}^{\frac{2(d+2)}{d}}([2^j, 2^{j+1}] \times \mathbf{R}^d)}^{1 - \frac{8}{d^2}} \left\| u_{\leq k\alpha - j/2 - N} \right\|_{L_{t,x}^2([2^j, 2^{j+1}] \times \mathbf{R}^d)}^{\frac{4}{d} - \frac{8}{d^2}}$$

$$\lesssim \left[(2^{k\beta - j/2 - N})^{-\frac{1}{2}} (2^{k\alpha - j/2 - N})^{\frac{d-1}{2}} \right]^{\frac{8}{d^2}} [\mathscr{M}(k\beta - N) + \mathscr{N}(k\beta - N)]^{\frac{8}{d^2}}$$

$$\times \mathscr{S}(k\beta - N)^{1 - \frac{8}{d^2}} 2^{j\left(\frac{2}{d} - \frac{4}{d^2}\right)}$$

$$\lesssim 2^{-k\frac{2\beta}{d^2}} [\mathscr{M}(k\beta - N) + \mathscr{N}(k\beta - N)].$$

This proves the lemma. □

Lemma 4.28 may be used to prove increasing regularity by induction, since the second and third terms in (4.100) are of order $1 + \frac{4}{d}$. Indeed, suppose that for some $\sigma > 0$,

$$\mathscr{S}(k) + \mathscr{N}(k) + \mathscr{M}(k) \lesssim 2^{-\sigma k}. \tag{4.101}$$

Expression (3.78) holds in L^2 for almost-periodic solutions to the mass-critical problem, so

$$\left\| P_{>k} u(1) \right\|_{L_x^2(\mathbf{R}^d)} \lesssim \sum_{j \geq 0} \left\| P_{>k} F(u) \right\|_{L_{t,x}^{\frac{2(d+2)}{d+4}}([2^j, 2^{j+1}] \times \mathbf{R}^d)},$$

and then by (4.92),

$$\mathscr{M}(k) \lesssim \sum_{j \geq 0} \mathscr{N}\left(k + \frac{j}{2} \right). \tag{4.102}$$

Then by (4.96) and (4.102),

$$\mathscr{M}(k) + \mathscr{S}(k) + \mathscr{N}(k) \lesssim \mathscr{M}(k) + \mathscr{N}(k) \lesssim \sum_{j \geq 0} \mathscr{N}\left(k + \frac{j}{2}\right).$$

By Lemma 4.28, with $\beta = 1 - \frac{1}{2d^2}$,

$$\sum_{j \geq 0} \mathscr{N}\left(k + \frac{j}{2}\right)$$

$$\lesssim_s \sum_{j \geq 0} \sum_{k_1 \leq \left(k + \frac{j}{2}\right)\beta - N} 2^{\left(k_1 - k - \frac{j}{2}\right)s} \mathscr{S}(k_1)$$

$$+ \sum_{j \geq 0} \left[\mathscr{S}\left(\frac{(k + \frac{j}{2})\beta}{2(d-1)} - N\right) + \mathscr{S}\left(\left(k + \frac{j}{2}\right)\beta - N\right)\right]^{\frac{4}{d}}$$

$$\times \mathscr{S}\left(\left(k + \frac{j}{2}\right)\beta - N\right)$$

$$+ \sum_{j \geq 0} 2^{-\left(k + \frac{j}{2}\right)\frac{2\beta}{d^2}} \left[\mathscr{M}\left(\left(k + \frac{j}{2}\right)\beta - N\right) + \mathscr{N}\left(\left(k + \frac{j}{2}\right)\beta - N\right)\right].$$

(4.103)

Plugging (4.101) into (4.103),

(4.103)

$$\lesssim \sum_{j \geq 0} \sum_{k_1 \leq \left(k + \frac{j}{2}\right)\beta - N} 2^{\left(k_1 - k - \frac{j}{2}\right)s} 2^{-\sigma k_1}$$

$$+ \sum_{j \geq 0} 2^{-\frac{(k+\frac{j}{2})\beta\sigma}{2(d-1)} \cdot \frac{4}{d}} 2^{-\left((k+\frac{j}{2})\beta - N\right)\sigma} + \sum_{j \geq 0} 2^{-\left(k+\frac{j}{2}\right) \cdot \frac{2\beta}{d^2}} 2^{-\left((k+\frac{j}{2})\beta - N\right)\sigma}$$

$$\lesssim_{N,s,\sigma} 2^{k\beta(s-\sigma)} 2^{-ks} + 2^{-\frac{2k\beta}{d(d-1)}\sigma} 2^{-k\beta\sigma} + 2^{-\frac{2k\beta}{d^2}} 2^{-k\beta\sigma}. \quad (4.104)$$

Iterating (4.103), starting with (4.99), proves that $\mathscr{M}(k) \lesssim_s 2^{-sk}$ for all $0 < s < 1 + \frac{4}{d}$.

When $d = 2$, by (4.23),

$$\|P_{>k-j/2} F(u)\|_{L^{4/3}_{t,x}([2^j, 2^{j+1}] \times \mathbf{R}^2)}$$

$$\lesssim \|F(u_{>k\beta - j/2 - N})\|_{L^{4/3}_{t,x}([2^j, 2^{j+1}] \times \mathbf{R}^2)}$$

4.3 Radial Mass-Critical Problem in Dimensions $d \geq 2$

$$+ \left\| O\left(\left(u_{k\alpha - j/2 - N \leq \cdot \leq k\beta - j/2 - N} \right)^2 \left(u_{>k\beta - j/2 - N} \right) \right) \right\|_{L_{t,x}^{4/3}([2^j, 2^{j+1}] \times \mathbf{R}^2)}$$

$$+ \left\| O\left(\left(u_{\leq k\alpha - j/2 - N} \right)^2 \left(u_{>k\beta - j/2 - N} \right) \right) \right\|_{L_{t,x}^{4/3}([2^j, 2^{j+1}] \times \mathbf{R}^2)}$$

$$\lesssim \mathscr{S}(k\alpha - N)^2 \mathscr{S}(k\beta - N) + [\mathscr{M}(k\beta - N) + \mathscr{N}(k\beta - N)] 2^{\frac{k(\alpha - \beta)}{2}}.$$

Take $\beta = \frac{12}{13}$, $\alpha = \frac{6}{13}$. If

$$\mathscr{M}(k) + \mathscr{N}(k) + \mathscr{S}(k) \lesssim 2^{-\sigma k},$$

then

$$\mathscr{M}(k) + \mathscr{N}(k) + \mathscr{S}(k) \lesssim 2^{-\frac{24}{13}\sigma k} + 2^{-\frac{12k}{13}\sigma} 2^{-\frac{3k}{13}},$$

and therefore, when $s < 3$, by induction,

$$\mathscr{M}(k) + \mathscr{N}(k) + \mathscr{S}(k) \lesssim_s 2^{-sk}.$$

Finally, for $d = 1$, by (4.23),

$$\left\| P_{>k-j/2} F(u) \right\|_{L_{t,x}^{6/5}([2^j, 2^{j+1}] \times \mathbf{R})}$$

$$\lesssim \left\| u_{> \frac{20k}{21} - j/2} \right\|_{L_{t,x}^6([2^j, 2^{j+1}] \times \mathbf{R})} \left\| u_{> \frac{10k}{21} - j/2} \right\|_{L_{t,x}^6([2^j, 2^{j+1}] \times \mathbf{R})}^4$$

$$+ 2^{-\frac{5k}{21}} \left(\mathscr{M}\left(\frac{20k}{21}\right) + \mathscr{N}\left(\frac{20k}{21}\right) \right),$$

which implies $\mathscr{M}(k) \lesssim_s 2^{-5k}$ when $s < 5$. This completes the proof of Theorem 4.26. $\square$

Proof of Theorem 4.27 Theorem 4.27 is proved using the double Duhamel method. As in (3.89),

$$\langle P_{>k} u(1), P_{>k} u(1) \rangle = \int_1^\infty \int_0^1 \langle e^{-it\Delta} P_{>k} F(u(t)), e^{-i\tau\Delta} P_{>k} F(u(\tau)) \rangle \, dt \, d\tau.$$

Recall from linear algebra that if $A + B = A' + B'$,

$$\langle A + B, A' + B' \rangle \lesssim |A|^2 + |A'|^2 + \langle B, B' \rangle. \tag{4.105}$$

For some $T > 2$, set

$$A = 0, \qquad B = \int_0^1 e^{-it\Delta} P_{>k} F(u(t)) \, dt,$$

$$A' = \int_1^T e^{-i\tau\Delta} P_{>k} F(u(\tau)) \, d\tau, \qquad B' = \int_T^\infty e^{-i\tau\Delta} P_{>k} F(u(\tau)) \, d\tau.$$

First consider $d \geq 4$. By Strichartz estimates,

$$\|A'\|_{L^2(\mathbf{R}^d)}$$
$$\lesssim \|P_{>k} F(u)\|_{N^0([1,T]\times\mathbf{R}^d)}$$
$$\lesssim \|P_{>k} F(u_{\leq k})\|_{N^0([1,T]\times\mathbf{R}^d)}$$
$$+ \left\| |u_{>k}| |u_{\leq \frac{k}{2(d-1)}}|^{\frac{4}{d}} \right\|_{L_t^1 L_x^2([1,T]\times\mathbf{R}^d)} + \left\| |u_{>k}| |u_{> \frac{k}{2(d-1)}}|^{\frac{4}{d}} \right\|_{L_{t,x}^{\frac{2(d+2)}{d+4}}([1,T]\times\mathbf{R}^d)}. \tag{4.106}$$

By (4.23) and Hölder's inequality,

$$\left\| |u_{>k}| |u_{\leq \frac{k}{2(d-1)}}|^{\frac{4}{d}} \right\|_{L_t^1 L_x^2([1,T]\times\mathbf{R}^d)}$$
$$\lesssim T^{1-\frac{2}{d}} \left\| |u_{>k}| |u_{\leq \frac{k}{2(d-1)}}|^{\frac{4}{d}} \right\|_{L_{t,x}^2([1,T]\times\mathbf{R}^d)} \|u_{>k}\|_{L_t^\infty L_x^2([1,T]\times\mathbf{R}^d)}^{1-\frac{4}{d}}$$
$$\lesssim_T \left(2^{-\frac{k}{2}} 2^{\frac{k}{4}}\right)^{4/d} \lesssim 2^{-k/d}. \tag{4.107}$$

Next, since $N(t) \leq 1$ for $t \geq 1$,

$$\|u_{>k}\|_{L_{t,x}^{\frac{2(d+2)}{d}}([1,T]\times\mathbf{R}^d)} \|u_{>\frac{k}{2(d-1)}}\|_{L_{t,x}^{\frac{2(d+2)}{d}}([1,T]\times\mathbf{R}^d)}^{4/d}$$
$$\lesssim_T \mathscr{S}(k) \mathscr{S}\left(\frac{k}{2(d-1)}\right)^{4/d}.$$

Finally, by Bernstein's inequality and the product rule,

$$\|P_{>k} F(u_{\leq k})\|_{N^0([1,T]\times\mathbf{R}^d)} \lesssim 2^{-k} \|\nabla F(u_{\leq k})\|_{N^0([1,T]\times\mathbf{R}^d)}$$
$$\lesssim 2^{-k} \left\| |\nabla u_{\leq k}| |u_{\leq k}|^{\frac{4}{d}} \right\|_{N^0([1,T]\times\mathbf{R}^d)}.$$

By the triangle inequality,

$$|\nabla u_{\leq k}| |u_{\leq k}|^{\frac{4}{d}} \lesssim \sum_{k_1 \leq k} |\nabla u_{k_1}| |u_{\leq k}|^{\frac{4}{d}}$$
$$\lesssim \sum_{k_1 \leq k} |\nabla u_{k_1}| |u_{\frac{k_1}{2(d-1)} \leq \cdot \leq k}|^{\frac{4}{d}} + \sum_{k_1 \leq k} |\nabla u_{k_1}| |u_{\leq \frac{k_1}{2(d-1)}}|^{\frac{4}{d}}. \tag{4.108}$$

4.3 Radial Mass-Critical Problem in Dimensions $d \geq 2$

Therefore, by (4.108), Hölder's inequality, and (4.23),

$$2^{-k} \left\| |\nabla u_{\leq k}| |u_{\leq k}|^{\frac{4}{d}} \right\|_{N^0([1,T] \times \mathbf{R}^d)}$$

$$\lesssim 2^{-k} \sum_{k_1 \leq k} \|\nabla u_{k_1}\|_{L_{t,x}^{\frac{2(d+2)}{d}}([1,T] \times \mathbf{R}^d)} \|u_{\geq \frac{k_1}{2(d-1)}}\|_{L_{t,x}^{\frac{2(d+2)}{d}}([1,T] \times \mathbf{R}^d)}^{4/d}$$

$$+ 2^{-k} T^{1-\frac{2}{d}} \sum_{k_1 \leq k} \left\| |\nabla u_{k_1}| |u_{\leq \frac{k_1}{2(d-1)}}|^{4/d} \right\|_{L_{t,x}^2([1,T] \times \mathbf{R}^d)} \|\nabla u_{k_1}\|_{L_t^\infty L_x^2([1,T] \times \mathbf{R}^d)}^{1-\frac{4}{d}}$$

$$\lesssim_T \sum_{k_1 \leq k} 2^{k_1-k} \left(2^{\frac{k_1}{4}} 2^{-\frac{k_1}{2}} \right)^{\frac{4}{d}} + \sum_{k_1 \leq k} 2^{(k_1-k)} \mathscr{S}(k_1) \mathscr{S}\left(\frac{k_1}{2(d-1)} \right)^{\frac{4}{d}}$$

$$\lesssim_T 2^{-\frac{k}{d}} + \mathscr{S}\left(\frac{k}{4(d-1)} \right)^{\frac{4}{d}} \sum_{k_1 \leq k} 2^{(k_1-k)} \mathscr{S}(k_1). \tag{4.109}$$

Next,

$$\langle B, B' \rangle_{L^2}$$

$$= \int_0^1 \int_T^\infty \langle e^{-it\Delta} P_{>k} F(u(t)), e^{-i\tau\Delta} F(u(\tau)) \rangle d\tau$$

$$= \int_0^1 \int_T^\infty \langle P_{>k} F(u(t)), e^{i(t-\tau)\Delta} P_{>k} F(u(\tau)) \rangle dt \, d\tau$$

$$\lesssim \int_0^1 \|P_{>k} F(u(t))\|_{L_x^{\frac{2d}{d+4}}(\mathbf{R}^d)} \left\| \int_T^\infty e^{i(t-\tau)\Delta} P_{>k} F(u(\tau)) d\tau \right\|_{L_x^{\frac{2d}{d-4}}(\mathbf{R}^d)} dt. \tag{4.110}$$

By Bernstein's inequality,

$$\|P_{>k} F(u)\|_{L_x^{\frac{2d}{d+4}}(\mathbf{R}^d)} \lesssim \|P_{>k} F(u_{\leq k})\|_{L_x^{\frac{2d}{d+4}}(\mathbf{R}^d)} + \left\| |u_{>k}| |u|^{\frac{4}{d}} \right\|_{L_x^{\frac{2d}{d+4}}(\mathbf{R}^d)}$$

$$\lesssim 2^{-k} \|\nabla F(u_{\leq k})\|_{L_x^{\frac{2d}{d+4}}(\mathbf{R}^d)} + \left\| |u_{>k}| |u|^{\frac{4}{d}} \right\|_{L_x^{\frac{2d}{d+4}}(\mathbf{R}^d)}$$

$$\lesssim \sum_{k_1 \leq k} 2^{k_1-k} \|P_{>k_1} u\|_{L_x^2(\mathbf{R}^d)}. \tag{4.111}$$

Since $N(t) \leq 1$ when $t \geq 1$, the dispersive estimate implies that for $T > 2$,

$$\left\| \int_T^\infty e^{i(t-\tau)\Delta} P_{>k} F(u(\tau)) d\tau \right\|_{L_t^\infty L_x^{\frac{2d}{d-4}}([0,1] \times \mathbf{R}^d)}$$

$$\lesssim \sup_{t \in [0,1]} \int_T^\infty \frac{1}{(t-\tau)^2} \sum_{k_1 \leq k} 2^{k_1-k} \mathscr{M}(k_1) dt \lesssim \frac{1}{T} \sum_{k_1 \leq k} 2^{k_1-k} \mathscr{M}(k_1).$$

Therefore,

$$(4.110) \lesssim \frac{1}{T}\left(\sum_{k_1 \le k} 2^{(k_1-k)} \mathscr{M}(k_1)\right) \|P_{>k} F(u(t))\|_{L_t^1 L_x^{\frac{2d}{d+4}}([0,1] \times \mathbf{R}^d)}. \quad (4.112)$$

Next, by (4.92) and (4.111), for any $j \le 0$,

$$\|P_{>k} F(u(t))\|_{L_t^1 L_x^{\frac{2d}{d+4}}([2^j, 2^{j+1}] \times \mathbf{R}^d)} \lesssim 2^j \sum_{k_1 \le k} 2^{(k_1-k)} \mathscr{M}\left(k_1 + \frac{j}{2}\right),$$

and so

$$\langle B, B' \rangle \lesssim (4.112)$$

$$\lesssim \frac{1}{T}\left(\sum_{k_1 \le k} \mathscr{M}(k_1) 2^{(k_1-k)}\right)\left(\sum_{j \le 0} 2^j \sum_{k_1 \le k} 2^{(k_1-k)} \mathscr{M}\left(k_1 + \frac{j}{2}\right)\right)$$

$$\lesssim \frac{1}{T}\left(\sum_{k_1 \le k} \mathscr{M}(k_1) 2^{(k_1-k)}\right)^2. \quad (4.113)$$

Therefore, by (4.109) and (4.113), there exists a constant $C(T)$ such that

$$\|P_{>k} u(1)\|_{L_x^2(\mathbf{R}^d)} \lesssim \frac{1}{T^{1/2}} \sum_{k_1 \le k} 2^{(k_1-k)} \mathscr{M}(k_1) + C(T) 2^{-\frac{k}{d}}$$

$$+ C(T) \mathscr{S}\left(\frac{k}{4(d-1)}\right)^{\frac{4}{d}} \sum_{k_1 \le k} 2^{(k_1-k)} \mathscr{S}(k_1). \quad (4.114)$$

The rescaling $u(t,x) \mapsto T^{d/4} u(Tt, T^{1/2}x)$ maps a self-similar solution to a self-similar solution, so in fact, (4.114) implies

$$\mathscr{M}(k) \lesssim \frac{1}{T^{1/2}} \sum_{k_1 \le k} 2^{(k_1-k)} \mathscr{M}(k_1) + C(T) 2^{-\frac{k}{d}}$$

$$+ C(T) \mathscr{S}\left(\frac{k}{4(d-1)}\right)^{\frac{4}{d}} \sum_{k_1 \le k} 2^{(k_1-k)} \mathscr{S}(k_1). \quad (4.115)$$

By Strichartz estimates and rescaling, $\mathscr{S}(k) \lesssim \mathscr{M}(k) + (4.106)$, so by (4.107)–(4.115),

$$\mathscr{M}(k) + \mathscr{S}(k) \lesssim \frac{1}{T^{1/2}} \sum_{k_1 \le k} 2^{(k_1-k)} [\mathscr{M}(k_1) + \mathscr{S}(k_1)] + C(T) 2^{-\frac{k}{d}}$$

$$+ C(T)\left[\mathscr{M}\left(\frac{k}{4(d-1)}\right)^{\frac{4}{d}} + \mathscr{S}\left(\frac{k}{4(d-1)}\right)^{\frac{4}{d}}\right]$$

$$\times \sum_{k_1 \le k} 2^{(k_1-k)} [\mathscr{M}(k_1) + \mathscr{S}(k_1)]. \quad (4.116)$$

4.3 Radial Mass-Critical Problem in Dimensions $d \geq 2$

Setting

$$\alpha(k) = \sup_j 2^{-\frac{1}{2d}|j-k|}[\mathscr{M}(j) + \mathscr{S}(j)]$$

and remembering (4.93) gives

$$\alpha(k) \lesssim \frac{1}{T^{1/2}}\alpha(k) + C(T) 2^{-\frac{k}{2d}} + C(T)\alpha\left(\frac{k}{4(d-1)}\right)^{\frac{4}{d}} \alpha(k). \quad (4.117)$$

The estimate (4.98) implies $\alpha(k) \searrow 0$ as $k \nearrow \infty$, so choosing T large, but fixed, for $k \geq K_0$, K_0 is a large, fixed constant, (4.117) implies

$$\alpha(k) \lesssim C(T) 2^{-\frac{k}{2d}}. \quad (4.118)$$

This completes the proof of Theorem 4.27 in the case when $d \geq 4$.

When $d = 3$ the proof is quite similar, save for some technical modifications. For any $j \geq 0$, by Bernstein's inequality,

$$\left\|P_{>k}F(u)\right\|_{L_t^4 L_x^1([2^j, 2^{j+1}] \times \mathbf{R}^3)}$$

$$\lesssim \left\|P_{>k}F(u_{\leq k})\right\|_{L_t^4 L_x^1([2^j, 2^{j+1}] \times \mathbf{R}^3)}$$

$$+ \left\|u_{>k}\right\|_{L_t^\infty L_x^2([2^j, 2^{j+1}] \times \mathbf{R}^3)} \left\|u\right\|_{L_t^{16/3} L_x^8([2^j, 2^{j+1}] \times \mathbf{R}^3)}^{4/3}$$

$$\lesssim 2^{-k} \left\|\nabla F(u_{\leq k})\right\|_{L_t^4 L_x^1([2^j, 2^{j+1}] \times \mathbf{R}^3)} + \mathscr{M}(k)$$

$$\lesssim \sum_{k_1 \leq k} 2^{k_1 - k} \left\|u_{k_1}\right\|_{L_{t,x}^{10/3}([2^j, 2^{j+1}] \times \mathbf{R}^3)}^{5/6} \left\|u_{k_1}\right\|_{L_t^\infty L_x^2([2^j, 2^{j+1}] \times \mathbf{R}^3)}^{1/6}$$

$$\times \left\|u\right\|_{L_t^\infty L_x^2([2^j, 2^{j+1}] \times \mathbf{R}^3)}^{4/3} + \mathscr{M}(k)$$

$$\lesssim \sum_{k_1 \leq k} 2^{k_1 - k} [\mathscr{M}(k_1) + \mathscr{S}(k_1)]. \quad (4.119)$$

Since $\left\|\frac{1}{t^{3/2}}\right\|_{L_t^{4/3}([2^j, 2^{j+1}])} \lesssim 2^{-\frac{3j}{4}}$, when $T > 2$ we find

$$\left\|\int_T^\infty e^{i(t-\tau)\Delta} P_{>k} F(u(\tau)) d\tau\right\|_{L_x^\infty(\mathbf{R}^3)} \lesssim \frac{1}{T^{3/4}} \sum_{k_1 \leq k} 2^{k_1 - k}[\mathscr{M}(k_1) + \mathscr{S}(k_1)].$$

$$(4.120)$$

Also, using (4.113) and (4.119), we have

$$\sum_{j \leq 0} \left\| P_{>k} F(u) \right\|_{L^1_{t,x}([2^j, 2^{j+1}] \times \mathbf{R}^3)}$$

$$\lesssim \sum_{j \leq 0} 2^{3j/4} \left\| P_{>k} F(u) \right\|_{L^4_t L^1_x ([2^j, 2^{j+1}] \times \mathbf{R}^3)}$$

$$\lesssim \sum_{j \leq 0} 2^{3j/4} \sum_{k_1 \leq k} 2^{k_1 - k} \left[\mathscr{M}\left(k_1 + \frac{j}{2}\right) + \mathscr{S}\left(k_1 + \frac{j}{2}\right) \right]$$

$$\lesssim \sum_{k_1 \leq k} 2^{k_1 - k} [\mathscr{M}(k_1) + \mathscr{S}(k_1)]. \tag{4.121}$$

Then, by (4.120) and (4.121), we have

$$\langle B, B' \rangle_{L^2} \lesssim \frac{1}{T^{3/4}} \left(\sum_{k_1 \leq k} 2^{k_1 - k} [M(k_1) + \mathscr{S}(k_1)] \right)^2.$$

To estimate A', consider

$$\left\| P_{>k} F(u) \right\|_{N^0([1,T] \times \mathbf{R}^3)}$$

$$\lesssim \left\| P_{>k} F(u_{\leq k}) \right\|_{N^0([1,T] \times \mathbf{R}^3)} + \left\| |u_{>k}| |u_{\leq \frac{k}{4}}|^{\frac{4}{3}} \right\|_{L^1_t L^2_x([1,T] \times \mathbf{R}^3)}$$

$$+ \left\| |u_{>k}| |u_{\geq \frac{k}{4}}|^{\frac{4}{3}} \right\|_{L^{\frac{10}{7}}_{t,x}([1,T] \times \mathbf{R}^3)}.$$

Strichartz estimates and $N(t) \leq 1$ for $t \geq 1$ imply

$$\left\| |u_{>k}| |u_{\geq \frac{k}{4}}|^{\frac{4}{3}} \right\|_{L^{10/7}_{t,x}([1,T] \times \mathbf{R}^3)} \lesssim_T \mathscr{S}(k) \mathscr{S}\left(\frac{k}{4}\right)^{\frac{4}{3}}.$$

Next, by (4.23) and (4.94),

$$\left\| |u_{>k}| |u_{\leq \frac{k}{4}}|^{\frac{4}{3}} \right\|_{L^1_t L^2_x([1,T] \times \mathbf{R}^3)}$$

$$\lesssim T^{1/6} \left\| |u_{>k}| |u_{\leq \frac{k}{4}}| \right\|^{2/3}_{L^2_{t,x}([1,T] \times \mathbf{R}^3)} \|u\|_{L^2_t L^6_x([1,T] \times \mathbf{R}^3)}$$

$$\lesssim_T \left(2^{-\frac{k}{2}} 2^{\frac{k}{4}}\right)^{2/3} \lesssim 2^{-\frac{k}{6}}.$$

4.3 Radial Mass-Critical Problem in Dimensions $d \geq 2$

Therefore, following (4.106), by Bernstein's inequality, (4.23), and (4.94),

$$\|A'\|_{L^2} \lesssim \|P_{>k}F(u_{\leq k})\|_{N([1,T]\times \mathbf{R}^3)} \lesssim 2^{-k}\|\nabla F(u_{\leq k})\|_{N([1,T]\times \mathbf{R}^3)}$$

$$\lesssim 2^{-k}\sum_{k_1 \leq k}\|\nabla u_{k_1}\|_{L_{t,x}^{10/3}([1,T]\times \mathbf{R}^3)}\|u_{\geq \frac{k_1}{4}}\|_{L_{t,x}^{10/3}([1,T]\times \mathbf{R}^3)}^{4/3}$$

$$+ 2^{-k}\sum_{k_1 \leq k}\left\| |\nabla u_{k_1}| |u_{\leq \frac{k_1}{4}}| \right\|_{L_{t,x}^2([1,T]\times \mathbf{R}^3)}^{2/3}$$

$$\times \|\nabla u_{k_1}\|_{L_t^2 L_x^6([1,T]\times \mathbf{R}^3)}^{1/3}\|u\|_{L_t^2 L_x^6([1,T]\times \mathbf{R}^3)}^{2/3}$$

$$\lesssim_T \sum_{k_1 \leq k} 2^{k_1 - k}\mathscr{S}(k_1)\mathscr{S}\left(\frac{k_1}{4}\right)^{4/3} + 2^{-k/6}.$$

Following the argument in (4.113)–(4.118), $\mathscr{M}(k) + \mathscr{S}(k) \lesssim 2^{-k/6}$.

Next take $d = 2$. In this case the nonlinearity is algebraic, so for $j \geq 0$,

$$\|P_{>k}F(u)\|_{L_t^2 L_x^1([2^j, 2^{j+1}]\times \mathbf{R}^2)}$$

$$\lesssim \|u_{>k-3}\|_{L_t^\infty L_x^2([2^j, 2^{j+1}]\times \mathbf{R}^2)}\|u\|_{L_{t,x}^4([2^j, 2^{j+1}]\times \mathbf{R}^2)}^2 \lesssim \mathscr{M}(k-3).$$

Then for $T > 2$,

$$\left\|\int_T^\infty e^{i(t-\tau)\Delta}F(u(\tau))d\tau\right\|_{L_{t,x}^\infty([0,1]\times \mathbf{R}^2)} \lesssim \frac{1}{T^{1/2}}\mathscr{M}(k-3).$$

When $j \leq 0$,

$$\|P_{>k}F(u)\|_{L_{t,x}^1([2^j, 2^{j+1}]\times \mathbf{R}^2)}$$

$$\lesssim 2^{j/2}\|u_{>k-3}\|_{L_t^\infty L_x^2([2^j, 2^{j+1}]\times \mathbf{R}^2)}\|u\|_{L_{t,x}^4([2^j, 2^{j+1}]\times \mathbf{R}^2)}^2 \lesssim 2^{j/2}\mathscr{M}\left(k + \frac{j}{2}\right),$$

so

$$\langle B, B'\rangle_{L^2} \lesssim \frac{1}{T^{1/2}}\mathscr{M}(k-3)\sum_{k_1 \leq k} 2^{\frac{(k_1-k)}{2}}\mathscr{M}(k_1).$$

Also,

$$\|P_{>k}F(u)\|_{L_{t,x}^{4/3}([1,T]\times \mathbf{R}^2)}$$

$$\lesssim \left\| |u_{>k-3}| |u_{\leq \frac{k}{2}}| \right\|_{L_{t,x}^2([1,T]\times \mathbf{R}^2)}\|u\|_{L_{t,x}^{4/3}([1,T]\times \mathbf{R}^2)}$$

$$+ \|u_{>k-3}\|_{L_{t,x}^4([1,T]\times \mathbf{R}^2)}\|u_{>\frac{k}{2}}\|_{L_{t,x}^4([1,T]\times \mathbf{R}^2)}\|u\|_{L_{t,x}^4([1,T]\times \mathbf{R}^2)}$$

$$\lesssim_T 2^{-\frac{k}{4}} + \mathscr{S}(k-3)\mathscr{S}\left(\frac{k}{2}\right).$$

Again following the argument in (4.113)–(4.118), we obtain $\mathscr{M}(k) + \mathscr{S}(k) \lesssim 2^{-k/4}$.

When $d=1$ the nonlinearity is also algebraic, so if $j \geq 0$,

$$\left\| P_{>k} F(u) \right\|_{L_t^{4/3} L_x^1 ([2^j, 2^{j+1}] \times \mathbf{R})}$$
$$\lesssim \left\| P_{>k-3} u \right\|_{L_t^\infty L_x^2 ([2^j, 2^{j+1}] \times \mathbf{R})} \left\| u \right\|^4_{L_t^{16/3} L_x^8 ([2^j, 2^{j+1}] \times \mathbf{R})} \lesssim \mathcal{M}(k-3),$$

so for $T > 2$,

$$\left\| \int_T^\infty e^{i(t-\tau)\Delta} F(u(\tau)) d\tau \right\|_{L_{t,x}^\infty ([0,1] \times \mathbf{R})} \lesssim \frac{1}{T^{1/4}} \mathcal{M}(k-3).$$

When $j \leq 0$,

$$\left\| P_{>k} F(u) \right\|_{L_{t,x}^1 ([2^j, 2^{j+1}] \times \mathbf{R})}$$
$$\lesssim 2^{j/4} \left\| P_{>k-3} u \right\|_{L_t^\infty L_x^2 ([2^j, 2^{j+1}] \times \mathbf{R})} \left\| u \right\|^4_{L_t^{16/3} L_x^8 ([2^j, 2^{j+1}] \times \mathbf{R})} \lesssim 2^{j/4} \mathcal{M}\left(k + \frac{j}{2}\right),$$

so

$$\langle B, B' \rangle_{L^2} \lesssim \frac{1}{T^{1/4}} \mathcal{M}(k-3) \sum_{k_1 \leq k} 2^{\frac{(k_1-k)}{4}} \mathcal{M}(k_1).$$

Also,

$$\left\| P_{>k} F(u) \right\|_{L_{t,x}^{6/5}(\theta)}$$
$$\lesssim \left\| |u_{>k-3}| |u_{\leq \frac{k}{2}}| \right\|^{1/2}_{L_{t,x}^2(\theta)} \left\| u_{>k-3} \right\|^{1/2}_{L_{t,x}^6(\theta)} \left\| u_{\leq \frac{k}{2}} \right\|^{1/2}_{L_{t,x}^\infty(\theta)} \left\| u \right\|^3_{L_{t,x}^6(\theta)}$$
$$+ \left\| u_{>k-3} \right\|_{L_{t,x}^6(\theta)} \left\| u_{>\frac{k}{2}} \right\|_{L_{t,x}^6([1,T] \times \mathbf{R}^2)} \left\| u \right\|^3_{L_{t,x}^6([1,T] \times \mathbf{R}^2)}$$
$$\lesssim_T 2^{-\frac{k}{4}} + \mathscr{S}(k-3) \mathscr{S}\left(\frac{k}{2}\right),$$

where we have written $\theta = [1,T] \times \mathbf{R}$ for clarity. Again following the argument in (4.113)–(4.118), it follows that $\mathcal{M}(k) + \mathscr{S}(k) \lesssim 2^{-k/4}$. This completes the proof of Theorem 4.27, which in turn completes the proof of Theorem 4.26. □

Having shown that an almost-periodic, self-similar solution does not exist, it only remains to prove that the only almost-periodic solution in the form of (4.88) is $u \equiv 0$. Killip *et al.* (2008, 2009) and Tao *et al.* (2007b) proved this for $d \geq 2$ when u is radial.

Theorem 4.29 *The defocusing, mass-critical nonlinear Schrödinger equation*

$$iu_t + \Delta u = F(u) = |u|^{\frac{4}{d}} u, \qquad u(0) = u_0 \in L^2(\mathbf{R}^d) \tag{4.122}$$

is globally well posed and scattering for all $u_0 \in L^2(\mathbf{R}^d)$ radial.

4.3 Radial Mass-Critical Problem in Dimensions $d \geq 2$

Proof This also follows from proving additional regularity.

Theorem 4.30 (Additional regularity) *If $u(t,x)$ is a radial, almost-periodic global solution to the mass-critical problem with $N(t) \leq 1$ for all $t \in \mathbf{R}$, then for $0 \leq s < 1 + \frac{4}{d}$, $d \geq 2$,*

$$u(t,x) \in L_t^\infty \dot{H}_x^s(\mathbf{R} \times \mathbf{R}^d).$$

Killip et al. (2008, 2009) and Tao et al. (2007b) used the in/out decomposition to prove this. If f is a radial function, then standard Fourier analysis implies

$$f(x) = C(d) \int \hat{f}(|\xi|) \int_{-\pi/2}^{\pi/2} e^{i|x||\xi|\sin(\theta)} (\cos(\theta))^{\frac{d-2}{2}} d\theta \, d|\xi|$$

and

$$\hat{f}(\xi) = C(d) \int f(|y|) \int_{-\pi/2}^{\pi/2} e^{i|y||\xi|\sin(\theta)} (\cos(\theta))^{\frac{d-2}{2}} d\theta \, d|y|.$$

The Bessel function

$$\int_{-\frac{\pi}{2}}^{\frac{\pi}{2}} e^{i|y||\xi|\sin(\theta)} (\cos(\theta))^{\frac{d-2}{2}} d\theta \qquad (4.123)$$

is then split into two pieces, the incoming piece and the outgoing piece,

$$[P^+ + P^-]f = f. \qquad (4.124)$$

Remark When $d > 4$, P^+ and P^- need not be bounded operators in $L_x^2(\mathbf{R}^d)$.

By time-translation invariance, assume without loss of generality that $t_1 = 0$. When calculating the norm of $P^\pm u(0)$ in some Hilbert space, it suffices to calculate the norms of

$$P^+ u(0) = -i \int_0^\infty e^{-it\Delta} P^+ F(u(t)) dt \qquad (4.125)$$

and

$$P^- u(0) = i \int_{-\infty}^0 e^{-it\Delta} P^- F(u(t)) dt. \qquad (4.126)$$

Equation (4.125) pulls back outgoing waves, while (4.126) pushes forward incoming waves, so it is possible to utilize the decay in space of a radially symmetric function to estimate (4.125) and (4.126). □

Remark When $d = 1$, the radial Sobolev embedding does not give any decay as $|x| \to \infty$ for radially symmetric functions. This is why Theorem 4.30 is restricted to $d \geq 2$.

Here, the proof will use the double Duhamel method, which is similar, although in this case the solution is not partitioned into two pieces that do not lie in $L_x^2(\mathbf{R}^d)$, as in (4.124).

Recall that Theorem 1.7 implies that the solution to $e^{it\Delta}P_k$ propagates with speed $\sim 2^k$. The radial Sobolev embedding theorem gives good estimates outside the cone $|x| = c2^k|t|$ for some $0 < c < 1$. The propagation speed $\sim 2^k$ also implies that when $e^{i(\tau-t)\Delta}\widetilde{P}_k$ acts on the part of $F(u(t))$ with support restricted to inside the cone, say when $t < 0$, all waves will become outgoing waves in positive times, and must lie outside the cone $|x| = c2^k|\tau|$ when $\tau > 0$. A similar argument can be made for waves inside the cone when $t > 0$, and $\tau < 0$. This fact can be utilized to prove additional regularity for an almost-periodic solution to (4.122).

Turning to the details, modifying (4.92), define

$$\mathscr{M}(k) = \sup_{t \in \mathbf{R}} \|P_{>k}u(t)\|_{L^2(\mathbf{R}^d)},$$

$$\mathscr{S}(k) = \sup_{T \in \mathbf{R}} \|u_{>k}\|_{L_{t,x}^{\frac{2(d+2)}{d}}([T,T+1]\times\mathbf{R}^d)},$$

and

$$\mathscr{N}(k) = \sup_{T \in \mathbf{R}} \|P_{>k}F(u)\|_{L_{t,x}^{\frac{2(d+2)}{d+4}}([T,T+1]\times\mathbf{R}^d)}.$$

As in (4.93), (4.95), and (4.96),

$$\mathscr{M}(k) + \mathscr{S}(k) + \mathscr{N}(k) \lesssim 1, \quad \lim_{k \to \infty} \mathscr{M}(k) = 0, \qquad (4.127)$$

and

$$\mathscr{S}(k) \lesssim \mathscr{M}(k) + \mathscr{N}(k).$$

In fact, $\mathscr{S}(k)$ is controlled by the frequency envelope controlling $\mathscr{N}(k)$.

Lemma 4.31 *For any $d \geq 1$, $0 < s < 1 + \frac{4}{d}$, and $\delta > 0$,*

$$\|u_{>k}\|_{L_{t,x}^{\frac{2(d+2)}{d}}([T,T+\delta]\times\mathbf{R}^d)}$$
$$\lesssim \mathscr{M}(k) + \eta(\delta) \sum_{k_1 \leq k} 2^{(k_1-k)s} \|u_{>k_1}\|_{L_{t,x}^{\frac{2(d+2)}{d}}([T,T+\delta]\times\mathbf{R}^d)},$$

where $\eta(\delta) \searrow 0$ as $\delta \searrow 0$.

Proof By (4.127), the Sobolev embedding theorem, Strichartz estimates, Hölder's inequality in time, and interpolation, it follows that

$$\lim_{\delta \searrow 0} \sup_{T \in \mathbf{R}} \|u\|_{L_{t,x}^{\frac{2(d+2)}{d}}([T,T+\delta]\times\mathbf{R}^d)} = 0. \qquad (4.128)$$

4.3 Radial Mass-Critical Problem in Dimensions $d \geq 2$

Then by (4.128) and Bernstein's inequality, for any $0 < s < 1 + \frac{4}{d}$,

$$\left\| P_{>k} F(u) \right\|_{L_{t,x}^{\frac{2(d+2)}{d+4}}([T,T+\delta]\times \mathbf{R}^d)}$$

$$\lesssim \left\| u_{>k} \right\|_{L_{t,x}^{\frac{2(d+2)}{d}}([T,T+\delta]\times \mathbf{R}^d)} \left\| u \right\|_{L_{t,x}^{\frac{2(d+2)}{d}}([T,T+\delta]\times \mathbf{R}^d)}^{4/d}$$

$$+ \left\| P_{>k} F(u_{\leq k}) \right\|_{L_{t,x}^{\frac{2(d+2)}{d+4}}([T,T+\delta]\times \mathbf{R}^d)}$$

$$\lesssim \eta(\delta)^{4/d} \left\| u_{>k} \right\|_{L_{t,x}^{\frac{2(d+2)}{d}}([T,T+\delta]\times \mathbf{R}^d)}$$

$$+ \eta(\delta)^{4/d} \sum_{k_1 \leq k} 2^{(k_1-k)s} \left\| u_{>k_1} \right\|_{L_{t,x}^{\frac{2(d+2)}{d}}([T,T+\delta]\times \mathbf{R}^d)}. \quad (4.129)$$

$\square$

Proof of Theorem 4.30 Let $\chi(t,y) \in C_0^{\infty}(\mathbf{R}^d)$ be a smooth function, $\chi(t,y) = 1$ on $|y| \leq c 2^k |t|$ for some small constant $c > 0$, and $\chi(t,y)$ be supported on $|y| \leq c 2^{k+1} |t|$. Following (4.105), let

$$A = \int_0^{\delta} e^{-it\Delta} P_{>k} F(u(t)) \, dt + \int_{\delta}^{\infty} e^{-it\Delta} (1-\chi)(t,y) P_{>k} F(u(t)) \, dt,$$

$$B = \int_{\delta}^{\infty} e^{-it\Delta} \chi(t,y) P_{>k} F(u(t)) \, dt,$$

and let A', B' be the corresponding integrals in the negative time direction.

The dual to the radial Strichartz estimate

$$\left\| |x|^{\frac{d-1}{2}} e^{it\Delta} u_0 \right\|_{L_t^4 L_x^{\infty}(\mathbf{R}\times\mathbf{R}^d)} \lesssim \left\| u_0 \right\|_{L_x^2(\mathbf{R}^d)}$$

implies

$$\left\| \int_{\delta}^{\infty} e^{-it\Delta} (1-\chi)(t,y) P_{>k} F(u)(t) \, dt \right\|_{L_x^2(\mathbf{R}^d)}$$

$$\lesssim \left\| \frac{1}{(c 2^k t)^{\frac{d-1}{2}}} (1-\chi)(t,y) P_{>k} F(u)(t) \right\|_{L_t^{4/3} L_x^1 ([\delta,\infty)\times \mathbf{R}^d)}. \quad (4.130)$$

Also, (4.102) and (4.127) imply

$$\left\| |x|^{\frac{d-1}{2}} u \right\|_{L_t^4 L_x^{\infty}([n,n+1]\times \mathbf{R}^d)} \lesssim 1. \quad (4.131)$$

When $d=2$, by (4.131) and the support properties of $1-\chi$, we have

$$\frac{1}{c^{1/2}2^{k/2}}\left\|\frac{1}{t^{1/2}}(1-\chi)(t,y)P_{>k}F(u(t))\right\|_{L_t^{4/3}L_x^1([\delta,\infty)\times\mathbf{R}^2)}$$

$$\lesssim \frac{1}{c^{1/2}2^{k/2}}\|P_{>k-3}u\|_{L_t^\infty L_x^2}\|u\|_{L_t^\infty L_x^2}$$

$$\times\left(\sum_{n=1}^\infty\left(\frac{1}{n^{1/2}\delta^{1/2}}\|(1-\chi)u\|_{L_t^4 L_x^\infty([\delta n,\delta(n+1)]\times\mathbf{R}^2)}\right)^{4/3}\right)^{3/4}$$

$$\lesssim \frac{2^{-k/2}}{\delta^{3/4}}\mathscr{M}(k-3). \tag{4.132}$$

Therefore, for any $s<3$, (4.129) and (4.132) imply

$$\|A\|_{L^2},\ \|A'\|_{L^2}$$
$$\lesssim_s \frac{2^{-k/2}}{\delta^{3/4}}\mathscr{M}(k-3)+\eta(\delta)^2\sum_{k_1\le k}2^{(k_1-k)s}\|P_{>k_1}u\|_{L_{t,x}^4([0,\delta]\times\mathbf{R}^2)}. \tag{4.133}$$

When $d=3$, Bernstein's inequality implies

$$\|P_{>k}F(u)\|_{L_x^1}\lesssim 2^{-11k/6}\||\nabla|^{11/6}F(u_{\le k})\|_{L_x^1}+\|u_{>k}\|_{L_x^2}\|u\|_{L_x^{8/3}}^{4/3},$$

so

$$\|A\|_{L^2},\ \|A'\|_{L^2}$$
$$\lesssim \frac{1}{c2^k}\left\|\frac{1}{t}(1-\chi)(t,y)P_{>k}F(u(t))\right\|_{L_t^{4/3}L_x^1([\delta,\infty)\times\mathbf{R}^3)}$$

$$\lesssim \frac{1}{c2^k\delta^{1/3}}\|u\|_{L_t^\infty L_x^2}\|P_{>k}u\|_{L_t^\infty L_x^2}\left(\sup_{n\in\mathbf{Z}}\|(1-\chi)u\|_{L_t^4 L_x^\infty([n,n+1]\times\mathbf{R}^3)}^{1/3}\right)$$

$$+\frac{1}{c^217k/6\delta^{1/4}}\||\nabla|^{11/6}u_{\le k}\|_{L_t^\infty L_x^2}\|u_{\le k}\|_{L_t^\infty L_x^{8/3}}^{4/3}$$

$$\lesssim \frac{2^{-k}}{\delta^{1/3}}\mathscr{M}(k)+\frac{1}{2^{7k/3}\delta^{1/4}}\sum_{k_1\le k}2^{11k_1/6}\mathscr{M}(k_1). \tag{4.134}$$

When $d\ge 4$, interpolating (4.130) with the usual Strichartz estimates,

$$\|A\|_{L^2},\ \|A'\|_{L^2}\lesssim\left\|\int_\delta^\infty e^{-it\Delta}(1-\chi)(t,y)P_{>k}F(u(t))dt\right\|_{L_x^2(\mathbf{R}^d)}$$

$$\lesssim\left\|\frac{1}{(c2^kt)^{3/2}}P_{>k}F(u(t))\right\|_{L_t^{4/3}L_x^{\frac{2d}{d+4}}([\delta,\infty)\times\mathbf{R}^d)}$$

$$\lesssim\frac{1}{c^{3/2}2^{3k/2}\delta^{3/4}}\sum_{k_1\le k}2^{\frac{k_1-k}{2}}\mathscr{M}(k_1). \tag{4.135}$$

4.3 Radial Mass-Critical Problem in Dimensions $d \geq 2$

Now take $\langle B, B' \rangle_{L^2}$. The cutoff function $\chi \in C_0^\infty(\mathbf{R}^d)$ is a Schwartz function, so for any M,

$$\hat{\chi}(t,\xi) \lesssim_M \frac{(c2^k t)^d}{(1+c2^k t|\xi|)^M}. \tag{4.136}$$

Therefore, by the Parseval identity, (4.136), Strichartz estimates, and (4.127), for $k \geq 0$, we have, since $|\xi| \gtrsim 2^k$, for $k > 0$,

$$\langle B, B' \rangle_{L^2} = \left\langle \int_{-\infty}^{-\delta} e^{-i\tau\Delta} \chi(\tau,y) P_{>k} F(u(\tau)) d\tau, \int_\delta^\infty e^{-it\Delta} \chi(t,y) P_{>k} F(u(t)) dt \right\rangle$$

$$= \left\langle \int_{-\infty}^{-\delta} e^{i(t-\tau)\Delta} \chi(\tau,y) P_{>k} F(u(\tau)) d\tau, \int_\delta^\infty \chi(t,y) P_{>k} F(u(t)) dt \right\rangle$$

$$\lesssim_d \left(\int_\delta^\infty \frac{(c2^k t)^d}{(1+c2^k t|\xi|)^{d+6}} dt \right)^2 \lesssim \delta^{-10} 2^{-10k}. \tag{4.137}$$

Take $d = 2$. By (4.132), (4.133), (4.137), and time-translation invariance,

$$\mathscr{M}(k) \lesssim_s \delta^{-5} 2^{-5k} + \frac{1}{2^k \delta^{3/4}} \mathscr{M}(k-3)$$

$$+ \eta(\delta)^2 \sum_{k_1 \leq k} 2^{(k_1-k)s} \left(\sup_{T \in \mathbf{R}} \|P_{>k_1} u\|_{L^4_{t,x}([T,T+\delta] \times \mathbf{R}^2)} \right). \tag{4.138}$$

Now, for any $s' < s$, define the frequency envelopes

$$\mathscr{M}'(k) = \sum_{k_1 \leq k} 2^{(k_1-k)s'} \mathscr{M}(k_1)$$

and $\quad \mathscr{S}'(k) = \sum_{k_1 \leq k} 2^{(k_1-k)s'} \left(\sup_{T \in \mathbf{R}} \|P_{>k_1} u\|_{L^{\frac{2(d+2)}{d}}_{t,x}([T,T+\delta] \times \mathbf{R}^d)} \right).$

Then by Lemma 4.31 and (4.138), choosing $\delta(s,s') > 0$ small enough yields

$$\mathscr{M}'(k) \lesssim_{s,s'} \frac{\delta^{-5}}{2^{s'k}} + \frac{1}{\delta^{3/4} 2^k} \mathscr{M}'(k) + \eta(\delta)^2 \mathscr{S}'(k)$$

and $\quad \mathscr{S}'(k) \lesssim_{s,s'} \mathscr{M}'(k) + \eta(\delta)^2 \mathscr{S}'(k).$

So when $d = 2$, for any $s' < 3$, we have $\mathscr{M}(k) \leq \mathscr{M}'(k) \lesssim_{s'} 2^{-s'k}$. Similarly, when $d = 3$, Lemma 4.31, (4.134), and (4.137) imply that for $s' < s < \frac{7}{3}$,

$$\mathscr{M}(k) \leq \mathscr{M}'(k) \lesssim_{s,s'} 2^{-s'k}.$$

Finally, when $d \geq 4$, Lemma 4.31, (4.135), and (4.137) imply that for $s' < s < 1 + \frac{4}{d}$,

$$\mathscr{M}(k) \leq \mathscr{M}'(k) \lesssim_{s,s'} 2^{-s'k}.$$

This completes the proof of Theorem 4.30. $\square$

Proof of Theorem 4.29 Theorem 4.30 directly implies that if u is an almost-periodic solution to the defocusing, mass-critical nonlinear Schrödinger equation of the form (4.88), then $u \equiv 0$.

Since u is radial, $\xi(t) = 0$. So for an almost-periodic solution in the form of (4.88), by the interaction Morawetz estimate in (4.32), then

$$\left\| |\nabla|^{\frac{3-d}{2}} |u|^2 \right\|_{L^2_{t,x}(\mathbf{R} \times \mathbf{R}^d)} < \infty. \tag{4.139}$$

When $d = 2$, the Sobolev embedding theorem implies

$$\|u\|_{L^4_t L^8_x(\mathbf{R} \times \mathbf{R}^d)} < \infty. \tag{4.140}$$

If $N(t) \sim 1$ and u is radial, then $x(t) = 0$, so $u(t)$ lies in a compact subset of $L^2_x(\mathbf{R}^d)$, which implies that if u is a nonzero solution to (4.122), then $\|u\|_{L^8_x} \geq \delta(u) > 0$, contradicting (4.140), forcing $u \equiv 0$. When $d \geq 3$, interpolating (4.139) with the Sobolev embedding theorem and product rule implies

$$\|u\|_{L^{d+1}_t L^{\frac{2(d+1)}{d-1}}_x (\mathbf{R} \times \mathbf{R}^d)} < \infty. \tag{4.141}$$

Once again, since $u(t)$ must lie in a compact subset of $L^2(\mathbf{R}^d)$ for all t, by Hölder's inequality, $\|u(t)\|_{L^{2(d+1)/(d-1)}_x} \geq \delta(u) > 0$ if u is a nonzero solution to (4.122), which contradicts (4.141).

Now suppose that $\liminf_{t \to \pm \infty} N(t) = 0$. In this case, (4.80), $\xi(t) = 0$, the Sobolev embedding, and Theorem 4.30 imply that for any $\eta > 0$,

$$E(u(t)) \lesssim C(\eta) N(t) + \eta^{\varepsilon(d)},$$

for some $\varepsilon(d) > 0$. Since $\eta > 0$ is arbitrary and $\liminf_{t \to \infty} N(t) = 0$, then by conservation of energy we have $E(u(t)) \equiv 0$ for all $t \in \mathbf{R}$, and thus $u \equiv 0$. □

4.4 Nonradial Mass-Critical Problem in Dimensions $d \geq 3$

For a nonradial, almost-periodic solution to the mass-critical problem,

$$iu_t + \Delta u = F(u) = |u|^{4/d} u, \quad u(0,x) = u_0 \in L^2_x(\mathbf{R}^d), \tag{4.142}$$

satisfying $N(t) \leq 1$ for all $t \in \mathbf{R}$, instead of proving additional regularity, a solution will be estimated using a frequency-localized interaction Morawetz estimate. This truncation will introduce some error terms, which will be estimated using long-time Strichartz estimates.

Long-time Strichartz estimates were first introduced in Dodson (2012) for the mass-critical nonlinear Schrödinger equation with $d \geq 3$, although they

4.4 Nonradial Mass-Critical Problem in Dimensions $d \geq 3$

were inspired by the frequency-localized interaction Morawetz estimates of Colliander et al. (2008) and the almost-Morawetz estimates used in the I-method (Colliander et al. (2007), Colliander and Roy (2011), Dodson (2011)). Dodson (2016a,b) extended the work of Dodson (2012) to dimensions one and two, Killip and Visan (2012) and Visan (2012) applied these estimates to simplify scattering proofs of scattering for the energy-critical problem for dimensions three and four, and Murphy (2014, 2015) used the long-time Strichartz estimates for intercritical and energy supercritical problems. The heuristic behind these estimates is that at frequencies far away from $N(t)$, the solution to (4.142) will be dominated by the free solution for long periods of time.

Let u be an almost-periodic solution to (4.142). Define three constants

$$0 < \eta_3 \ll \eta_2 \ll \eta_1 < 1, \quad (4.143)$$

according to the following rule. First, by (4.86) and (4.87) there exists some $\eta_1 > 0$ such that

$$|\xi'(t)| + |N'(t)| \leq \frac{N(t)^3}{\eta_1}. \quad (4.144)$$

Writing $\Phi = 2^{-20}\eta_3^{-1/4d}$ for clarity, by (4.80), choose η_2 and η_3 such that

$$\int_{|x-x(t)| \geq \frac{\Phi}{N(t)}} |u(t,x)|^2 dx + \int_{|\xi-\xi(t)| > \Phi N(t)} |\hat{u}(t,\xi)|^2 d\xi \leq \eta_2^2 \quad (4.145)$$

and

$$\eta_3 < \eta_2^{10}.$$

Let k_0 be a positive integer and let $[a,b]$ be a compact interval satisfying

$$\int_a^b \int |u(t,x)|^{2(d+2)/d} dx\, dt = 2^{k_0}. \quad (4.146)$$

Rescaling using (1.81), it is possible to have $N(t)$ satisfying

$$\int_a^b N(t)^3 dt = \eta_3 2^{k_0}. \quad (4.147)$$

Remark Strictly speaking, after rescaling using (1.81), the compact interval would rescale to $[\frac{a}{\lambda^2}, \frac{b}{\lambda^2}]$. However, since a, b, and λ are constants, there is no problem with calling $\frac{a}{\lambda^2}$ the new a and $\frac{b}{\lambda^2}$ the new b.

Definition 4.32 (Galilean Littlewood–Paley projection)   For $j > 0$, let P_j be the Littlewood–Paley operator defined by Definition 1.5. Let

$$P_{\xi_0,j} f = e^{ix\cdot\xi_0} P_j\left(e^{-ix\cdot\xi_0} f\right).$$

Abusing notation slightly, let $P_{\xi_0,0} = P_{\xi_0,\leq 0}$, and for $j < 0$, let $P_{\xi_0,j} = 0$. For $1 \leq p \leq \infty$ and $1 \leq q \leq \infty$, define the norm

$$\left\|P_{\xi(t),j}f\right\|_{L_t^p L_x^q(I\times\mathbf{R}^d)} = \left\|\left\|P_{\xi(t),j}f(t,x)\right\|_{L_x^q(\mathbf{R}^d)}\right\|_{L_t^p(I)}. \tag{4.148}$$

The long-time Strichartz estimate in Dodson (2012) proved that for all $0 \leq j \leq k_0$,

$$\left\|P_{\xi(t),>j}u\right\|_{L_t^2 L_x^{\frac{2d}{d-2}}([a,b]\times\mathbf{R}^d)} \lesssim 2^{\frac{k_0-j}{2}},$$

where crucially the implicit constant did not depend on k_0.

Definition 4.33 (Long-time Strichartz seminorm) When $d \geq 3$, define

$$\|u\|_{X([a,b]\times\mathbf{R}^d)}^2 = \sup_{0\leq j\leq k_0} 2^{j-k_0} \left\|P_{\xi(t),\geq j}u\right\|_{L_t^2 L_x^{\frac{2d}{d-2}}([a,b]\times\mathbf{R}^d)}^2. \tag{4.149}$$

Theorem 4.34 (Long-time Strichartz estimate) *If u is an almost-periodic solution to (4.142) for $d \geq 3$, and u satisfies (4.146) and (4.147), then*

$$\|u\|_{X([a,b]\times\mathbf{R}^d)} \lesssim 1, \tag{4.150}$$

with constant independent of k_0.

Proof Partition $[a,b]$ in two different but related ways.

Definition 4.35 (Small intervals) Divide $[a,b]$ into consecutive, disjoint intervals J_l, $l = 0,\ldots,2^{k_0}-1$, such that

$$\int_{J_l} \|u(t)\|_{L_x^{2(d+2)/d}(\mathbf{R}^d)}^{2(d+2)/d} dt = 1.$$

Remark Let $N(J_l) = \sup_{t\in J_l} N(t)$. By (4.81) and (4.86),

$$\int_{J_l} N(t)^3\, dt \sim \sup_{t\in J_l} N(t) = N(J_l) \sim \inf_{t\in J_l} N(t). \tag{4.151}$$

Definition 4.36 Divide $[a,b]$ into 2^{k_0} consecutive, disjoint intervals J_α, $\alpha = 0,\ldots,2^{k_0}-1$ such that

$$\int_{J_\alpha} \left(N(t)^3 + \eta_3\|u(t)\|_{L_x^{2(d+2)/d}(\mathbf{R}^d)}^{2(d+2)/d}\right) dt = 2\eta_3. \tag{4.152}$$

For any integer $0 \leq j < k_0$, let

$$G_k^j = \bigcup_{\alpha=k2^j}^{(k+1)2^j-1} J_\alpha, \tag{4.153}$$

4.4 Nonradial Mass-Critical Problem in Dimensions $d \geq 3$

and for $j \geq k_0$, let
$$G_k^j = [a,b]. \tag{4.154}$$

If $G_\alpha^i = [\tilde{a}, \tilde{b}]$, let $\xi\left(G_\alpha^i\right) = \xi(\tilde{a})$. Define $\xi(J_l)$ in a similar manner.

Remark For $N(t)$ constant, there is an upper bound on the number of J_l intervals that intersect any one J_α, and an upper bound on the number of J_α intervals that intersect any one J_l. For a general $N(t)$, the partitions J_α and J_l can be quite different.

Remark By (4.144) and (4.152), for all $t \in G_\alpha^i$, we have
$$\left|\xi(t) - \xi\left(G_\alpha^i\right)\right| \leq \int_{G_\alpha^i} \eta_1^{-1} N(t)^3 \, dt \leq \eta_3 \eta_1^{-1} 2^i,$$

so for all $t \in G_\alpha^i$,
$$\{\xi : 2^{i-1} \leq |\xi - \xi(t)| \leq 2^{i+1}\} \subset \{\xi : 2^{i-2} \leq |\xi - \xi(G_\alpha^i)| \leq 2^{i+2}\}. \tag{4.155}$$

Now fix some j and $G_k^j \subset [a,b]$ according to (4.153) and (4.154). For any $t, t_0 \in G_k^j$,
$$P_{\xi(t), \geq j} u(t) = e^{i(t-t_0)\Delta} P_{\xi(t), \geq j} u(t_0) - i \int_{t_0}^t e^{i(t-\tau)\Delta} P_{\xi(t), j} F(u(\tau)) \, d\tau.$$

Then by the appropriate choice of t_0^k for each G_k^j, conservation of mass, the number of G_k^j intervals, Strichartz estimates, and (4.155),

$$\left\|P_{\xi(t), \geq j} u(t)\right\|_{L_t^2 L_x^{\frac{2d}{d-2}}([a,b] \times \mathbf{R}^d)}^2$$
$$\lesssim 2^{k_0 - j} + \inf_{t \in [a,b]} \left\|P_{\xi(t), \geq j-1} u(t)\right\|_{L_x^2(\mathbf{R}^d)}^2$$
$$+ \sum_{G_k^j \subset [a,b]} \left\|P_{\xi(t), \geq j} \int_{t_0^k}^t e^{i(t-\tau)\Delta} F(u(\tau)) \, d\tau\right\|_{L_t^2 L_x^{\frac{2d}{d-2}}(G_k^j \times \mathbf{R}^d)}^2. \tag{4.156}$$

Also by (4.155), Strichartz estimates, and the fact that P_j is given by the convolution with a function whose L^1-norm is uniformly bounded,

$$\sum_{G_k^j \subset [a,b]} \left\|P_{\xi(t), \geq j} \int_{t_0^k}^t e^{i(t-\tau)\Delta} F(u(\tau)) \, d\tau\right\|_{L_t^2 L_x^{\frac{2d}{d-2}}(G_k^j \times \mathbf{R}^d)}^2$$
$$\lesssim \left\|P_{\xi(\tau), \geq j-1} F(u(\tau))\right\|_{L_t^2 L_x^{\frac{2d}{d+2}}([a,b] \times \mathbf{R}^d)}^2. \tag{4.157}$$

Let n_1 be an integer satisfying $\eta_1 \sim 2^{n_1}$. Consistent with Definition 4.32, $P_{\xi(t), \geq a}$ refers to $P_{\xi(t), \geq \sup(a, 0)}$, and so on.

Remark Remember that a lowercase Latin letter refers to (1.24) and an uppercase Latin letter refers to (1.25).

By Bernstein's inequality, Hölder's inequality, and u having finite mass, let

$$\sigma = \begin{cases} \frac{1}{2} & \text{if } j \leq k_0, \\ 0 & \text{if } j > k_0. \end{cases}$$

Then

$$\left\| P_{\xi(\tau), \geq j-1} F\left(P_{\xi(\tau), \leq j+n_1} u(\tau)\right) \right\|_{L_t^2 L_x^{\frac{2d}{d+2}}([a,b] \times \mathbf{R}^d)}$$

$$\lesssim 2^{-j} \left\| (\nabla - i\xi(\tau)) P_{\xi(\tau), \leq j+n_1} u \right\|_{L_t^2 L_x^{\frac{2d}{d-2}}([a,b] \times \mathbf{R}^d)}$$

$$\lesssim \eta_1^{1/2} 2^{\sigma(k_0 - j)} \|u\|_{X([a,b] \times \mathbf{R}^d)}. \tag{4.158}$$

Next, by (4.145), since $F(u) - F(u_l) \lesssim |u_h| \left(|u_h|^{\frac{4}{d}} + |u_l|^{\frac{4}{d}} \right)$ when $u = u_l + u_h$,

$$\left\| F(u) - F\left(P_{\xi(\tau), \leq j+n_1} u \right) \right\|_{L_t^2 L_x^{\frac{2d}{d+2}}([a,b] \times \mathbf{R}^d)}$$

$$\lesssim \left\| P_{\xi(\tau), > j+n_1} u \right\|_{L_t^2 L_x^{\frac{2d}{d-2}}([a,b] \times \mathbf{R}^d)} \left\| P_{\xi(\tau), \geq 2^{-20} \eta_3^{-\frac{1}{4d}} N(\tau)} u \right\|_{L_t^\infty L_x^2([a,b] \times \mathbf{R}^d)}^{4/d}$$

$$+ \left\| \left(P_{\xi(\tau), > j+n_1} u \right) \left(P_{\xi(\tau), \leq 2^{-20} \eta_3^{-\frac{1}{4d}} N(\tau)} u \right)^{4/d} \right\|_{L_t^2 L_x^{\frac{2d}{d+2}}([a,b] \times \mathbf{R}^d)}$$

$$\lesssim 2^{\sigma(k_0 - j)} \eta_1^{-\frac{1}{2}} \eta_2^{\frac{4}{d}} \|u\|_{X([a,b] \times \mathbf{R}^d)}$$

$$+ \left\| \left(P_{\xi(\tau), > j+n_1} u \right) \left(P_{\xi(\tau), \leq 2^{-20} \eta_3^{-\frac{1}{4d}} N(\tau)} u \right)^{4/d} \right\|_{L_t^2 L_x^{\frac{2d}{d+2}}([a,b] \times \mathbf{R}^d)}. \tag{4.159}$$

Now let $\chi(t,x)$ be a smooth function. We have

$$\chi(t,x) = \begin{cases} 1 & \text{if } |x - x(t)| \leq 2^{-20} \eta_3^{-1/4d}/N(t), \\ 0 & \text{if } |x - x(t)| \geq 2^{-19} \eta_3^{-1/4d}/N(t). \end{cases}$$

Again by (4.145),

$$\left\| \left(P_{\xi(\tau), > j+n_1} u \right) \left(P_{\xi(\tau), \leq 2^{-20} \eta_3^{-1/4d} N(\tau)} u \right)^{4/d} \right\|_{L_t^2 L_x^{\frac{2d}{d+2}}([a,b] \times \mathbf{R}^d)}$$

$$\lesssim \left\| \left(P_{\xi(\tau), > j+n_1} u \right) \left(P_{\xi(\tau), \leq 2^{-20} \eta_3^{-1/4d} N(\tau)} u \right)^{4/d} \chi(\tau, x) \right\|_{L_t^2 L_x^{\frac{2d}{d+2}}([a,b] \times \mathbf{R}^d)}$$

$$+ \left\| P_{\xi(\tau), > j+n_1} u \right\|_{L_t^2 L_x^{\frac{2d}{d-2}}([a,b] \times \mathbf{R}^d)}$$

4.4 Nonradial Mass-Critical Problem in Dimensions $d \geq 3$

$$\times \left\| \left(P_{\xi(\tau), \leq 2^{-20}\eta_3^{-1/4d}N(\tau)} u \right) (1 - \chi(\tau, x)) \right\|_{L_t^\infty L_x^2([a,b] \times \mathbf{R}^d)}^{4/d}$$

$$\lesssim \left\| \left(P_{\xi(\tau), > j+n_1} u \right) \left(P_{\xi(\tau), \leq 2^{-20}\eta_3^{-1/4d}N(\tau)} u \right)^{4/d} \chi(\tau, x) \right\|_{L_t^2 L_x^{\frac{2d}{d+2}}([a,b] \times \mathbf{R}^d)}$$

$$+ \eta_1^{-1/2} \eta_2^{4/d} 2^{\sigma(k_0-j)} \|u\|_{X([a,b] \times \mathbf{R}^d)}. \tag{4.160}$$

Finally,

$$\left\| \chi(\tau, x) \left| P_{\xi(\tau), > j+n_1} u \right| \left| P_{\xi(\tau), \leq 2^{-20}\eta_3^{-1/4d}N(\tau)} u \right|^{4/d} \right\|_{L_t^2 L_x^{\frac{2d}{d+2}}([a,b] \times \mathbf{R}^d)}^2$$

$$= \sum_{J_l \subset [a,b]} \left\| \chi(\tau, x) \left| P_{\xi(\tau), > j+n_1} u \right| \left| P_{\xi(\tau), \leq 2^{-20}\eta_3^{-1/4d}N(\tau)} u \right|^{4/d} \right\|_{L_t^2 L_x^{\frac{2d}{d+2}}(J_l \times \mathbf{R}^d)}^2. \tag{4.161}$$

When $d = 3$, (4.23) implies

$$\left\| \chi(\tau, x) \left| P_{\xi(\tau), > j+n_1} u \right| \left| P_{\xi(\tau), \leq 2^{-20}\eta_3^{-1/12}N(\tau)} u \right|^{4/3} \right\|_{L_t^2 L_x^{6/5}(J_l \times \mathbf{R}^3)}$$

$$\lesssim \left\| \left(P_{\xi(\tau), > j+n_1} u \right) \left(P_{\xi(\tau), \leq 2^{-20}\eta_3^{-1/12}N(\tau)} u \right) \right\|_{L_{t,x}^2(J_l \times \mathbf{R}^3)} \| \chi(\tau, x) \|_{L_t^\infty L_x^6(J_l \times \mathbf{R}^3)}$$

$$\lesssim \frac{\eta_1^{-1/2} \eta_3^{-1/8} N(J_l)^{1/2}}{2^{j/2}}.$$

Summing over intervals J_l in l^2, by (4.152) and (4.153), we get

$$(4.161) \lesssim \eta_1^{-1/2} \eta_3^{3/8} 2^{(k_0-j)/2}. \tag{4.162}$$

When $d \geq 4$, (4.23) and Hölder's inequality imply

$$\left\| \chi(\tau, x) \left| P_{\xi(\tau), > j+n_1} u \right| \left| P_{\xi(\tau), \leq 2^{-20}\eta_3^{-\frac{1}{4d}}N(\tau)} u \right|^{4/d} \right\|_{L_t^2 L_x^{2d/(d+2)}(J_l \times \mathbf{R}^d)}$$

$$\lesssim \left\| \left(P_{\xi(\tau), > j+n_1} u \right) \left(P_{\xi(\tau), \leq 2^{-20}\eta_3^{-\frac{1}{4d}}N(\tau)} u \right) \right\|_{L_{t,x}^2(J_l \times \mathbf{R}^d)}^{4/d}$$

$$\times \left\| P_{\xi(\tau), > j+n_1} u \right\|_{L_t^2 L_x^{2d/(d-2)}(J_l \times \mathbf{R}^d)}^{1-(4/d)} \| \chi(\tau, x) \|_{L_t^\infty L_x^{d^2/2(d-2)}}$$

$$\lesssim \frac{2^{-40(d-1)/d} \eta_1^{-2/d} N(J_l)^{2(d-1)/d} \eta_3^{-(d-1)/2d^2}}{2^{2j/d}}$$

$$\times \left\| P_{\xi(\tau), > j+n_1} u \right\|_{L_t^2 L_x^{2d/(d-2)}(J_l \times \mathbf{R}^d)}^{1-4/d} \frac{2^{-40(d-2)/d} \eta_3^{-(d-2)/2d^2}}{N(J_l)^{2(d-2)/d}}. \tag{4.163}$$

When $d = 4$,

$$\left(\sum_{J_l} (4.163)^2\right)^{1/2} \lesssim \eta_1^{-\frac{1}{2}} \eta_3^{\frac{11}{32}} 2^{\frac{k_0-j}{2}}. \qquad (4.164)$$

Applying Hölder's inequality in time when $d \geq 5$,

$$\left(\sum_{J_l \subset [a,b]} \frac{N(J_l)^{4/d}}{2^{4j/d}} \|P_{\xi(\tau), \geq j+n_1} u\|_{L_t^2 L_x^{2d/(d-2)}(J_l \times \mathbf{R}^d)}^{2-8/d} \eta_3^{-(2d-3)/d^2} \eta_1^{-4/d}\right)^{1/2}$$

$$\lesssim 2^{\sigma(k_0-j)} \eta_3^{\frac{d^2-2d+3}{2d^2}} \eta_1^{-2/d} \|u\|_{X([a,b] \times \mathbf{R}^d)}^{1-4/d}. \qquad (4.165)$$

Therefore, by (4.156), (4.158), (4.159), (4.160), (4.162), (4.164), and (4.165),

$$\|u\|_{X([a,b] \times \mathbf{R}^d)} \lesssim 1.$$

This completes the proof of Theorem 4.34. $\square$

Long-time Strichartz estimates may be used to prove scattering for the non-radial, mass-critical problem.

Theorem 4.37 *The defocusing, mass-critical initial value problem* (4.142) *is globally well posed and scattering for all $u_0 \in L^2(\mathbf{R}^d)$, $d \geq 3$.*

Proof To prove this it suffices to prove that if u is an almost-periodic solution to (4.142) satisfying $N(t) \leq 1$ for all $t \in \mathbf{R}$, then $u \equiv 0$. Because u lies in L^2 but need not lie in $\dot{H}^1$, in order to use (3.120) and (3.121), a frequency-localized interaction Morawetz estimate will be utilized. Let $[-T, T]$ be an interval such that $\int_{-T}^T N(t)^3 \, dt = K$ and $\int_{-T}^T \int |u(t,x)|^{2(d+2)/d} dx \, dt = 2^{k_0}$. Let $\lambda = \eta_3 2^{k_0}/K$ and rescale:

$$u_\lambda(t,x) = \lambda^{d/2} u(\lambda^2 t, \lambda x), \quad \int_{-T/\lambda^2}^{T/\lambda^2} N_\lambda(t)^3 \, dt = \eta_3 2^{k_0}. \qquad (4.166)$$

Then by Theorem 4.34,

$$\|u_\lambda\|_{X\left(\left[-\frac{T}{\lambda^2}, \frac{T}{\lambda^2}\right] \times \mathbf{R}^d\right)} \lesssim 1.$$

Recall from (4.149) that the X-norm is constructed from $L_t^2 L_x^{\frac{2d}{d-2}}$-norms, which are invariant under (4.166).

Define the Fourier multiplier

$$I: L^2(\mathbf{R}^d) \to H^1(\mathbf{R}^d),$$

4.4 Nonradial Mass-Critical Problem in Dimensions $d \geq 3$

given by $m\left(\frac{\eta_3 \xi}{K}\right)$, where $m \in C^\infty\left(\mathbf{R}^d\right)$ is a radial function:

$$m(\xi) = \begin{cases} 1, & |\xi| \leq 1, \\ 0, & |\xi| > 2. \end{cases}$$

This Fourier multiplier contains the frequency support of most of u, since by (4.144),

$$\int_{-T}^{T} |\xi'(t)| \, dt \leq \eta_1^{-1} K \ll \eta_3^{-1} K.$$

Following (3.102), let

$$\partial_t M(t) = \int \frac{(x-y)_j}{|x-y|} \operatorname{Im} \left[\overline{Iu}(t,x) \partial_j Iu(t,x)\right] |Iu(t,y)|^2 dx \, dy.$$

Remark This is a slight abuse of notation since a truncated interaction Morawetz potential need not be a time derivative of a function in the form of (3.101).

Relation (4.80) gives sufficiently good decay for $u(t,x)$ at high frequencies to bound the endpoints of the interaction Morawetz estimate as we now show.

Lemma 4.38 *If $N(t) \leq 1$ on* **R**,

$$|\partial_t M(t)| \lesssim o(K).$$

Here $o(K)$ is a quantity that satisfies $\frac{o(K)}{K} \to 0$ as $K \nearrow \infty$.

Proof By (4.80),

$$\int \frac{(x-y)_j}{|x-y|} |Iu(t,y)|^2 \operatorname{Im} \left[\overline{Iu}(t,x) (\partial_j - i\xi_j(t)) Iu(t,x)\right] dx \, dy$$

$$\lesssim \|Iu\|_{L_t^\infty L_x^2(\mathbf{R} \times \mathbf{R}^d)}^3 \|(\nabla - i\xi(t)) Iu(t,x)\|_{L_t^\infty L_x^2(\mathbf{R} \times \mathbf{R}^d)}$$

$$\lesssim \sum_{2^k \leq \eta_3^{-1} K} 2^k \|P_{\xi(t),k} u(t,x)\|_{L_t^\infty L_x^2(\mathbf{R} \times \mathbf{R}^d)}$$

$$+ \eta_3^{-1} K \|P_{|\xi - \xi(t)| \geq \eta_3^{-1} K} u(t,x)\|_{L_t^\infty L_x^2(\mathbf{R} \times \mathbf{R}^d)} \lesssim o(K).$$

Also, following (4.19),

$$\int \frac{(x-y)_j}{|x-y|} |Iu(t,y)|^2 \operatorname{Im} \left[|Iu(t,x)|^2 i\xi_j(t)\right] dx \, dy$$

$$= \int \frac{(x-y)_j}{|x-y|} |Iu(t,x)|^2 |Iu(t,y)|^2 \xi_j(t) \, dx \, dy = 0,$$

which proves Lemma 4.38. □

Lemma 4.39 *For any d, if u satisfies* (4.80),

$$N(t)^3 \lesssim \left\| |\nabla|^{\frac{3-d}{2}} |Iu(t,x)|^2 \right\|_{L_x^2(\mathbf{R}^d)}^2. \tag{4.167}$$

Proof First suppose $d \leq 3$. By conservation of mass and the Sobolev embedding theorem,

$$\left\| P_{\leq \eta N(t)} \left(|Iu(t,x)|^2 \right) \right\|_{L_x^2(\mathbf{R}^d)}^2 \lesssim \eta^d N(t)^d.$$

By Hölder's inequality, $|\xi(t)| \ll \eta_3^{-1} K$, and (4.80), for $N(t) \leq 1$, there exists $C(m_0, d)$ sufficiently large so that

$$\frac{m_0^4}{4} \leq \left(\int_{|x-x(t)| \leq \frac{C(m_0,d)}{N(t)}} |Iu(t,x)|^2 dx \right)^2 \lesssim \frac{C(m_0,d)^d}{N(t)^d} \|Iu(t,x)\|_{L_x^4(\mathbf{R}^d)}^4,$$

where m_0 is the L^2-norm $\|u(t,x)\|_{L^2} = m_0$ and $m_0 > 0$.

Then by Bernstein's inequality,

$$\frac{C(m_0,d)^d}{N(t)^d} \|Iu(t,x)\|_{L_x^4(\mathbf{R}^d)}^4$$

$$\lesssim \frac{C(m_0,d)^d}{N(t)^d} \left(\|P_{> \eta N(t)} |Iu|^2 \|_{L_x^2}^2 + \|P_{\leq \eta N(t)} |Iu|^2 \|_{L_x^2}^2 \right)$$

$$\lesssim \frac{C(m_0,d)^d}{N(t)^d} \left(\eta^{d-3} N(t)^{d-3} \left\| |\nabla|^{\frac{3-d}{2}} |Iu|^2 \right\|_{L_x^2}^2 + \eta^d N(t)^d \right)$$

$$\lesssim C(m_0,d)^d \eta^d + \frac{C(m_0,d)^d}{N(t)^3} \left\| |\nabla|^{\frac{3-d}{2}} |Iu|^2 \right\|_{L_x^2}^2.$$

Taking $\eta(u)$ sufficiently small implies the lemma for $d \leq 3$.

For $d > 3$, (3.120) implies

$$\left\| |\nabla|^{\frac{3-d}{2}} |Iu(t,x)|^2 \right\|_{L_x^2(\mathbf{R}^d)}^2$$

$$\sim \int \frac{1}{|x-y|^3} |Iu(t,x)|^2 |Iu(t,y)|^2 dx dy$$

$$\gtrsim N(t)^3 \int_{\substack{|x-x(t)| \\ \leq C(m_0,d)N(t)}} \int_{\substack{|y-x(t)| \\ \leq C(m_0,d)N(t)}} |Iu(t,x)|^2 |Iu(t,y)|^2 dx dy \sim N(t)^3.$$

$\square$

Inequality (4.167) implies a lower bound for $\frac{d^2}{dt^2} M(t)$ as follows.

4.4 Nonradial Mass-Critical Problem in Dimensions $d \geq 3$

Theorem 4.40 *When $\mu = +1$, $\int_{-T}^{T} N(t)^3 \, dt = K$, and $d \geq 3$, we have*

$$\int_{-T}^{T} \frac{d^2}{dt^2} M(t) \, dt \gtrsim K.$$

Proof Plugging the Fourier multiplier I into (4.142) yields

$$\partial_t Iu = i\Delta Iu - iF(Iu) + iF(Iu) - iIF(u).$$

If

$$\partial_t Iu = i\Delta Iu - iF(Iu),$$

then (4.167) combined with the proof of Lemma 3.23 would imply

$$\int_{-T}^{T} \frac{d^2}{dt^2} M(t) \, dt \gtrsim \left\| |\nabla|^{\frac{3-d}{2}} |Iu(t,x)|^2 \right\|^2_{L^2_{t,x}([0,T] \times \mathbf{R}^d)} \gtrsim K.$$

Therefore, it only remains to control the errors arising from

$$iF(Iu) - iIF(u). \tag{4.168}$$

Let $\mathscr{E}(t)$ denote such an error. Then

$$\mathscr{E}(t) = \int_{-T}^{T} \int \frac{(x-y)_j}{|x-y|} |Iu(t,y)|^2 \operatorname{Re}\left[\left(IF(\bar{u}) - F(\overline{Iu})\right) \cdot \partial_j Iu\right](t,x) \, dx \, dy \, dt$$

$$+ \int_{-T}^{T} \int \frac{(x-y)_j}{|x-y|} |Iu(t,y)|^2 \operatorname{Re}\left[\overline{Iu} \cdot \partial_j (F(Iu) - IF(u))\right](t,x) \, dx \, dy \, dt$$

$$+ 2 \int_{-T}^{T} \int \frac{(x-y)_j}{|x-y|} \operatorname{Im}\left[\overline{Iu} \cdot IF(u)\right](t,y) \operatorname{Im}[\overline{Iu} \cdot \partial_j Iu](t,x) \, dx \, dy \, dt. \tag{4.169}$$

Similarly to (4.18), the error $\mathscr{E}(t)$ is Galilean invariant. Indeed,

$$\int \frac{(x-y)_j}{|x-y|} |Iu(t,y)|^2 \xi_j(t) \operatorname{Im}\left[\left(F(\overline{Iu}) - IF(\bar{u})\right) \cdot Iu\right](t,x) \, dx \, dy$$

$$+ \int \frac{(x-y)_j}{|x-y|} |Iu(t,y)|^2 \xi_j(t) \operatorname{Im}\left[\overline{Iu} \cdot (IF(u) - F(Iu))\right](t,x) \, dx \, dy$$

$$= 2 \int \frac{(x-y)_j}{|x-y|} |Iu(t,y)|^2 \xi_j(t) \operatorname{Im}\left[\overline{Iu} \cdot IF(u)\right](t,x) \, dx \, dy. \tag{4.170}$$

Then by the antisymmetry of $\frac{(x-y)_j}{|x-y|}$ we have

$$2\int\int \frac{(x-y)_j}{|x-y|}|Iu(t,y)|^2 \xi_j(t)\, \mathrm{Im}\left[\overline{Iu}IF(u)\right](t,x)\,dxdy$$
$$+2\int \frac{(x-y)_j}{|x-y|}\xi_j(t)\, \mathrm{Im}\left[\overline{Iu}IF(u)\right](t,y)\,|Iu(t,x)|^2\,dxdy = 0.$$

This implies

$$\mathscr{E}(t)$$
$$= \int_{-T}^{T}\int \frac{(x-y)_j}{|x-y|}|Iu(t,y)|^2 \mathrm{Re}\left[\left(IF(\bar{u}) - F(\overline{Iu})\right)(\partial_j - i\xi_j(t))Iu\right](t,x)\,dxdydt \tag{4.171}$$

$$+\int_{-T}^{T}\int \frac{(x-y)_j}{|x-y|}|Iu(t,y)|^2 \mathrm{Re}\left[\overline{Iu}(\partial_j - i\xi_j(t))(F(Iu) - IF(u))\right](t,x)\,dxdydt \tag{4.172}$$

$$+2\int_{-T}^{T}\int \frac{(x-y)_j}{|x-y|}\mathrm{Im}\left[\overline{Iu}IF(u)\right](t,y)\,\mathrm{Im}\left[\overline{Iu}(\partial_j - i\xi_j(t))Iu\right](t,x)\,dxdydt. \tag{4.173}$$

First take (4.171).

Lemma 4.41 *Since* $\left|\frac{(x-y)_j}{|x-y|}\right| \leq 1$, *it follows that*

$$\int_{-T}^{T}\int \frac{(x-y)_j}{|x-y|}|Iu(t,y)|^2 \mathrm{Re}\left[\left(IF(\bar{u}) - F(\overline{Iu})\right)(\partial_j - \xi_j(t))Iu\right](t,x)\,dxdydt$$
$$\lesssim o(K). \tag{4.174}$$

Proof To simplify notation, let

$$\widetilde{\nabla}(t) = \nabla - i\xi(t) \quad \text{and} \quad \widetilde{\overline{\nabla}}(t) = \nabla + i\xi(t).$$

Combining (4.150) with the scale invariance of $L_t^2 L_x^{\frac{2d}{d-2}}$ implies that for $j \leq 0$,

$$\left\|P_{|\xi - \xi(t)| > 2^j \eta_3^{-1} K}u\right\|_{L_t^2 L_x^{\frac{2d}{d-2}}([-T,T]\times\mathbf{R}^d)} \lesssim 2^{-j/2}. \tag{4.175}$$

4.4 Nonradial Mass-Critical Problem in Dimensions $d \geq 3$

Let

$$u_l = P_{|\xi - \xi(t)| \leq \eta_3^{-1} K/4} u,$$

and let

$$u = u_l + u_h.$$

Then

$$IF(u_l) - F(Iu_l) = IF(u_l) - F(u_l) = (I - 1)F(u_l).$$

By Bernstein's inequality, since $I - 1$ is supported on $|\xi| \geq \eta_3^{-1} K$,

$$\left\| (I - 1) F(u_l) \right\|_{L_t^2 L_x^{\frac{2d}{d+2}}([-T,T] \times \mathbf{R}^d)}$$

$$\lesssim \frac{\eta_3}{K} \left\| \widetilde{\nabla} u_l \right\|_{L_t^2 L_x^{\frac{2d}{d-2}}([-T,T] \times \mathbf{R}^d)} \left\| P_{>\frac{\eta_3^{-1}K}{4}} |u_l|^{4/d} \right\|_{L_t^\infty L_x^{d/2}([-T,T] \times \mathbf{R}^d)}. \quad (4.176)$$

Remark It is true that

$$\widetilde{\nabla}\left(|u|^{4/d} u\right) = |u|^{4/d} \left(\widetilde{\nabla} u\right) + u \nabla |u|^{4/d}$$

$$= \left(1 + \frac{2}{d}\right) |u|^{4/d} \left(\widetilde{\nabla} u\right) + \frac{2}{d} |u|^{(4/d) - 2} u^2 \overline{\left(\widetilde{\nabla} u\right)}. \quad (4.177)$$

However, the only characteristic of $|u|^{4/d}$ that will be used is that $F(x) = |x|^{4/d}$ is a Hölder continuous function of order $4/d$ when $d \geq 4$, and is differentiable when $d = 3$. Since this is also true of $|x|^{(4/d)-2} x^2$, it will be enough to simply treat the first term in (4.177).

Again by (4.150),

$$\left\| \widetilde{\nabla} Iu \right\|_{L_t^2 L_x^{\frac{2d}{d-2}}([-T,T] \times \mathbf{R}^d)} \lesssim \eta_3^{-1} K \sum_{j \leq 0} 2^{j/2} \lesssim \eta_3^{-1} K. \quad (4.178)$$

Splitting,

$$IF(u) - F(Iu) = IF(u) - IF(u_l) + F(u_l) - F(Iu) + IF(u_l) - F(u_l). \quad (4.179)$$

Then by Taylor's theorem,

$$F(u) - F(u_l) = u_h O\left(|u|^{4/d}\right) \quad \text{and} \quad F(Iu) - F(u_l) = (Iu_h) O\left(|Iu|^{4/d}\right).$$

By the fundamental theorem of calculus, for $|\xi_2| \ll |\xi|$,

$$|m(\xi + \xi_2) - m(\xi)| \lesssim |\xi_2||\nabla m(\xi)| \lesssim \frac{|\xi_2|}{|\xi|}. \qquad (4.180)$$

Therefore, by Bernstein's inequality, (4.180), and the fractional chain rule of Visan (2007),

$$\begin{aligned}
&\left\|IF(u) - IF(u_l) + F(u_l) - F(Iu)\right\|_{L_t^2 L_x^{2d/(d+2)}([-T,T] \times \mathbf{R}^d)} \\
&= \left\|I(u_h|Iu|^{4/d}) - (Iu_h)|Iu|^{4/d}\right\|_{L_t^2 L_x^{\frac{2d}{(d+2)}}([-T,T] \times \mathbf{R}^d)} \\
&\lesssim \frac{\eta_3^{3/d}}{K^{3/d}} \|u_h\|_{L_t^2 L_x^{\frac{2d}{d-2}}([-T,T] \times \mathbf{R}^d)} \left\||\nabla|^{3/d}|Iu|^{4/d}\right\|_{L_t^\infty L_x^2([-T,T] \times \mathbf{R}^d)}. \qquad (4.181)
\end{aligned}$$

Combining (4.145), (4.176), (4.178), and (4.181) gives

(4.174)
$$\begin{aligned}
&\lesssim \eta_3^{-1} K \left(\frac{\eta_3}{K} \|\widetilde{\nabla} Iu\|_{L_t^2 L_x^{2d/(d-2)}([-T,T] \times \mathbf{R}^d)} + \|u_h\|_{L_t^2 L_x^{2d/(d-2)}([-T,T] \times \mathbf{R}^d)} \right) \\
&\quad \times \left(\frac{\eta_3^{3/d}}{K^{3/d}} \||\nabla|^{3/d}|Iu|^{4/d}\|_{L_t^\infty L_x^{d/2}([-T,T] \times \mathbf{R}^d)} + \|u_h\|_{L_t^\infty L_x^2([-T,T] \times \mathbf{R}^d)}^{4/d} \right) \\
&\lesssim \eta_3^{-1} K \left(\frac{\eta_3^{3/d}}{K^{3/d}} \||\nabla|^{3/d}|Iu|^{4/d}\|_{L_t^\infty L_x^{d/2}([-T,T] \times \mathbf{R}^d)} + \|u_h\|_{L_t^\infty L_x^2([-T,T] \times \mathbf{R}^d)}^{4/d} \right) \\
&\lesssim o(K).
\end{aligned}$$

The last inequality follows from the fact that $\eta_3(u) > 0$ is fixed. Letting $K \to \infty$, $N(t) \leq 1$, then the Hölder continuity of $|u|^{4/d}$ combined with Bernstein's inequality and (4.80) implies that, when $K \nearrow \infty$,

$$\left\|P_{|\xi - \xi(t)| > K^{1/2}} \right\|_{L_t^\infty L_x^2([-T,T] \times \mathbf{R}^d)} + K^{-3/2d} \|u\|_{L_t^\infty L_x^2([-T,T] \times \mathbf{R}^d)}^{4/d} \to 0. \qquad \square$$

Next consider (4.172).

Lemma 4.42

$$\int_{-T}^{T} \int \frac{(x-y)_j}{|x-y|} |Iu(t,y)|^2 \operatorname{Re}\left[\overline{Iu}(\partial_j - i\xi_j(t))(F(Iu) - IF(u)) \right](t,x) \, dx \, dy \, dt$$
$$\lesssim o(K). \qquad (4.182)$$

4.4 Nonradial Mass-Critical Problem in Dimensions $d \geq 3$

Proof Integrating by parts,

$$(4.182)$$

$$= -\int_{-T}^{T}\int\int \frac{1}{|x-y|}|Iu(t,y)|^2 \operatorname{Re}\left[(F(Iu) - IF(u))\overline{Iu}\right](t,x)\,dx\,dy\,dt \tag{4.183}$$

$$-\int_{-T}^{T}\int\int \frac{(x-y)_j}{|x-y|}|Iu(t,y)|^2$$
$$\times \operatorname{Re}\left[(F(Iu) - IF(u))(\partial_j + i\xi_j(t))\overline{Iu}\right](t,x)\,dx\,dy\,dt. \tag{4.184}$$

The last line, (4.184), is equal to the complex conjugate of (4.174). To show that (4.183) $\lesssim o(K)$, by the Hardy–Littlewood–Sobolev theorem,

$$(4.183) \lesssim \|F(Iu) - IF(u)\|_{L_t^2 L_x^{\frac{2d}{d+2}}([-T,T]\times\mathbf{R}^d)}$$

$$\times \|Iu\|_{L_t^4 L_x^{\frac{2d}{d-7/3}}([-T,T]\times\mathbf{R}^d)} \|Iu\|^2_{L_t^8 L_x^{\frac{2d}{d-5/6}}([-T,T]\times\mathbf{R}^d)}$$

$$\lesssim o(K). \tag{4.185}$$

The last estimate follows from the same argument estimating (4.179), which shows that

$$\|F(Iu) - IF(u)\|_{L_t^2 L_x^{\frac{2d}{d+2}}([-T,T]\times\mathbf{R}^d)} \leq o_K(1).$$

Also, by (4.145), $\|P_{\xi(t),j}u\|_{L_t^\infty L_x^2([-T,T]\times\mathbf{R}^d)} \to 0$ as $j \nearrow \infty$. Interpolating this fact with

$$\|P_{\xi(t),>j}u\|_{L_t^2 L_x^{\frac{2d}{d-2}}([-T,T]\times\mathbf{R}^d)} \lesssim \left(\frac{K\eta_3^{-1}}{2^j}\right)^{1/2}$$

implies that for any $2 < p < \infty$, $s > \frac{1}{p}$, if $(p,q) \in \mathscr{A}_d$ is an admissible pair, then

$$\||\nabla|^s Iu\|_{L_t^p L_x^q([-T,T]\times\mathbf{R}^d)} \lesssim o\left(K^s \eta_3^{-s}\right). \qquad \square$$

Finally consider (4.173).

Lemma 4.43 *For $d \geq 3$,*

$$\int_{-T}^{T}\int\int \frac{(x-y)_j}{|x-y|} \operatorname{Im}\left[\overline{Iu} \cdot IF(u)\right](t,y) \operatorname{Im}\left[\overline{Iu}(\partial_j - \xi_j(t))Iu\right](t,x)\,dx\,dy\,dt$$
$$\lesssim o(K).$$

Proof Because $\text{Im}(\bar{u} F(u)) = \text{Im}(|u|^{\frac{2(d+2)}{d}}) = 0$, it follows that

$$0 = \text{Im}\left[(1-I)\bar{u} \cdot (1-I) F(u) + (1-I)\bar{u} \cdot IF(u) + I\bar{u} \cdot (1-I) F(u) \right.$$
$$\left. + I\bar{u} \cdot IF(u) \right],$$

and therefore

$$\text{Im}\left[I\bar{u} \cdot IF(u)\right]$$
$$= \text{Im}\left[-(1-I)\bar{u} \cdot (1-I) F(u) - (1-I)\bar{u} \cdot IF(u) - I\bar{u} \cdot (1-I) F(u)\right].$$

By (4.175) and (4.176), we have

$$\int_{-T}^{T} \int \int \frac{(x-y)_j}{|x-y|} \text{Im}\left[(1-I)\bar{u} \cdot (1-I) F(u)\right](t,y)$$
$$\times \text{Im}\left[\overline{Iu}(\partial_j - \xi_j(t)) Iu\right](t,x) \, dx\,dy\,dt \lesssim o(K).$$

Next, integrating by parts, by the Fourier support of $(1-I)$,

$$\int_{-T}^{T} \int \int \frac{(x-y)_j}{|x-y|} \text{Im}\left[(1-I)\bar{u} \cdot IF(u) + I\bar{u} \cdot (1-I) F(u)\right](t,y)$$
$$\times \text{Im}\left[\overline{Iu}(\partial_j - \xi_j(t)) Iu\right](t,x) \, dx\,dy\,dt$$
$$= -\int_{-T}^{T} \int \int \frac{(x-y)_j}{|x-y|} \text{Im}\left[\frac{\widetilde{\nabla}_y}{\widetilde{\Delta}_y}(1-I)\bar{u} \cdot \widetilde{\nabla}_y IF(u)\right.$$
$$\left. + \widetilde{\nabla}_y I\bar{u} \cdot \frac{\widetilde{\nabla}_y}{\widetilde{\Delta}_y}(1-I) F(u)\right](t,y)$$
$$\times \text{Im}\left[\overline{Iu}(\partial_j - \xi_j(t)) Iu\right](t,x) \, dx\,dy\,dt$$
$$+ \int_{-T}^{T} \int \int \frac{1}{|x-y|} O\left(\left[\left|\frac{\widetilde{\nabla}_y}{\widetilde{\Delta}_y}(1-I)\bar{u}\right| |IF(u)|\right.\right.$$
$$\left.\left. + |I\bar{u}| \left|\frac{\widetilde{\nabla}_y}{\widetilde{\Delta}_y}(1-I) F(u)\right|\right](t,y)\right)$$
$$\times |Iu(t,x)| |\widetilde{\nabla} Iu(t,x)| \, dx\,dy\,dt. \qquad (4.186)$$

Then by (4.150), (4.175), (4.176), (4.185), the chain rule $\nabla F(Iu) = F'(Iu) \nabla Iu$,

4.4 Nonradial Mass-Critical Problem in Dimensions $d \geq 3$

and the Hardy–Littlewood–Sobolev inequality,

$$(4.186) \lesssim \frac{1}{K} \left\| P_{>\eta_3^{-1}K} u \right\|_{L_t^2 L_x^{\frac{2d}{d-2}}([-T,T] \times \mathbf{R}^d)} \left\| \widetilde{\nabla} I u \right\|_{L_t^2 L_x^{\frac{2d}{d-2}}([-T,T] \times \mathbf{R}^d)}$$

$$\times \|u\|_{L_t^\infty L_x^2([-T,T] \times \mathbf{R}^d)}^{1+4/d} \left\| \widetilde{\nabla} I u \right\|_{L_t^\infty L_x^2([-T,T] \times \mathbf{R}^d)}$$

$$+ \left\| P_{>\eta_3^{-1}K} u \right\|_{L_t^2 L_x^{\frac{2d}{d-2}}([-T,T] \times \mathbf{R}^d)} \left\| IF(u) - F(Iu) \right\|_{L_t^2 L_x^{\frac{2d}{d+2}}([-T,T] \times \mathbf{R}^d)}$$

$$\times \|u\|_{L_t^\infty L_x^2([-T,T] \times \mathbf{R}^d)} \left\| \widetilde{\nabla} I u \right\|_{L_t^\infty L_x^2([-T,T] \times \mathbf{R}^d)}$$

$$+ \frac{1}{K} \left\| P_{>\eta_3^{-1}K} F(u) \right\|_{L_t^2 L_x^{\frac{2d}{d+2}}([-T,T] \times \mathbf{R}^d)} \left\| \widetilde{\nabla} I u \right\|_{L_t^2 L_x^{\frac{2d}{d-2}}([-T,T] \times \mathbf{R}^d)}$$

$$\times \|u\|_{L_t^\infty L_x^2([-T,T] \times \mathbf{R}^d)} \left\| \widetilde{\nabla} I u \right\|_{L_t^\infty L_x^2([-T,T] \times \mathbf{R}^d)}$$

$$+ \frac{1}{K} \left\| P_{>\eta_3^{-1}K} u \right\|_{L_t^2 L_x^{\frac{2d}{d-2}}([-T,T] \times \mathbf{R}^d)} \left\| IF(u) \right\|_{L_t^4 L_x^{\frac{2d}{d+1}}([-T,T] \times \mathbf{R}^d)}$$

$$\times \|u\|_{L_t^\infty L_x^2([-T,T] \times \mathbf{R}^d)} \left\| \widetilde{\nabla} I u \right\|_{L_t^4 L_x^{\frac{2d}{d-1}}([-T,T] \times \mathbf{R}^d)}$$

$$+ \frac{1}{K} \left\| P_{>\eta_3^{-1}K} F(u) \right\|_{L_t^2 L_x^{\frac{2d}{d+2}}([-T,T] \times \mathbf{R}^d)} \left\| Iu \right\|_{L_t^4 L_x^{\frac{2d}{d-3}}([-T,T] \times \mathbf{R}^d)}$$

$$\times \|u\|_{L_t^\infty L_x^2([-T,T] \times \mathbf{R}^d)} \left\| \widetilde{\nabla} I u \right\|_{L_t^4 L_x^{\frac{2d}{d-1}}([-T,T] \times \mathbf{R}^d)}$$

$$\lesssim o(K).$$

This completes the proof of Lemma 4.43. □

Combining Lemmas 4.41–4.43 completes the proof of Theorem 4.40. □

Corollary 4.44 *There exists no nonzero almost-periodic solution to* (4.142) *satisfying* $N(t) \leq 1$ *for all* $t \in \mathbf{R}$, *and*

$$\int_{-\infty}^{\infty} N(t)^3 \, dt = \infty. \tag{4.187}$$

Proof By Theorem 4.40 and Lemmas 4.41–4.43, if $\int_{-T}^{T} N(t)^3 \, dt = K$, then

$$K \lesssim o(K).$$

This implies that there must be an upper bound on K. □

Now given the uniform bound $\int_{-T}^{T} N(t)^3 \, dt = K \leq K_0 < \infty$ for any $T < \infty$ and $\xi(0) = 0$, (4.144) implies that there exist $|\xi_+|, |\xi_-| \lesssim \eta_1^{-1} K$ such that

$$\lim_{t \to \pm\infty} \xi(t) = \xi_\pm.$$

Make a Galilean transformation mapping $\xi_+ \mapsto 0$. Then (4.144) and (4.187) also imply that $\lim_{t \to \pm\infty} N(t) = 0$, so by (4.80), for any j,

$$\lim_{t \to \pm\infty} \|P_j u(t)\|_{L^2_x(\mathbf{R}^d)} = 0. \qquad (4.188)$$

Therefore, for any $t \in [T, +\infty)$,

$$\|P_{>j} u(t)\|_{L^2_x(\mathbf{R}^d)} \lesssim \|P_{>j} F(u(t))\|_{L^2_t L^{\frac{2d}{d+2}}_x([T,\infty) \times \mathbf{R}^d)}.$$

Also by Theorem 4.34 and a scaling argument, if $\int_T^\infty N(t)^3 \, dt = \widetilde{K}$, then for any $2^j \leq \eta_3^{-1} \widetilde{K}$,

$$\|P_{\xi(t), >j} F(u(t))\|_{L^2_t L^{\frac{2d}{d-2}}_x([T,\infty) \times \mathbf{R}^d)} \lesssim \left(\frac{\widetilde{K} \eta_3^{-1}}{2^j}\right)^{1/2}.$$

Now let $j_0 \in \mathbf{Z}$ be the integer satisfying $2^{j_0} \sim \eta_3^{-1} K$. For $j \geq 0$, as in the proof of Theorem 4.34, (4.188) implies

$$\|P_{>j+j_0} F(u(t))\|_{L^2_t L^{\frac{2d}{d+2}}_x([T,\infty) \times \mathbf{R}^d)}$$

$$\lesssim \left\|\chi(t,x) (P_{>j+j_0+n_1} u) \left(P_{|\xi-\xi(t)| \leq \eta_3^{-1} N(t)} u\right)^{4/d}\right\|_{L^2_t L^{\frac{2d}{d+2}}_x([T,\infty) \times \mathbf{R}^d)}$$

$$+ \eta_2^{4/d} \|P_{>j+j_0+n_1} u\|_{L^2_t L^{\frac{2d}{d-2}}_x([T,\infty) \times \mathbf{R}^d)}$$

$$+ \sum_{k \leq j+j_0+n_1} \left(\frac{2^k}{2^j \eta_3^{-1} \widetilde{K}}\right)^{1/2} \|P_{\xi(t), >k} u\|_{L^2_t L^{\frac{2d}{d-2}}_x([T,\infty) \times \mathbf{R}^d)}. \qquad (4.189)$$

Then, arguing by induction,

$$\|P_{j+j_0} F(u(t))\|_{L^2_t L^{\frac{2d}{d+2}}_x([T,\infty) \times \mathbf{R}^d)} \lesssim 2^{-j/2}.$$

In particular, this implies that for any j,

$$\|P_{>j} u(t)\|_{L^2_x(\mathbf{R}^d)} \leq C(d, \eta_1, \eta_3) \widetilde{K}^{1/2} 2^{-j/2}. \qquad (4.190)$$

Applying the same calculation as in (4.97) to any small interval $J_l \subset [T, \infty)$,

$$\|P_{>j} F(u(t))\|_{L^{\frac{2(d+2)}{d+4}}_{t,x}(J_l \times \mathbf{R}^d)} \lesssim \widetilde{K}^{\frac{1}{2(d+2)}} 2^{-\frac{j}{2(d+2)}}. \qquad (4.191)$$

Plugging (4.190) and (4.191) back in to the bilinear estimate in (4.189),

$$\|u(t)\|_{L^\infty_t \dot{H}^{1/2}_x([T,\infty) \times \mathbf{R}^d)} \lesssim \widetilde{K}^{1/2}. \qquad (4.192)$$

4.4 Nonradial Mass-Critical Problem in Dimensions $d \geq 3$

The same is true for the interval $(-\infty, -T]$. By the dominated convergence theorem $\int_T^\infty N(t)^3 \, dt \to 0$, so by (4.192),

$$\lim_{T \to \pm \infty} \int \int |u(T,y)|^2 \frac{(x-y)}{|x-y|} \cdot \text{Im} \left[\bar{u} \nabla u\right](T,x) \, dx \, dy = 0,$$

and therefore by (3.121) and (4.19),

$$\left\| |\nabla|^{\frac{3-d}{2}} |u(t,x)|^2 \right\|_{L^2_{t,x}(\mathbf{R} \times \mathbf{R}^d)}^2 = 0,$$

and thus $u \equiv 0$. This completes the proof of Theorem 4.37. □

5

Low-Dimensional Well-Posedness Results

5.1 The Energy-Critical Problem in Dimensions Three and Four

Proof of scattering for the defocusing, energy-critical Schrödinger problem when $d = 3,4$ will also use frequency-localized interaction Morawetz estimates. In this case, $u \in \dot{H}^1$, but need not lie in L^2. In fact, the focusing, nonlinear Schrödinger equation

$$iu_t + \Delta u = -|u|^{\frac{4}{d-2}} u \qquad (5.1)$$

has a soliton solution

$$W(x) = \frac{1}{\left(1 + \frac{|x|^2}{d(d-2)}\right)^{\frac{d-2}{2}}},$$

and $W \in L^2(\mathbf{R}^d)$ if and only if $d \geq 5$. Examining the proof of Theorem 3.18, it is clear that the proof could be applied equally well to an almost-periodic solution to the focusing equation (5.1) as to the defocusing equation. In fact, Killip and Visan (2010) proved scattering for the focusing equation. Thus it is impossible to prove that an almost-periodic solution in dimensions $d = 3,4$ lies in $L^2(\mathbf{R}^d)$ using only linear methods.

The proof of scattering for the radially symmetric energy-critical problem used the Morawetz estimate (2.7), which was truncated in space. Then, since $u(t) \in L^{\frac{2d}{d-2}}(\mathbf{R}^d)$, it follows that $u(t)$ truncated in space lies in $L_x^2(\mathbf{R}^d)$. However, for the nonradial, energy-critical problem, this estimate is much less useful (see (3.60)), hence the use of a frequency-localized interaction Morawetz estimate.

At this point, a discussion on the chronology of the results presented in

5.1 The Energy-Critical Problem in Dimensions Three and Four

this book is in order. The global well-posedness and scattering results for the three-dimensional, energy-critical Schrödinger equation in Colliander et al. (2008), and also Kenig and Merle (2010), were the first scattering results for nonradial problems with critical regularity. Subsequently, the four-dimensional energy-critical result of Ryckman and Visan (2007) was proved, followed by the higher-dimensional energy-critical results and mass-critical results presented in Chapters 3 and 4.

Remark Scattering for the defocusing, energy-critical problem when $d \geq 5$ was proved in Visan (2007, 2006) and prior to Killip and Visan (2010), using frequency-localized interaction Morawetz estimates. The proof in Visan (2007, 2006) is quite technical, and will not be presented in this book.

The ordering of topics in this book is for expository reasons. Put simply, the proof in Colliander et al. (2008) was a very deep argument, introducing frequency-localized interaction Morawetz estimates, a version of long-time Strichartz estimates, and the double Duhamel argument to the study of nonlinear Schrödinger equations. In the author's opinion, it was better to introduce these concepts individually, for problems that only required one or two such tools. The reader should think of Sections 5.1 and 5.2 as the "final exam" for this book.

Theorem 5.1 *The nonlinear Schrödinger equation*

$$iu_t + \Delta u = F(u) = |u|^{\frac{4}{d-2}}u, \qquad u(0,x) = u_0, \tag{5.2}$$

is globally well posed and scattering for any $u_0 \in \dot{H}^1(\mathbf{R}^d)$, for $d = 3$ or $d = 4$.

Proof Recall from Theorems 3.14 and 3.15 that it suffices to show that if u is an almost-periodic solution to (5.2) on $\mathbf{R}$ with $N(t) \geq 1$ for all $t \in \mathbf{R}$, then $u \equiv 0$. The proof will follow five steps:

1. Show that the frequency-localized interaction Morawetz estimates must hold on some closed interval I with positive measure.
2. Prove Strichartz estimates on an interval $I \subset \mathbf{R}$ on which the bound on the interaction Morawetz estimate holds.
3. Show that there is an upper bound on $N(t)$ on the interval I. As $N(t) \nearrow \infty$, (3.60) and (3.61) imply that for any N_*, the mass at frequencies higher than N_* must go to zero. Therefore, to prove that $N(t)$ is bounded above it is enough to show that there is a lower bound on the mass at high frequencies.
4. Bootstrap Strichartz estimates on I with defocusing localized interaction Morawetz estimate bounds on I to prove that I must also be open and closed in $\mathbf{R}$, and therefore $I = \mathbf{R}$.

5. Make an argument similar to the proof of Theorem 3.21, proving that if u is an almost-periodic solution to (5.2) with $N(t) \sim 1$ and with frequency-localized interaction Morawetz estimates, then $u \equiv 0$.

Take the constants

$$0 < \eta_5 \ll \eta_4 \ll \eta_3 \ll \eta_2 \ll \eta_1 \ll 1.$$

The quantity η_j will depend on a power (that depends only on the dimension) of η_{j-1} of the form $\eta_j \ll \eta_{j-1}^{c(d)}$.

Proposition 5.2 (Frequency-localized interaction Morawetz estimate) *Suppose that u is an almost-periodic solution to (5.2) on an interval $I \subset \mathbf{R}$ and $N_* < \frac{\eta_3}{C(\eta_3)} \cdot \min_{t \in I} N(t)$, for $C(\eta_3) > 0$ defined in (3.60) and (3.61). Choose $j_* = \sup\{j \in \mathbf{Z} : 2^j \leq N_*\}$. Then, when $d = 3$,*

$$\int\!\!\int_I |u_{\geq j_*}(t,x)|^4 dx\,dt + \int\!\!\int_I \int \frac{|u_{\geq j_*}(t,y)|^2 |u_{\geq j_*}(t,x)|^6}{|x-y|} dx\,dy\,dt \lesssim \eta_1 N_*^{-3}, \tag{5.3}$$

and when $d = 4$,

$$\int\!\!\int_I \int \frac{|P_{\geq j_*} u(t,x)|^2 |P_{\geq j_*} u(t,y)|^2}{|x-y|^3} dx\,dy\,dt$$

$$+ \int\!\!\int_I \int \frac{|P_{\geq j_*} u(t,x)|^2 |P_{\geq j_*} u(t,y)|^4}{|x-y|} dx\,dy\,dt \lesssim \eta_1 N_*^{-3}. \tag{5.4}$$

Proof By scaling symmetry normalize $N_* = 1$ and $N(t) \geq \frac{C(\eta_3)}{\eta_3}$ for all $t \in I$, where $C(\eta_3)$ is defined in (3.60) and (3.61). Then

$$\left\|u_{<\frac{1}{\eta_3}}\right\|_{L_t^\infty \dot{H}^1(I \times \mathbf{R}^d)} + \left\|u_{<\frac{1}{\eta_3}}\right\|_{L_t^\infty L_x^{\frac{2d}{d-2}}(I \times \mathbf{R}^d)} \lesssim \eta_3. \tag{5.5}$$

Remark Here $u_{<\frac{1}{\eta_3}}$ denotes $u_{<j_3}$, where $j_3 = \sup\{j : 2^j \leq \frac{1}{\eta_3}\}$.

Then if $u_l = u_{\leq 0}$ and $u = u_l + u_h$, (5.5) implies

$$\|u_l\|_{L_t^\infty \dot{H}^1(I \times \mathbf{R}^d)} + \|u_l\|_{L_t^\infty L_x^{\frac{2d}{d-2}}(I \times \mathbf{R}^d)} \lesssim \eta_3, \tag{5.6}$$

and (5.5), Bernstein's inequality, and conservation of energy imply

$$\|u_h\|_{L_t^\infty L_x^2(I \times \mathbf{R}^d)} \lesssim \eta_3. \tag{5.7}$$

Since $u_h \in L_t^\infty L_x^2 \cap L_t^\infty L_x^{\frac{2d}{d-2}}$, then, when $d = 3, 4$, for any $t \in I$, it follows

5.1 The Energy-Critical Problem in Dimensions Three and Four

that $u_h \in L_x^2(\mathbf{R}^d) \cap L_x^4(\mathbf{R}^d)$. Therefore, for $d=3$, by the Hardy–Littlewood–Sobolev inequality,

$$\int |u_h(t,x)|^4 dx + \int \frac{|u_h(t,x)|^2 |u_h(t,y)|^6}{|x-y|} dx dy \lesssim 1,$$

and when $d=4$, we get

$$\int \frac{|u_h(t,x)|^2 |u_h(t,y)|^2}{|x-y|^3} dx dy + \int \frac{|u_h(t,x)|^2 |u_h(t,y)|^4}{|x-y|} dx dy \lesssim 1.$$

Therefore, there exists some $I \subset \mathbf{R}$ with positive measure and $0 \in I$ such that (5.3) and (5.4) hold.

Now make a bootstrap assumption. When $d=3$, assume that for some interval I,

$$\int_I \int |u_h(t,x)|^4 dx dt + \int_I \int \frac{|u_h(t,y)|^2 |u_h(t,x)|^6}{|x-y|} dx dy dt \leq C_0 \eta_1, \quad (5.8)$$

and when $d=4$ assume that for some interval $I \subset \mathbf{R}$,

$$\int_I \int \int \frac{|u_h(t,x)|^2 |u_h(t,y)|^2}{|x-y|^3} dx dy dt$$
$$+ \int_I \int \int \frac{|u_h(t,x)|^2 |u_h(t,y)|^4}{|x-y|} dx dy dt \leq C_0 \eta_1. \quad (5.9)$$

To prove Proposition 5.2 it suffices to show that if (5.8) and (5.9) hold on $I \subset \mathbf{R}$ then (5.8) and (5.9) hold on a slightly larger interval.

To estimate the error terms arising from the frequency-localized interaction Morawetz estimates, Colliander et al. (2008), Ryckman and Visan (2007), and Visan (2006, 2007) proved bounds on certain Strichartz estimates of u_h and u_l under the assumption that (5.8) and (5.9) hold on I. The Strichartz estimates will differ depending on the dimension, for various technical reasons. For example, when $d=4$, the Sobolev embedding theorem implies

$$\||\nabla|^{-1/2} f\|_{L_x^8(\mathbf{R}^4)} \lesssim \|f\|_{L_x^4(\mathbf{R}^4)}. \quad (5.10)$$

However, when $d=3$, $\||\nabla|^{-1/2} f\|_{L_x^\infty(\mathbf{R}^3)} \lesssim \|f\|_{L_x^6(\mathbf{R}^3)}$ does not hold, which impacts which Strichartz estimates may be used in step 2.

Theorem 5.3 *For $d=4$, if u is an almost-periodic solution to (5.2) on an interval I that satisfies (5.9), and $N(t) \geq \frac{C(\eta_3)}{\eta_3}$ on I, then*

$$\|u_l\|_{\dot{S}^1(I \times \mathbf{R}^4)} + \||\nabla|^{-1/2} u_h\|_{S^0(I \times \mathbf{R}^4)} \lesssim C_0^{1/2} \eta_1^{1/2}. \quad (5.11)$$

Proof Let
$$Z = \|u_l\|_{\dot S^1(I\times \mathbf{R}^4)} + \||\nabla|^{-1/2} u_h\|_{\dot S^0(I\times \mathbf{R}^4)}.$$

By interpolation and conservation of energy,
$$\|u_h\|_{L^3_{t,x}(I\times\mathbf{R}^4)} \lesssim \|\nabla u_h\|^{1/3}_{L^\infty_t L^2_x(I\times\mathbf{R}^4)} \||\nabla|^{-1/2} u_h\|^{2/3}_{L^2_t L^4_x(I\times\mathbf{R}^4)} \lesssim Z^{2/3}, \quad (5.12)$$
and by (5.6),
$$\|\nabla u_l\|_{L^3_{t,x}(I\times\mathbf{R}^4)} \lesssim \|\nabla u_l\|^{2/3}_{L^2_t L^4_x(I\times\mathbf{R}^4)} \|\nabla u_l\|^{1/3}_{L^\infty_t L^2_x(I\times\mathbf{R}^4)} \lesssim Z^{2/3} \eta_3^{1/3}.$$

Also by (3.118), the bootstrap assumption (5.9), conservation of energy, and interpolation, we have
$$\|u_h\|_{L^6_t L^{24/7}_x(I\times\mathbf{R}^4)} \lesssim \||\nabla|^{1/2} u_h\|^{1/3}_{L^\infty_t L^{8/3}_x(I\times\mathbf{R}^4)} \||\nabla|^{-1/4} u_h\|^{2/3}_{L^4_{t,x}(I\times\mathbf{R}^4)} \lesssim C_0^{1/6} \eta_1^{1/6}. \quad (5.13)$$

Then by Strichartz estimates and (5.6), for any $t_0 \in I$,
$$Z \lesssim \|u_l(t_0)\|_{\dot H^1_x(\mathbf{R}^4)} + \|u_h(t_0)\|_{L^2_x(\mathbf{R}^4)} + \|\nabla P_l F(u)\|_{L^2_t L^{4/3}_x(I\times\mathbf{R}^4)}$$
$$+ \||\nabla|^{-1/2} P_h F(u)\|_{L^2_t L^{4/3}_x(I\times\mathbf{R}^4)}$$
$$\lesssim \eta_3 + \|\nabla P_l F(u)\|_{L^2_t L^{4/3}_x(I\times\mathbf{R}^4)} + \||\nabla|^{-1/2} P_h F(u)\|_{L^2_t L^{4/3}_x(I\times\mathbf{R}^4)}.$$

Now make the following split:
$$F(u) = F(u_l) + O(u_h u^2).$$

Then
$$\|\nabla F(u_l)\|_{L^2_t L^{4/3}_x(I\times\mathbf{R}^4)} \lesssim \|\nabla u_l\|_{L^2_t L^4_x(I\times\mathbf{R}^4)} \|u_l\|^2_{L^\infty_t L^4_x(I\times\mathbf{R}^4)} \lesssim \eta_3^2 Z,$$
so by Bernstein's inequality,
$$\|\nabla P_l F(u_l)\|_{L^2_t L^{4/3}_x(I\times\mathbf{R}^4)} + \||\nabla|^{-1/2} P_h F(u_l)\|_{L^2_t L^{4/3}_x(I\times\mathbf{R}^4)} \lesssim \eta_3^2 Z. \quad (5.14)$$

Next, by the Sobolev embedding theorem and (5.12),
$$\||u_l|^2 u_h\|_{L^2_t L^{4/3}_x(I\times\mathbf{R}^4)} \lesssim \|u_h\|_{L^3_{t,x}(I\times\mathbf{R}^4)} \|u_l\|_{L^6_{t,x}(I\times\mathbf{R}^4)} \|u_l\|_{L^\infty_t L^4_x(I\times\mathbf{R}^4)}$$
$$\lesssim \|u_h\|_{L^3_{t,x}(I\times\mathbf{R}^4)} \|\nabla u_l\|^{1/3}_{L^2_t L^4_x(I\times\mathbf{R}^4)} \|u_l\|^{5/3}_{L^\infty_t \dot H^1_x(I\times\mathbf{R}^4)}$$
$$\lesssim \eta_3^{5/3} Z, \quad (5.15)$$

so by the Fourier support of P_l and P_h,
$$\|\nabla P_l(u_l^2 u_h)\|_{L^2_t L^{4/3}_x} + \||\nabla|^{-1/2} P_h(u_l^2 u_h)\|_{L^2_t L^{4/3}_x} \lesssim \eta_3^{5/3} Z, \quad (5.16)$$

5.1 The Energy-Critical Problem in Dimensions Three and Four

and by the Sobolev embedding theorem and (5.13),

$$\left\| |\nabla|^{-1/2} |u_h|^3 \right\|_{L_t^2 L_x^{4/3}(I \times \mathbf{R}^4)} \lesssim \left\| |u_h|^3 \right\|_{L_t^2 L_x^{8/7}(I \times \mathbf{R}^4)} \lesssim C_0^{1/2} \eta_1^{1/2}. \tag{5.17}$$

Therefore, combining (5.14)–(5.17),

$$Z \lesssim \eta_3 + \eta_3^2 Z + \eta_3^{5/3} Z + C_0^{1/2} \eta_1^{1/2},$$

which implies

$$Z \lesssim C_0^{1/2} \eta_1^{1/2},$$

which was to be proved. □

When $d = 3$, the estimate corresponding to (5.10) holds only for frequency-localized functions.

Theorem 5.4 *When $d = 3$,*

$$\|u_l\|_{\dot{S}^1(I \times \mathbf{R}^3)} + \left(\sup_{j \geq 0} 2^{-j/2} \|u_j\|_{L_t^2 L_x^6(I \times \mathbf{R}^3)} \right) \lesssim C_0^{1/2} \eta_1^{1/2}. \tag{5.18}$$

Proof Set

$$Z = \|u_l\|_{\dot{S}^1(I \times \mathbf{R}^3)} + \left(\sup_{j \geq 0} 2^{-j/2} \|u_j\|_{L_t^2 L_x^6(I \times \mathbf{R}^3)} \right).$$

For any $t_0 \in I$, Strichartz estimates imply

$$Z \lesssim \|u_l(t_0)\|_{\dot{H}_x^1(\mathbf{R}^3)} + \|u_h(t_0)\|_{\dot{H}_x^{-1/2}(\mathbf{R}^3)} + \|\nabla P_l F(u)\|_{L_t^2 L_x^{6/5}(I \times \mathbf{R}^3)}$$

$$+ \left(\sup_{j \geq 0} 2^{-j/2} \|P_j F(u)\|_{L_t^2 L_x^{6/5}(I \times \mathbf{R}^3)} \right). \tag{5.19}$$

Again by (5.6),

$$\left\| |\nabla|^{-1/2} u_h(t_0) \right\|_{L_x^2(\mathbf{R}^3)} + \|\nabla u_l(t_0)\|_{L_x^2(\mathbf{R}^3)} \lesssim \eta_3. \tag{5.20}$$

Taking any $j \geq 0$, by (5.6) and Bernstein's inequality, yields

$$\|\nabla F(u_l)\|_{N^0(I \times \mathbf{R}^3)} + 2^{-j/2} \|P_j F(u_l)\|_{N^0(I \times \mathbf{R}^3)}$$
$$\lesssim \|\nabla F(u_l)\|_{N^0(I \times \mathbf{R}^3)} \lesssim \|\nabla u_l\|_{L_t^2 L_x^6(I \times \mathbf{R}^3)} \|u_l\|_{L_t^\infty L_x^6(I \times \mathbf{R}^3)}^4$$
$$\lesssim \eta_3^4 Z. \tag{5.21}$$

Next, by interpolation and Bernstein's inequality, we find

$$\|O(u_l^4 u_h)\|_{N^0(I \times \mathbf{R}^3)} \lesssim \|u_h\|_{L_t^\infty L_x^2(I \times \mathbf{R}^3)} \|\nabla u_l\|_{L_t^8 L_x^{12/5}(I \times \mathbf{R}^3)}^4$$
$$\lesssim \eta_3 \|\nabla u_l\|_{L_t^2 L_x^6(I \times \mathbf{R}^3)} \|\nabla u_l\|_{L_t^\infty L_x^2(I \times \mathbf{R}^3)}^3 \lesssim \eta_3^4 Z. \tag{5.22}$$

Therefore,
$$\|\nabla P_l O(u_l^4 u_h)\|_{L_t^2 L_x^{6/5}} + \||\nabla|^{-1/2} P_h O(u_l^4 u_h)\|_{L_t^2 L_x^{6/5}} \lesssim \eta_3^4 Z. \quad (5.23)$$

Finally, by the Sobolev embedding theorem, for any j,
$$\|P_j(u_h^2 u^3)\|_{L_t^2 L_x^{6/5}(I \times \mathbf{R}^3)} \lesssim 2^{j/2} \|(u_h^2 u^3)\|_{L_t^2 L_x^1(I \times \mathbf{R}^3)}$$
$$\lesssim \|u_h\|_{L_{t,x}^4(I \times \mathbf{R}^3)}^2 \|u\|_{L_t^\infty L_x^6(I \times \mathbf{R}^3)}^3 \lesssim C_0^{1/2} \eta_1^{1/2}. \quad (5.24)$$

Therefore, by Strichartz estimates and (5.19)–(5.24),
$$Z \lesssim \eta_3 + \eta_3^4 Z + C_0^{1/2} \eta_1^{1/2},$$
which was to be proved. □

Moving right along to step 3 in the proof of Theorem 5.1, we derive an upper bound on $N(t)$ for all $t \in I$. To do this, we improve the bounds in Theorems 5.3 and 5.4 at very low frequencies.

Lemma 5.5 *For $j \le 0$ and $d = 3, 4$, for any $\delta > 0$, for $\eta_3(\delta) > 0$ sufficiently small,*
$$\|P_{\le j} u\|_{\dot{S}^1(I \times \mathbf{R}^d)} \lesssim_\delta \left(\inf_{t \in I} \|P_{\le 0} u(t)\|_{\dot{H}^1(\mathbf{R}^d)}\right) + 2^{j(\frac{3}{2} - \delta)} C_0^{1/2} \eta_1^{1/2}. \quad (5.25)$$

Proof By Strichartz estimates and Bernstein's inequality, for any $t_0 \in I$,
$$\|P_{\le j} u(t)\|_{\dot{S}^1(I \times \mathbf{R}^d)} \lesssim \|P_{\le j} u(t_0)\|_{\dot{H}^1(\mathbf{R}^d)} + \|F(u_{\le j})\|_{\dot{N}^1(I \times \mathbf{R}^d)}$$
$$+ 2^j \|P_{\le j}[F(u) - F(u_{\le j})]\|_{L_t^2 L_x^{\frac{2d}{d+2}}(I \times \mathbf{R}^d)}. \quad (5.26)$$

Now since $j \le 0$, $\|P_{\le j} u\|_{L_t^\infty \dot{H}^1(I \times \mathbf{R}^d)} \lesssim \eta_3$, so by interpolation,
$$\|F(u_{\le j})\|_{\dot{N}^1(I \times \mathbf{R}^d)} \lesssim \|\nabla u_{\le j}\|_{L_t^\infty L_x^2(I \times \mathbf{R}^d)}^{\frac{4}{d-2}} \|\nabla u_{\le j}\|_{L_t^2 L_x^{\frac{2d}{d-2}}(I \times \mathbf{R}^d)}$$
$$\lesssim \eta_3^{\frac{4}{d-2}} \|u_{\le j}\|_{\dot{S}^1(I \times \mathbf{R}^d)}. \quad (5.27)$$

Next, decompose:
$$F(u) - F(u_{\le j}) = O\left(|u_{\ge j}||u|^{\frac{4}{d-2}}\right) \lesssim O\left(|u_{\ge j}||u_{\le j}|^{\frac{4}{d-2}}\right) + O\left(|u_{\ge j}|^{\frac{d+2}{d-2}}\right)$$
$$\lesssim O\left(|u_{\ge j}||u_{\le j}|^{\frac{4}{d-2}}\right) + O\left(|u_{j \le \cdot \le 0}|^{\frac{d+2}{d-2}}\right) + O\left(|u_h|^{\frac{d+2}{d-2}}\right).$$
$$(5.28)$$

5.1 The Energy-Critical Problem in Dimensions Three and Four

For some $0 < \delta < \frac{1}{2}$, define the frequency envelope

$$\alpha_j = \sum_{j \leq k \leq 0} 2^{(\frac{3}{2}-\delta)(j-k)} \|P_{\leq k} u\|_{\dot{S}^1(I \times \mathbf{R}^d)}.$$

Then by (5.6), (5.26), (5.27), (5.28), interpolation, and the Sobolev embedding theorem, when $d = 3$ or $d = 4$,

$$\alpha_j \lesssim \sum_{j \leq k \leq 0} 2^{(\frac{3}{2}-\delta)(j-k)} \left(\inf_{t \in I} \|u_{\leq 0}(t)\|_{\dot{H}^1(\mathbf{R}^d)} \right)$$

$$+ \eta_3^{\frac{4}{d-2}} \sum_{j \leq k \leq 0} 2^{(\frac{3}{2}-\delta)(j-k)} \|P_{\leq k} u\|_{\dot{S}^1(I \times \mathbf{R}^d)}$$

$$+ \sum_{j \leq k \leq 0} 2^{(\frac{3}{2}-\delta)(j-k)} 2^k \|P_{\leq k} [F(u) - F(u_{\leq k})]\|_{L_t^2 L_x^{\frac{2d}{d+2}}(I \times \mathbf{R}^d)}. \quad (5.29)$$

By the Sobolev embedding theorem and (5.28),

$$\sum_{j \leq k \leq 0} 2^{(\frac{3}{2}-\delta)(j-k)} 2^k \|P_{\leq k}[F(u) - F(u_{\leq k})]\|_{L_t^2 L_x^{\frac{2d}{d+2}}(I \times \mathbf{R}^d)}$$

$$\lesssim \sum_{j \leq k \leq 0} 2^{(\frac{3}{2}-\delta)(j-k)} 2^k \|u_{\geq k}\|_{L_t^\infty L_x^2(I \times \mathbf{R}^d)} \|\nabla u_{\leq k}\|_{L_t^2 L_x^{\frac{2d}{d-2}}(I \times \mathbf{R}^d)} \|u_{\leq k}\|_{L_t^\infty \dot{H}^1(I \times \mathbf{R}^d)}^{\frac{4}{d-2}-1}$$

$$+ \sum_{j \leq k \leq 0} 2^{(\frac{3}{2}-\delta)(j-k)} 2^{\frac{3k}{2}} \|u_{k \leq \cdot \leq 0}\|_{L_t^3 L_x^{\frac{6d}{3d-4}}(I \times \mathbf{R}^d)}^{\frac{3}{2}} \|u_{k \leq \cdot \leq 0}\|_{L_t^\infty L_x^{\frac{2d}{d-2}}(I \times \mathbf{R}^d)}^{\frac{4}{d-2}-\frac{1}{2}}$$

$$+ \sum_{j \leq k \leq 0} 2^{(\frac{3}{2}-\delta)(j-k)} 2^k \|P_{\leq k}\left(|u_h|^{\frac{d+2}{d-2}}\right)\|_{L_t^2 L_x^{\frac{2d}{d+2}}(I \times \mathbf{R}^d)}.$$

Bernstein's inequality, (5.6), and interpolation yield

$$\sum_{j \leq k \leq 0} 2^{\frac{3k}{2}} 2^{(\frac{3}{2}-\delta)(j-k)} \|u_{k \leq \cdot \leq 0}\|_{L_t^3 L_x^{\frac{6d}{3d-4}}}^{\frac{3}{2}}$$

$$\lesssim \eta_3^{\frac{1}{2}} \sum_{j \leq k \leq 0} 2^{(\frac{3}{2}-\delta)j} 2^{\delta k} \sum_{k \leq k_1 \leq 0} 2^{-\frac{k_1}{2}} \|u_{k_1}\|_{L_t^2 L_x^{\frac{2d}{d-2}}}$$

$$\lesssim_\delta \eta_3^{\frac{1}{2}} \sum_{j \leq k_1 \leq 0} 2^{(\frac{3}{2}-\delta)(j-k_1)} \|\nabla u_{k_1}\|_{L_t^2 L_x^{\frac{2d}{d-2}}}.$$

Therefore,

$$(5.29) \lesssim_\delta \left(\inf_{t \in I} \|u_{\leq 0}(t)\|_{\dot{H}^1(\mathbf{R}^d)} \right) + \eta_3^{\frac{4}{d-2}-1} \alpha_j + \eta_3^{\frac{4}{d-2}} \alpha_j$$

$$+ \sum_{j \leq k \leq 0} 2^{(j-k)(\frac{3}{2}-\delta)} 2^k \|P_{\leq k}\left(|u_h|^{\frac{d+2}{d-2}}\right)\|_{L_t^2 L_x^{\frac{2d}{d+2}}(I \times \mathbf{R}^d)}. \quad (5.30)$$

For $\eta_3(\delta) > 0$ sufficiently small, the second and third terms on the right-hand side of (5.30) can be absorbed into the left-hand side. Finally, by the Sobolev embedding theorem, (5.8), (5.9), and (5.12), we get

$$2^k\left\|P_{\leq k}\left(|u_h|^{\frac{d+2}{d-2}}\right)\right\|_{L_t^2 L_x^{\frac{2d}{d+2}}(I\times\mathbf{R}^d)} \lesssim 2^{\frac{3}{2}k}\left\||u_h|^{\frac{d+2}{d-2}}\right\|_{L_t^2 L_x^{\frac{2d}{d+3}}(I\times\mathbf{R}^d)} \lesssim 2^{\frac{3}{2}k}C_0^{1/2}\eta_1^{1/2}. \tag{5.31}$$

Plugging (5.31) into (5.30) gives

$$\alpha_j \lesssim \left(\inf_{t\in I}\|u_{\leq 0}(t)\|_{H^1(\mathbf{R}^d)}\right) + 2^{j(\frac{3}{2}-\delta)}C_0^{1/2}\eta_1^{1/2}, \tag{5.32}$$

which proves Lemma 5.5. $\square$

Lemma 5.5 implies an upper bound on $N(t)$ when $d = 4$.

Proposition 5.6 *Suppose u is an almost-periodic solution to the energy-critical initial value problem with $N(t) \geq C(\eta_3)/\eta_3$ for all $t \in \mathbf{R}$ and $d = 4$. Then for all $t \in I$, we have $N(t) \lesssim C(\eta_5) N_{\min}$.*

Proof Since $N(t)$ was normalized so that $N(t) \geq C(\eta_3)/\eta_3$ on I, there exists some $t_{\min} \in I$ such that $N(t_{\min}) \leq 2C(\eta_3)/\eta_3$. Then by Bernstein's inequality and (3.60), if u is nonzero, for some $\eta > 0$ small,

$$\left\|P_{N(t_{\min})c(\eta)<\cdot<N(t_{\min})C(\eta)} u(t_{\min})\right\|_{L^2(\mathbf{R}^d)} \gtrsim \eta\frac{\eta_3}{C(\eta_3)}. \tag{5.33}$$

To prove the proposition it suffices to show that there does not exist $t_{\text{evac}} \in I$ such that $N(t_{\text{evac}}) \gg C(\eta_5)$. By time reversal symmetry, suppose that $t_{\min} < t_{\text{evac}}$. Choosing $C(\eta_5)$ sufficiently large, (3.60) implies

$$\left\|P_{<\eta_5^{-1}} u(t_{\text{evac}})\right\|_{\dot{H}^1(\mathbf{R}^d)} \leq \eta_5. \tag{5.34}$$

Therefore, if one can show that (5.33) holds for all $t \in I$, then (5.34) implies a contradiction when $N(t) \gg C(\eta_5)$, proving that $N(t) \lesssim C(\eta_5)$ for all $t \in I$.

Modifying notation, let $u_{h'} = P_{\geq j} u$ for some $j \leq 0$ to be specified later, and let $u_{h'} + u_{l'} = u$. Then by (1.112),

$$\frac{d}{dt}\int |u_{h'}(t,x)|^2 dx = \int \{P_{h'}F(u), u_{h'}\}_m dx$$
$$= \int \{P_{h'}(F(u) - F(u_{h'}) - F(u_{l'})), u_{h'}\}_m dx$$
$$- \int \{P_{l'}F(u_{h'}), u_{h'}\}_m dx + \int \{P_{h'}F(u_{l'}), u_{h'}\}_m dx.$$

5.1 The Energy-Critical Problem in Dimensions Three and Four 173

By Bernstein's inequality, (5.5), (5.6), and (5.25),

$$\int_I \int |\{P_{h'} F(u_{l'}), u_{h'}\}_m | \, dx \, dt$$

$$\lesssim 2^{-j} \|\nabla u_{l'}\|_{L_t^2 L_x^4 (I \times \mathbf{R}^4)} \|u_{l'}\|_{L_t^4 L_x^8 (I \times \mathbf{R}^4)}^2 \|u_{h'}\|_{L_t^\infty L_x^2 (I \times \mathbf{R}^4)}$$

$$\lesssim \eta_3 2^{-2j} \left(\left(\inf_{t \in I} \|u_{\leq 0}(t)\|_{\dot{H}^1(\mathbf{R}^4)} \right) + C_0^{1/2} \eta_1^{1/2} 2^{j(\frac{3}{2} - \delta)} \right)^3. \quad (5.35)$$

Also by the Sobolev embedding theorem, (5.6), (5.11), (5.25), and Bernstein's inequality,

$$\int_I \int |\{P_{l'} F(u_{h'}), u_{h'}\}_m | \, dx \, dt$$

$$\lesssim 2^j \|u_{h'}\|_{L_{t,x}^3 (I \times \mathbf{R}^4)}^3 \|u\|_{L_t^\infty \dot{H}^1 (I \times \mathbf{R}^4)} \lesssim 2^j C_0 \eta_1 + 2^j \|u_{j \leq \cdot \leq 0}\|_{L_{t,x}^3 (I \times \mathbf{R}^d)}^3$$

$$\lesssim 2^j C_0 \eta_1 + 2^j \left(\sum_{j \leq k \leq 0} 2^{-k} \left(\inf_{t \in I} \|u_{\leq 0}(t)\|_{\dot{H}^1(\mathbf{R}^d)} \right) + 2^{-k} 2^{k(\frac{3}{2} - \delta)} C_0^{1/2} \eta_1^{1/2} \right)^3.$$

$$(5.36)$$

Finally, make the splitting

$$|F(u) - F(u_{h'}) - F(u_{l'})| \lesssim |u_{h'}||u_{l'}||u|.$$

Then by (5.11), (5.25), and Bernstein's inequality,

$$\int_I \int |\{P_{h'}(F(u) - F(u_{h'}) - F(u_{l'})), u_{h'}\}| \, dx \, dt$$

$$\lesssim \int_I \int |u_{h'}|^3 |u_{l'}| + |u_{h'}|^2 |u_{l'}|^2 \, dx \, dt$$

$$\lesssim \|u_{h'}\|_{L_{t,x}^3 (I \times \mathbf{R}^4)}^3 \|u_{l'}\|_{L_{t,x}^\infty (I \times \mathbf{R}^d)}$$

$$+ \|u_{l'}\|_{L_{t,x}^6 (I \times \mathbf{R}^4)}^2 \left[\|u_{\geq 0}\|_{L_{t,x}^3 (I \times \mathbf{R}^4)}^2 + \|u_{j \leq \cdot \leq 0}\|_{L_{t,x}^3 (I \times \mathbf{R}^4)}^2 \right]$$

$$\lesssim 2^j C_0 \eta_1 + 2^j \left(\sum_{j \leq k \leq 0} \left(2^{-k} \inf_{t \in I} \|u_{\leq 0}(t)\|_{\dot{H}^1(\mathbf{R}^d)} \right) + 2^{-k} 2^{k(3/2 - \delta)} C_0^{1/2} \eta_1^{1/2} \right)^3$$

$$+ \left(\inf_{t \in I} \|u_{\leq 0}(t)\|_{\dot{H}^1} + C_0^{1/2} \eta_1^{1/2} 2^{j(\frac{3}{2} - \delta)} \right)^2$$

$$\times \left(\sum_{j \leq k \leq 0} \left(2^{-k} \inf_{t \in I} \|u_{\leq 0}(t)\|_{\dot{H}^1(\mathbf{R}^d)} \right) + 2^{-k} 2^{k(3/2 - \delta)} C_0^{1/2} \eta_1^{1/2} \right)^2.$$

$$(5.37)$$

Therefore, if (5.34) holds, then taking j such that $2^j = \eta_4$, and fixing $\delta > 0$

(for example, $\delta = \frac{1}{4}$), relations (5.34), (5.35), (5.36), and (5.37) imply

$$\int_I \int |\{P_{h'} F(u) - F(u_{h'}), u_{h'}\}_m| \, dx \, dt \ll \eta \frac{\eta_3}{C(\eta_3)}. \tag{5.38}$$

Therefore, (5.34), (5.38), and the fundamental theorem of calculus contradict (5.33), proving that $N(t) \leq C(\eta_1, \eta_3, \eta_4)$ for all $t_{\min} \leq t \leq t_{\text{evac}}$, contradicting $N(t_{\text{evac}}) \gg C(\eta_5)$, hence $N(t) \leq C(\eta_1, \eta_3, \eta_4)$ for all $t \in I$. □

The proof of the corresponding three-dimensional result is slightly different from Proposition 5.6 and will be postponed.

Proposition 5.7 *Suppose u is an almost-periodic solution to the energy-critical initial value problem with $N(t) \geq 1$ for all $t \in \mathbf{R}$ and $d = 3$. Then for all $t \in I$, $N(t) \lesssim C(\eta_5) N_{\min}$.*

Let us proceed to step 4 of the proof of Theorem 5.1 when $d = 4$.

Proposition 5.8 *Let u_h and u_l be as in Proposition 5.2, with $N_* = 1$. Then when $d = 4$,*

$$\int_I \int \int \frac{|u_h(t,y)|^2 |u_h(t,x)|^2}{|x-y|^3} dx \, dy \, dt + \int_I \int \int \frac{|u_h(t,y)|^2 |u_h(t,x)|^4}{|x-y|} dx \, dy \, dt \tag{5.39}$$

$$\lesssim \|u_h\|_{L_t^\infty L_x^2(I \times \mathbf{R}^4)}^3 \|u_h\|_{L_t^\infty \dot H_x^1(I \times \mathbf{R}^4)} \tag{5.40}$$

$$+ \|u_h\|_{L_t^\infty L_x^2(I \times \mathbf{R}^4)} \|u_h\|_{L_t^\infty \dot H_x^1(I \times \mathbf{R}^4)} \int_I \int |\{P_h F(u) - F(u_h), u_h\}_m| \, dx \, dt \tag{5.41}$$

$$+ \|u_h\|_{L_t^\infty L_x^2(I \times \mathbf{R}^4)}^2 \int_I \int |u_h(t,x)| |u_l(t,x)|^2 |\nabla u_l(t,x)| \, dx \, dt \tag{5.42}$$

$$+ \|u_h\|_{L_t^\infty L_x^2(I \times \mathbf{R}^4)}^2 \int_I \int |\nabla u_l(t,x)| |u_h(t,x)|^3 \, dx \, dt \tag{5.43}$$

$$+ \|u_h\|_{L_t^\infty L_x^2(I \times \mathbf{R}^4)}^2 \int_I \int |\nabla P_l(F(u))(t,x)| |u_h(t,x)| \, dx \, dt \tag{5.44}$$

$$+ \int_I \int \int \frac{|u_h(t,y)|^2 |u_l(t,x)|^3 |u_h(t,x)|}{|x-y|} dx \, dy \, dt \tag{5.45}$$

$$+ \int_I \int \int \frac{|u_h(t,y)|^2 |P_l F(u)(t,x)| |u_h(t,x)|}{|x-y|} dx \, dy \, dt \tag{5.46}$$

$$+ \int_I \int \int \frac{|u_h(t,y)|^2 |u_l(t,x)| |u_h(t,x)|^3}{|x-y|} dx \, dy \, dt. \tag{5.47}$$

Proof Recall the interaction Morawetz quantity (3.101):

$$M^{\text{int}}(t) = \int |u(t,y)|^2 h(x-y) |u(t,x)|^2 dx \, dy.$$

5.1 The Energy-Critical Problem in Dimensions Three and Four

Then by (1.112), (3.102), and integrating by parts, if $h(x) = |x|$,

$$\frac{1}{2}\frac{d}{dt}M^{\text{int}}(t) = -\int |u(t,y)|^2 \frac{x-y}{|x-y|} \cdot \text{Im}[\bar{u}\nabla u](t,x)\,dx\,dy.$$

Splitting,

$$(i\partial_t + \Delta)u_h = \mathcal{N}_1 + \mathcal{N}_2 = F(u_h) + P_h F(u) - F(u_h), \quad u_h(0,x) = P_h u(0,x),$$

the contribution of $\mathcal{N}_1 = F(u_h)$ may be inserted directly into (3.116)–(3.117), yielding (5.39) and (5.40).

Relations (5.41)–(5.47) are precisely the error terms that arise from $\mathcal{N}_2$. First take the mass bracket in (3.117). Then

$$\int_I \int\int |\{F(u_h) - P_h F(u), u_h\}_m(t,x)| \, |u_h(t,y)| \, |\nabla u_h(t,y)| \, dx\,dy\,dt \lesssim (5.41).$$

Now take the momentum bracket in (3.116). Split

$$\{\mathcal{N}_2, u_h\}_p = \{F(u_h) - P_h F(u), u_h\}_p = \{F(u_h) - F(u) + P_l F(u), u_h\}_p.$$

By definition of $\{\cdot,\cdot\}_p$ and the product rule,

$$\{f,g\}_p = \nabla \text{Re}(f\bar{g}) + O(|\nabla f||g|), \tag{5.48}$$

so integrating by parts gives

$$\int_I \int\int |u_h(t,y)|^2 \frac{(x-y)}{|x-y|} \cdot \{P_l F(u), u_h\}_p(t,x)\,dx\,dy\,dt$$

$$\lesssim \int_I \int\int |u_h(t,y)|^2 \frac{1}{|x-y|} |P_l F(u)(t,x)| \, |u_h(t,x)| \, dx\,dy\,dt$$

$$+ \int_I \int\int |u_h(t,y)|^2 |\nabla P_l F(u)(t,x)| \, |u_h(t,x)| \, dx\,dy\,dt$$

$$\lesssim (5.44) + (5.46).$$

Expanding $\{F(u_h) - F(u), u_h\}_p$, (1.118) implies

$$\{F(u_h) - F(u), u_h\}_p$$
$$= \{F(u_h), u_h\}_p - \{F(u), u\}_p + \{F(u), u_l\}_p$$
$$= -\frac{1}{2}\nabla|u_h|^4 + \frac{1}{2}\nabla|u|^4 + \{F(u_l), u_l\}_p + \{F(u) - F(u_l), u_l\}_p$$
$$= -\frac{1}{2}\nabla|u_h|^4 - \frac{1}{2}\nabla|u_l|^4 + \frac{1}{2}\nabla|u|^4 + \{F(u) - F(u_l), u_l\}_p.$$

Integrating by parts,

$$\frac{1}{2}\int_I \int\int |u_h(t,y)|^2 \frac{(x-y)}{|x-y|} \cdot \nabla\big[|u|^4 - |u_h|^4 - |u_l|^4\big](t,x)\,dx\,dy\,dt$$

$$\lesssim (5.45) + (5.47).$$

By (5.48), we have $|F(u) - F(u_l)| \lesssim |u_l|^2|u_h| + |u_h|^3$. Finally, integrating by parts yields

$$\int_I \int \int |u_h(t,y)|^2 \frac{(x-y)}{|x-y|} \cdot \{F(u) - F(u_l), u_l\}_p (t,x) \, dx \, dy \, dt$$
$$\lesssim (5.42) + (5.43) + (5.45) + (5.47).$$

This completes the proof of Proposition 5.8. $\square$

We are now ready to complete the proof of Proposition 5.2 when $d = 4$. First, (5.7) implies that $(5.40) \lesssim \eta_3^3$.

Next, taking (5.35), (5.36), and (5.37) with $j = 0$, combined with (5.6), yields

$$\int_I \int |\{u_h, P_h F(u) - F(u_h)\}_m| \, dx \, dt \lesssim C_0 \eta_1,$$

and therefore $(5.41) \lesssim C_0 \eta_1 \eta_3$.

Now consider (5.42)–(5.44). By (5.6), (5.7), (5.11), the Sobolev embedding theorem, and interpolation we have

$$(5.42) \lesssim \eta_3^2 \|u_h\|_{L_t^\infty L_x^2} \|u_l\|_{L_t^4 L_x^8}^2 \|\nabla u_l\|_{L_t^2 L_x^4} \lesssim \eta_3^4 (C_0 \eta_1).$$

Also by (5.6), (5.7), (5.11), (5.12), and the Sobolev embedding theorem,

$$(5.43) \lesssim \eta_3^2 \|u_h\|_{L_{t,x}^3}^3 \|u_l\|_{L_{t,x}^\infty} \lesssim \eta_3^3 (C_0 \eta_1).$$

Make the following split:

$$P_l F(u) = P_l F(u_l) + P_l O\left(|u_h| |u_l|^2\right) + P_l O\left(|u_h|^3\right).$$

Bernstein's inequality, (5.6), (5.7), (5.11), (5.12), and interpolation imply

$$\|\nabla P_l F(u)\|_{L_t^1 L_x^2} \lesssim \|\nabla u_l\|_{L_t^2 L_x^4}^2 \|\nabla u_l\|_{L_t^\infty L_x^2}$$
$$+ \|u_h\|_{L_{t,x}^3} \|\nabla u_l\|_{L_t^2 L_x^4}^{4/3} \|\nabla u_l\|_{L_t^\infty L_x^2}^{2/3} + \|u_h\|_{L_{t,x}^3}^3$$
$$\lesssim C_0 \eta_1 \eta_3 + C_0 \eta_1 \eta_3^{2/3} + C_0 \eta_1 \lesssim C_0 \eta_1. \tag{5.49}$$

Then by (5.6),

$$(5.44) \lesssim \eta_3^2 C_0 \eta_1. \tag{5.50}$$

Now, by Hardy's inequality,

$$\sup_y \left\| \frac{1}{|x-y|} |u_l(t,x)|^3 \right\|_{L_x^2} \lesssim \left\| |\nabla |u_l|^3 \right\|_{L_x^2}, \tag{5.51}$$

so the first term on the right-hand side of (5.49) and (5.50) imply

$$(5.45) \lesssim \eta_3^3 C_0 \eta_1.$$

Inequalities (5.49) and (5.51) also imply

$$(5.46) \lesssim \|u_h\|_{L_t^\infty L_x^2}^3 \|\nabla P_l F(u_h)\|_{L_t^1 L_x^2} \lesssim \eta_3^3 C_0 \eta_1.$$

Finally, by the Hardy–Littlewood–Sobolev theorem, along with (5.6), (5.7), (5.11), and (5.12), we get

$$(5.47) \lesssim \|u_h\|_{L_{t,x}^3}\|u_h\|_{L_t^\infty L_x^2}\|u_l\|_{L_{t,x}^\infty}\|u_h\|_{L_t^\infty L_x^4}\|u_h\|_{L_{t,x}^3}^2 \lesssim \eta_3^2 C_0 \eta_1.$$

This implies the bootstrap estimate that

$$(5.39) \lesssim C_0 \eta_1 \quad \Rightarrow \quad (5.39) \ll C_0 \eta_1.$$

This finally completes the proof of Proposition 5.2. $\square$

Step 5 of the proof of Theorem 5.1 in dimension $d = 4$ is relatively straightforward. Indeed, by Proposition 5.6,

$$N(t) \sim 1 \quad \text{for all } t \in \mathbf{R},$$

and by Proposition 5.2,

$$\int_\mathbf{R} \int \int \frac{|u_h(t,y)|^2 |u_h(t,x)|^2}{|x-y|^3} dx\,dy\,dt + \int_\mathbf{R} \int \int \frac{|u_h(t,y)|^2 |u_h(t,x)|^4}{|x-y|} dx\,dy\,dt$$
$$\lesssim 1.$$

Then by (3.118),

$$\||\nabla|^{-1/4} u_h\|_{L_{t,x}^4(\mathbf{R} \times \mathbf{R}^4)} \lesssim 1. \tag{5.52}$$

Now by Strichartz estimates and the fact that $N(t) \sim 1$ for all $t \in \mathbf{R}$, for any $T < \infty$,

$$\|\nabla u_h\|_{L_t^2 L_x^4([-T,T] \times \mathbf{R}^4)} \lesssim T^{1/2}. \tag{5.53}$$

Interpolating (5.52) and (5.53),

$$\|u_h\|_{L_t^{20/9} L_x^4([-T,T] \times \mathbf{R}^4)} \lesssim T^{2/5}.$$

But since T may be arbitrarily large this shows that there exists a sequence $t_n \in \mathbf{R}$ such that $u(t_n) \to 0$ in $L_x^4(\mathbf{R}^4)$. Then almost-periodicity implies that $u(t_n) \to 0$ in $\dot{H}^1$, from which we deduce that $u \equiv 0$. This finally completes the proof of Theorem 5.1 when $d = 4$. $\square$

5.2 Three-Dimensional Energy-Critical Problem

Returning to the three-dimensional problem, recall that the proof of Theorem 5.1 left off with the statement, but not the proof, of Proposition 5.7.

Proposition 5.9 *Suppose $d = 3$. If u is an almost-periodic solution on the maximal interval I, then $N(t) \leq C(\eta_1, \eta_3, \eta_4)$ for all $t \in I$.*

Proof If J is an interval with $N(t) \leq C_1(\eta_1, \eta_3, \eta_4)$ and $J \subset I$,

$$\int_J N(t)^2 \, dt \lesssim_{\eta_1, \eta_3, \eta_4} \int_J N(t)^{-1} \, dt \lesssim \|u_h\|^4_{L^4_{t,x}(I \times \mathbf{R}^3)} \lesssim C_0 \eta_1. \tag{5.54}$$

Indeed, suppose I_0 is an interval satisfying

$$\|u\|_{L^{10}_{t,x}(I_0 \times \mathbf{R}^3)} = 1. \tag{5.55}$$

By (3.60), (3.61), the Sobolev embedding theorem, Hölder's inequality, interpolation, and the Littlewood–Paley theorem, for nonzero u and $\eta > 0$ small and fixed, we have

$$\left\|P_{\leq C(\eta)N(t)} u_h(t)\right\|^6_{L^6_x(\mathbf{R}^3)} \lesssim \left\|P_{\leq C(\eta)N(t)} u_h\right\|^2_{L^\infty_x(\mathbf{R}^3)} \left\|P_{\leq C(\eta)N(t)} u_h\right\|^4_{L^4_x(\mathbf{R}^3)}$$
$$\lesssim C(\eta) N(t) \|u_h(t)\|^4_{L^4_x(\mathbf{R}^3)}.$$

By (3.60), (3.61), concentration compactness, and $N(t_1) \sim_{I_0} N(t_2)$ for any $t_1, t_2 \in I_0$,

$$\int_{I_0} N(t)^{-1} \, dt \lesssim \|u_h\|^4_{L^4_{t,x}(I_0 \times \mathbf{R}^3)}.$$

Then by (3.64),

$$\|u\|_{\dot{S}^1(J \times \mathbf{R}^3)} \lesssim C(\eta_1, \eta_3, \eta_4). \tag{5.56}$$

Next, recall the estimate on low frequencies in Lemma 5.5.

Lemma 5.10 *If (5.8) holds, then for $j \leq 0$ and any $\delta > 0$,*

$$\|P_{\leq j} u\|_{\dot{S}^1(I \times \mathbf{R}^3)} \lesssim_\delta \left(\min_{t \in I} \|P_{\leq 0} u(t)\|_{\dot{H}^1(\mathbf{R}^3)}\right) + 2^{j(\frac{3}{2} - \delta)} C_0^{1/2} \eta_1^{1/2}.$$

Suppose there exist $t_{\min}, t_{\text{evac}} \in I$ such that $N(t_{\min}) \leq \frac{2C(\eta_3)}{\eta_3}$, which by (5.33) implies

$$\|u_h(t_{\min})\|_{L^2} \geq \eta \frac{\eta_3}{C(\eta_3)}, \tag{5.57}$$

and $N(t_{\text{evac}}) \geq C(\eta_5)$. Then (5.34) holds at t_{evac} and thus by the continuity of $N(t)$, there exists $t_{\min} < t_1 < t_{\text{evac}}$ such that $N(t_1) = C_1(\eta_1, \eta_3, \eta_4)$ and $N(t) \leq C_1(\eta_1, \eta_3, \eta_4)$ for all $t_{\min} \leq t \leq t_1$.

5.2 Three-Dimensional Energy-Critical Problem

Again let $u_{h'} = P_{\geq j}u$, $u = u_{h'} + u_{l'}$ for some $j \leq 0$. By (1.112),

$$\frac{d}{dt}\int |u_{h'}(t,x)|^2 dx = \int \{P_{h'}F(u), u_{h'}\}_m dx$$

$$= \int \{P_{h'}(F(u) - F(u_{h'}) - F(u_{l'})), u_{h'}\}_m dx$$

$$- \int \{P_{l'}F(u_{h'}), u_{h'}\}_m dx + \int \{P_{h'}F(u_{l'}), u_{h'}\}_m dx. \tag{5.58}$$

Let $J = [t_{\min}, t_1]$. By Bernstein's inequality and Lemma 5.10, for any $\delta > 0$,

$$\int_J \int |\{P_{h'}F(u_{l'}), u_{h'}\}_m| \, dx \, dt$$

$$\lesssim_\delta 2^{-j} \|u_{l'}\|_{S^1(I \times \mathbf{R}^3)}^5 \|u_{h'}\|_{L_t^\infty L_x^2(I \times \mathbf{R}^3)}$$

$$\lesssim 2^{-2j}\left(\left(\inf_{t \in I}\|u_{\leq 0}(t)\|_{\dot{H}^1(\mathbf{R}^3)}\right) + C_0^{1/2}\eta_1^{1/2} 2^{j(3/2-\delta)}\right)^5. \tag{5.59}$$

Next,

$$\int_J \int |\{P_{h'}(F(u) - F(u_{h'}) - F(u_{l'})), u_{h'}\}_m| \, dx \, dt$$

$$\lesssim \int_J \int |u_{h'}|^2 |u_{l'}||u|^3 \, dx \, dt$$

$$\lesssim \int_J \int |u_{h'}|^5 |u_{l'}| \, dx \, dt + \int_J \int |u_{h'}|^2 |u_{l'}|^4 \, dx \, dt. \tag{5.60}$$

Choose j such that $2^j \sim \eta_4$ and j_5 such that $2^{j_5} \sim \eta_5^{1/2}$. Splitting $u_{l'} = u_{<j_5} + u_{j_5 \leq \cdot \leq j}$ and $u_{h'} = u_{j < \cdot < 0} + u_{\geq 0}$, by the Sobolev embedding theorem, Lemma 5.10, and Bernstein's inequality, since $|u_{h'}|^5 \lesssim |u_{j < \cdot < 0}|^5 + |u_{\geq 0}|^5$,

$$(5.60) \lesssim \|u_{h'}\|_{L_t^\infty L_x^2(J \times \mathbf{R}^3)}^2 \|u_{l'}\|_{L_t^4 L_x^\infty(J \times \mathbf{R}^3)}^4$$

$$+ \|u_h\|_{L_t^2 L_x^6(J \times \mathbf{R}^3)} \|u_h\|_{L_{t,x}^4(J \times \mathbf{R}^3)}^2 \|u\|_{L_t^\infty L_x^6(J \times \mathbf{R}^3)}^2 \|u_{<j_5}\|_{L_{t,x}^\infty(J \times \mathbf{R}^3)}$$

$$+ \|u_h\|_{L_{t,x}^4(J \times \mathbf{R}^3)}^2 \|u\|_{L_t^\infty L_x^6(J \times \mathbf{R}^3)}^3 \|u_{j_5 \leq \cdot \leq j}\|_{L_t^2 L_x^\infty(J \times \mathbf{R}^3)}$$

$$+ \|u_{j \leq \cdot \leq 0}\|_{L_t^6 L_x^{18/7}(J \times \mathbf{R}^3)}^2 \|u_{j \leq \cdot \leq 0}\|_{L_t^6 L_x^{18}(J \times \mathbf{R}^3)}^3 \|u_{l'}\|_{L_t^6 L_x^{18}(J \times \mathbf{R}^3)}. \tag{5.61}$$

By the Sobolev embedding theorem, Bernstein's inequality, the Cauchy–Schwarz inequality, and the Littlewood–Paley theorem,

$$\|u(t)\|_{L_x^\infty}^4$$
$$\lesssim \sum_{j_1 \le j_2 \le j_3 \le j_4} 2^{(j_1-j_3)/2} 2^{(j_2-j_4)/2} \left(2^{-j_1/2}\|P_{j_1}u(t)\|_{L_x^\infty}\right)\left(2^{-j_2/2}\|P_{j_2}u(t)\|_{L_x^\infty}\right)$$
$$\times \left(2^{j_3/2}\|P_{j_3}u(t)\|_{L_x^\infty}\right)\left(2^{j_4/2}\|P_{j_4}u(t)\|_{L_x^\infty}\right)$$
$$\lesssim \sum_j \|P_j u(t)\|_{\dot{H}^1(\mathbf{R}^3)}^2 \left(\sup_j 2^j \|P_j u(t)\|_{L_x^6(\mathbf{R}^3)}\right)^2.$$

Therefore,
$$\|u\|_{L_t^4 L_x^\infty(I\times\mathbf{R}^3)}^4 \lesssim \|u\|_{L_t^\infty \dot{H}^1(I\times\mathbf{R}^3)}^2 \|u\|_{\dot{S}^1(I\times\mathbf{R}^3)}^2, \tag{5.62}$$

and by Bernstein's inequality, Lemma 5.10, and (5.56),

$$(5.61) \lesssim \frac{1}{\eta_4^2}\left(\inf_{t\in I}\|u_{\le 0}(t)\|_{\dot{H}^1(\mathbf{R}^3)} + C_0^{1/2}\eta_1^{1/2}\eta_4^{3/2-\delta}\right)^4$$
$$+ \eta_5^{1/4} C(\eta_1,\eta_3,\eta_4) C_0^{1/2}\eta_1^{1/2}$$
$$+ C_0^{1/2}\eta_1^{1/2}\left(\eta_5^{-1/4}\left(\inf_{t\in I}\|u_{\le 0}(t)\|_{\dot{H}^1(\mathbf{R}^3)}\right) + \eta_4^{1-\delta}C_0^{1/2}\eta_1^{1/2}\right)$$
$$+ \sum_{j\le k\le 0} 2^{-2k}\left(\left(\inf_{t\in I}\|u_{\le 0}(t)\|_{\dot{H}^1(\mathbf{R}^3)}\right) + C_0^{1/2}\eta_1^{1/2} 2^{k(\frac{3}{2}-\delta)}\right)^5$$
$$\times \left(\left(\inf_{t\in I}\|u_{\le 0}(t)\|_{\dot{H}^1(\mathbf{R}^3)}\right) + C_0^{1/2}\eta_1^{1/2}\eta_4^{3/2-\delta}\right). \tag{5.63}$$

Finally,
$$\int_J\int \{P_{l'}F(u_{h'}), u_{h'}\}_m\, dx\, dt$$
$$= \int_J\int \{F(u_{h'}), P_{l'}u_{h'}\}_m\, dx\, dt = \int_J\int \{F(u_{h'}), P_{h'}u_{l'}\}_m\, dx\, dt$$
$$\lesssim \int_J\int |u_{h'}|^5 |u_{l'}|\, dx\, dt. \tag{5.64}$$

The estimate of (5.64) may be controlled by some of the terms in (5.61). Then (5.58), (5.59), (5.63), and (5.64) imply that for all $t \in J$,

$$\|u_{h'}(t,x)\|_{L_x^2(\mathbf{R}^3)} \ge \frac{\eta}{2}\frac{\eta_3}{C(\eta_3)}. \tag{5.65}$$

By Bernstein's inequality and conservation of energy, (5.65) implies that for all $t \in J$,

$$\left\|P_{\eta_4 < \cdot \le \eta_4^{-1}} u(t)\right\|_{L_x^2(\mathbf{R}^3)} \ge \frac{1}{4} c(\eta_3). \tag{5.66}$$

5.2 Three-Dimensional Energy-Critical Problem

Then by (3.60) and (3.61), (5.66) implies that $N(t) \leq C_2(\eta_3, \eta_4)$. Crucially, $C_2(\eta_3, \eta_4)$ does not depend on $C_1(\eta_1, \eta_3, \eta_4)$, so choosing $C_1(\eta_1, \eta_3, \eta_4) \gg C_2(\eta_3, \eta_4)$ and making a bootstrap argument completes the proof of Proposition 5.9. $\square$

Moving on to step 4 when $d = 3$, we have the following result.

Proposition 5.11 *Suppose u is an almost-periodic solution to (5.2) when $d = 3$. Then since $N_* = 1$,*

$$\int_I \int |u_h(t,x)|^4 \, dx\, dt + \int_I \int \int_{|x-y| \leq 2R} \frac{|u_h(t,y)|^2 |u_h(t,x)|^6}{|x-y|} \, dx\, dy\, dt$$

$$\lesssim \|u_h\|_{L_t^\infty L_x^2(I \times \mathbf{R}^d)}^3 \|u_h\|_{L_t^\infty \dot{H}_x^1(I \times \mathbf{R}^d)} \tag{5.67}$$

$$+ \int_I \int \int_{|x-y| \leq 2R} \frac{|u_h(t,y)|^2 |u_l(t,x)|^5 |u_h(t,x)|}{|x-y|} \, dx\, dy\, dt \tag{5.68}$$

$$+ \sum_{j=0}^{4} \int_I \int \int_{|x-y| \leq 2R} |u_h(t,y)| \, |P_h O(u_h^j u_l^{5-j})(t,y)| \, |u_h(t,x)| \, |\nabla u_h(t,x)| \, dx\, dy\, dt \tag{5.69}$$

$$+ \int_I \int \int_{|x-y| \leq 2R} |u_h(t,y)| \, |P_l O(u_h^5)(t,y)| \, |u_h(t,x)| \, |\nabla u_h(t,x)| \, dx\, dy\, dt \tag{5.70}$$

$$+ \|\nabla u_h\|_{L_t^\infty L_x^2(I \times \mathbf{R}^3)} \Big(\|u_h\|_{L_t^\infty L_x^6(I \times \mathbf{R}^3)}^3 \|u_h\|_{L_t^\infty L_x^2(I \times \mathbf{R}^3)} \|u_l\|_{L_t^\infty L_x^6(I \times \mathbf{R}^3)}$$

$$+ \|u_l\|_{L_t^\infty L_x^6(I \times \mathbf{R}^3)}^5 + \|u_h\|_{L_t^\infty L_x^6(I \times \mathbf{R}^3)}^{9/2} \|u_h\|_{L_t^\infty L_x^2(I \times \mathbf{R}^3)}^{1/2} \Big)$$

$$\times \int_I \left(\sup_{x \in \mathbf{R}^3} \int_{B(x,2R)} |u_h(t,y)|^2 \, dy \right) dt \tag{5.71}$$

$$+ \frac{1}{R^3} \|u_h\|_{L_t^\infty L_x^2(I \times \mathbf{R}^d)}^2 \int_I \left(\sup_{x \in \mathbf{R}^3} \int_{B(x,2R)} |u_h(t,y)|^2 \, dy \right) dt \tag{5.72}$$

$$+ \frac{1}{R} \int_I \int_{|x-y| \leq 2R} |u_h(t,y)|^2 \left(|\nabla u_h(t,x)|^2 + |u_h(t,x)|^6 \right) dx\, dy\, dt. \tag{5.73}$$

Proof The function u_h solves the initial value problem

$$(i\partial_t + \Delta) u_h = P_h F(u) = P_h \left(|u|^4 u \right) = \mathcal{N}.$$

By (3.104)–(3.108) and the fundamental theorem of calculus, if $h(x)$ is a function such that $|\nabla h(x)|$ is uniformly bounded, then

$$\int_I \int (-\Delta\Delta h(x-y))|u_h(t,x)|^2|u_h(t,y)|^2 \, dx \, dy \, dt \tag{5.74}$$

$$+4\int_I \int h_{jk}(x-y)\operatorname{Re}(\partial_j \bar{u}_h \partial_k u_h)(t,x)|u_h(t,y)|^2 \, dx \, dy \tag{5.75}$$

$$-4\int h_{jk}(x-y)\operatorname{Im}(\bar{u}_h \partial_j u_h)(t,x)\operatorname{Im}(\bar{u}_h \partial_k u_h)(t,y) \, dx \, dy \, dt \tag{5.76}$$

$$+4\int h_j(x-y)\operatorname{Im}(\bar{u}_h \partial_j u_h)(t,x)\{\mathcal{N}, u_h\}_m(t,y) \, dx \, dy \, dt \tag{5.77}$$

$$+2\int h_j(x-y)\{\mathcal{N}, u_h\}^j_p(t,x)|u_h(t,y)|^2 \, dx \, dy \, dt$$

$$\lesssim \|u_h\|^3_{L_t^\infty L_x^2(I\times\mathbb{R}^3)} \|u_h\|_{L_t^\infty H^1(I\times\mathbb{R}^3)}.$$

Rather than taking $h(x) = |x|$, as was done in the case $d = 4$, choose $h(x) = |x|\chi\left(\frac{|x|}{R}\right)$, where χ is a smooth, nonnegative bump function supported on $0 \leq |x| \leq 2$ and $\chi(|x|) = 1$ for $0 \leq |x| \leq 1$. Let $\tilde{\chi}$ denote some other bump function with the same properties as χ. The following are straightforward to calculate:

$$h_j(x) = \frac{x^j}{|x|}\tilde{\chi}(|x|), \tag{5.78}$$

$$h_{jk}(x) = \frac{1}{|x|}\left(\delta_{jk} - \frac{x_j}{|x|}\frac{x_k}{|x|}\right)\tilde{\chi}(|x|) + \frac{x^j}{|x|}\frac{x^k}{|x|}\tilde{\chi}'(|x|), \tag{5.79}$$

$$\Delta h(x) = \frac{2}{|x|}\tilde{\chi}(|x|) + \tilde{\chi}'(|x|), \tag{5.80}$$

$$-\Delta\Delta h(x) = 8\pi\delta(x) - \frac{4}{|x|}\tilde{\chi}''(|x|) - \tilde{\chi}'''(|x|). \tag{5.81}$$

By (5.78)–(5.81) and the support properties of χ,

$$(5.74) = \int_I \int \int (-\Delta\Delta h(x-y))|u_h(t,x)|^2|u_h(t,y)|^2 \, dx \, dy \, dt$$

$$= 8\pi \int_I \int |u_h(t,x)|^4 \, dx \, dt$$

$$+ O\left(\frac{1}{R^3}\int_I \int \int_{R\leq|x-y|\leq 2R}|u_h(t,x)|^2|u_h(t,y)|^2 \, dx \, dy \, dt\right)$$

$$= 8\pi \int_I \int |u_h(t,x)|^4 \, dx \, dt + (5.72).$$

Next, (3.109)–(3.110) still imply that (5.79) for h_{jk} is positive definite, so that

$$(5.75) + (5.76) \geq -\frac{1}{R}\int_I \int \int_{R\leq|x-y|\leq 2R}|u_h(t,y)|^2|\nabla u_h(t,x)|^2 \, dx \, dy \, dt$$

$$\geq -(5.73).$$

5.2 Three-Dimensional Energy-Critical Problem

Moreover, by (1.113),

$$\{\mathcal{N}, u_h\}_m = \{P_h F(u) - F(u_h), u_h\}_m$$
$$= \{P_h(F(u) - F(u_h))\}_m + \{P_l F(u), u_h\}_m. \quad (5.82)$$

Since $h_j(x)$ is uniformly bounded, (5.82) implies

$$|(5.77)| \lesssim (5.69) + (5.70).$$

By (1.118),

$$\{\mathcal{N}, u_h\}_p$$
$$= \{P_h F(u), u_h\}_p = \{F(u), u_h\}_p - \{P_l F(u), u_h\}_p$$
$$= \{F(u), u\}_p - \{F(u), u_l\}_p - \{P_l F(u), u_h\}_p$$
$$= \{F(u), u\}_p - \{F(u_l), u_l\}_p - \{F(u) - F(u_l), u_l\}_p - \{P_l F(u), u_h\}_p$$
$$= -\frac{2}{3}\nabla(|u|^6 - |u_l|^6) - \{F(u) - F(u_l), u_l\}_p - \{P_l F(u), u_h\}_p.$$

Integrating by parts, by (5.78)–(5.81), Hölder's inequality, and the support properties of χ and $\widetilde{\chi}$, we have

$$-\frac{2}{3}\int\int h_j(x-y)|u_h(t,y)|^2 \partial_j(|u(t,x)|^6 - |u_l(t,x)|^6)dxdy$$
$$= \frac{2}{3}\iint\left(\frac{2}{|x-y|}\widetilde{\chi}(x-y) + \widetilde{\chi}'(x-y)\right)|u_h(t,y)|^2\left(|u(t,x)|^6 - |u_l(t,x)|^6\right)dxdy$$
$$\geq \frac{2}{3}\int\int_{|x-y|\leq 2R} \frac{|u_h(t,y)|^2|u_h(t,x)|^6}{|x-y|}dxdy$$
$$- C\int\int \frac{|u_h(t,y)|^2|u_l(t,x)|^5|u_h(t,x)|}{|x-y|}dxdy$$
$$- \frac{C}{R}\int\int_{|x-y|\leq 2R}|u_h(t,y)|^2|u_h(t,x)|^6 dxdy.$$

Now take $\{P_l F(u), u_h\}_p$. By Bernstein's inequality, we get

$$\int_I \int\int |u_h(t,y)|^2 h_j(x-y)\{P_l F(u), u_h\}_p^j(t,x)dxdydt$$
$$\lesssim \|\{P_l F(u), u_h\}_p\|_{L_T^\infty L_x^1(I\times\mathbf{R}^3)}\left(\int_I \left(\sup_{x\in\mathbf{R}^3}\int_{B(x,2R)}|u_h(t,y)|^2 dy\right)dt\right)$$
$$\lesssim \|\nabla u_h\|_{L_T^\infty L_x^2(I\times\mathbf{R}^3)}\|P_l F(u)\|_{L_T^\infty L_x^2(I\times\mathbf{R}^3)}\left(\int_I \left(\sup_{x\in\mathbf{R}^3}\int_{B(x,2R)}|u_h(t,y)|^2 dy\right)dt\right)$$
$$+ \|u_h\|_{L_T^\infty L_x^2(I\times\mathbf{R}^3)}\|\nabla P_l F(u)\|_{L_T^\infty L_x^2(I\times\mathbf{R}^3)}\left(\int_I \left(\sup_{x\in\mathbf{R}^3}\int_{B(x,2R)}|u_h(t,y)|^2 dy\right)dt\right)$$
$$\lesssim \|\nabla u_h\|_{L_T^\infty L_x^2(I\times\mathbf{R}^3)}\|P_l F(u)\|_{L_T^\infty L_x^2(I\times\mathbf{R}^3)}\left(\int_I \left(\sup_{x\in\mathbf{R}^3}\int_{B(x,2R)}|u_h(t,y)|^2 dy\right)dt\right).$$
$$(5.83)$$

By the Sobolev embedding theorem,

$$\sum_{j=0}^{4} \|P_l O(u_h^j u_l^{5-j})\|_{L^2(\mathbf{R}^3)} \lesssim \|u_h^4\|_{L^1(\mathbf{R}^3)} \|u_l\|_{L^\infty(\mathbf{R}^3)} + \|u_l\|_{L^{10}(\mathbf{R}^3)}^5$$

$$\lesssim \|u_h\|_{L_x^6(\mathbf{R}^3)}^3 \|u_h\|_{L_x^2(\mathbf{R}^3)} \|u_l\|_{L_x^6(\mathbf{R}^3)} + \|u_l\|_{L_x^6(\mathbf{R}^3)}^5 \tag{5.84}$$

and

$$\|P_l O(u_h^5)\|_{L^2(\mathbf{R}^3)} \lesssim \|u_h\|_{L^6(\mathbf{R}^3)}^{9/2} \|u_h\|_{L^2(\mathbf{R}^3)}^{1/2}.$$

Therefore, $(5.83) \lesssim (5.71)$.

Finally, take

$$\{F(u) - F(u_l), u_l\}_P. \tag{5.85}$$

By (5.48),

$$\{F(u) - F(u_l), u_l\}_P = O((F(u) - F(u_l))\nabla u_l) + \nabla O((F(u) - F(u_l))u_l). \tag{5.86}$$

Make the split

$$F(u) - F(u_l) = \sum_{j=1}^{5} O(u_h^j u_l^{5-j}).$$

By the Sobolev embedding theorem, Bernstein's inequality, and analysis similar to (5.84)–(5.85),

$$\sum_{j=1}^{5} \int_I \int \int \tilde{\chi}(|x-y|) \frac{(x-y)}{|x-y|} \cdot O(u_h^j u_l^{5-j} \nabla u_l)(t,x) |u_h(t,y)|^2 dx dy dt \lesssim (5.71).$$

Observe that by the Sobolev embedding theorem and the support properties of u_l and (5.6),

$$\|\nabla u_l\|_{L^\infty} \lesssim \|\nabla u_l\|_{L^2} \leq \|\nabla u_h\|_{L^2} \quad \text{and} \quad \|\nabla u_l\|_{L^6} \lesssim \|u_l\|_{L^6}. \tag{5.87}$$

Taking the second term in (5.86) and integrating by parts, for any $\varepsilon > 0$,

$$\int_I \int \int \tilde{\chi}(|x-y|) \frac{(x-y)}{|x-y|} \cdot \nabla O((F(u)-F(u_l))u_l)(t,x) |u_h(t,y)|^2 dx dy dt$$

$$\leq C(\varepsilon)(5.68) + \varepsilon \int_I \int \int_{|x-y| \leq 2R} \frac{|u_h(t,x)|^6 |u_h(t,y)|^2}{|x-y|} dx dy dt. \tag{5.88}$$

Absorbing the second term in (5.88) into

$$\int_I \int \int_{|x-y| \leq 2R} \frac{|u_h(t,x)|^6 |u_h(t,y)|^2}{|x-y|} dx dy dt$$

completes the proof of Proposition 5.11. $\square$

5.2 Three-Dimensional Energy-Critical Problem

Using this estimate, it is possible to prove Proposition 5.2, the frequency-localized interaction Morawetz estimate, in dimension $d = 3$.

Proof of Proposition 5.2 Proposition 5.2 is proved by showing that the average over dyadic $1 \leq R \leq C(\eta_1, \eta_2) 2^{J(\eta_2)}$ is $\ll \eta_1$ under the bootstrap assumption (5.8). First take (5.67). As usual (5.67) $\lesssim \eta_3^3$, so the average is $\ll \eta_1$.

Next take (5.68). By (5.5), (5.18), and (5.62),

$$\|u_l\|_{L_t^4 L_x^\infty(I \times \mathbf{R}^3)} \lesssim \|\nabla u_l\|_{L_t^2 L_x^6(I \times \mathbf{R}^3)}^{1/2} \|\nabla u_l\|_{L_t^\infty L_x^2(I \times \mathbf{R}^3)}^{1/2} \lesssim (C_0^{1/4} \eta_1^{1/4}) \eta_3^{1/2}, \tag{5.89}$$

and therefore by Young's inequality and (5.18),

$$\int_I \int \int_{|x-y| \leq 2R} \frac{|u_h(t,y)|^2 |u_l(t,x)|^5 |u_h(t,x)|}{|x-y|} dx\,dy\,dt$$

$$\lesssim R^2 \|u_h\|_{L_{t,x}^4(I \times \mathbf{R}^3)}^2 \|u_h\|_{L_t^\infty L_x^2(I \times \mathbf{R}^3)} \|u_l\|_{L_t^4 L_x^\infty(I \times \mathbf{R}^3)}^2 \|u_l\|_{L_{t,x}^\infty(I \times \mathbf{R}^3)}^3$$

$$\lesssim R^2 (C_0 \eta_1) \eta_3^4. \tag{5.90}$$

Averaging (5.90) over $1 \leq R \leq C(\eta_1, \eta_2) 2^{J(\eta_2)}$ implies

$$\frac{1}{J} \sum_{\substack{1 \leq R \leq \\ C(\eta_1,\eta_2) 2^{J(\eta_2)}}} (5.68) \ll \eta_1.$$

Next take (5.69). By (5.18), (5.89), and Young's inequality, we get

$$\sum_{j=1}^{4} \int_I \int \int_{|x-y| \leq 2R} |u_h(t,y)| \left|P_h O\big(u_h^j u_l^{5-j}\big)(t,y)\right| |u_h(t,x)| |\nabla u_h(t,x)|\,dx\,dy\,dt$$

$$\lesssim R^{3/4} \|\nabla u_h\|_{L_t^\infty L_x^2} \|u_h\|_{L_{t,x}^4}^3 \|u_l\|_{L_t^4 L_x^\infty} \|u\|_{L_t^\infty L_x^6}^3 \lesssim R^{3/4} (C_0 \eta_1) \eta_3^{1/2}.$$

Bernstein's inequality, (5.18), and (5.89), yield

$$\int_I \int \int_{|x-y| \leq 2R} |u_h(t,y)| \left|P_h O\big(u_l^5\big)(t,y)\right| |u_h(t,x)| |\nabla u_h(t,x)|\,dx\,dy\,dt$$

$$\lesssim \|u_h\|_{L_t^\infty L_x^2} \|\nabla u_h\|_{L_t^\infty L_x^2} \|\nabla u_l\|_{L_t^2 L_x^6} \|u_l\|_{L_t^4 L_x^\infty}^2 \|u_l\|_{L_t^\infty L_x^6}^2 \|u_h\|_{L_t^\infty L_x^2}$$

$$\lesssim (C_0 \eta_1) \eta_3^5.$$

Therefore,

$$\frac{1}{J} \sum_{\substack{1 \leq R \leq \\ C(\eta_1,\eta_2) 2^{J(\eta_2)}}} (5.69) \lesssim \frac{1}{J} \sum_{\substack{1 \leq R \leq \\ C(\eta_1,\eta_2) 2^{J(\eta_2)}}} (C_0 \eta_1) \left(\eta_3^5 + R^{3/4} \eta_3^{1/2}\right) \ll \eta_1.$$

Next take (5.70). By Bernstein's inequality and (5.18),

$$\|u_j\|_{L^4_t L^3_x(I\times\mathbf{R}^3)} \lesssim \|u_j\|^{1/2}_{L^\infty_t L^2_x(I\times\mathbf{R}^3)} \|u_j\|^{1/2}_{L^2_t L^6_x(I\times\mathbf{R}^3)}$$

$$\lesssim 2^{-j/2} \|\nabla u_j\|^{1/2}_{L^\infty_t L^2_x(I\times\mathbf{R}^3)} (C_0\eta_1)^{1/4},$$

which by (5.5) implies

$$\|u_h\|_{L^4_t L^3_x(I\times\mathbf{R}^3)} \lesssim (C_0\eta_1)^{1/4} \eta_3^{1/2}.$$

Then by the Sobolev embedding theorem and Young's inequality,

$$\int_I \int \int_{|x-y|\leq 2R} |u_h(t,y)| |P_l O(u_h^5)(t,y)| |u_h(t,x)| |\nabla u_h(t,x)| \, dx\, dy\, dt$$

$$\lesssim R^{1/2} \|\nabla u_h\|_{L^\infty_t L^2_x} \|u_h\|_{L^4_t L^3_x} \|u_h\|^3_{L^4_{t,x}} \|u_h\|^3_{L^\infty_t L^6_x}$$

$$\lesssim R^{1/2} (C_0\eta_1) \eta_3^{1/2}.$$

Averaging,

$$\frac{1}{J} \sum_{\substack{1\leq R\leq \\ C(\eta_1,\eta_2)2^{J(\eta_2)}}} (5.70) \ll \eta_1.$$

It remains to prove

$$\frac{1}{J} \sum_{\substack{1\leq R\leq \\ C(\eta_1,\eta_2)2^{J(\eta_2)}}} (5.71) + (5.72) + (5.73) \ll \eta_1.$$

To prove this let

$$Y = \frac{1}{J} \sum_{1\leq R\leq 2^J} \int_I \int \int_{|x-y|\leq 2R} \frac{|u_h(t,y)|^2 |u_h(t,x)|^6}{|x-y|} \, dx\, dy\, dt \qquad (5.91)$$

and

$$W = \sup_{\substack{1\leq R\leq \\ C(\eta_1,\eta_2)2^J}} \frac{1}{R} \int_I \left(\sup_{x\in\mathbf{R}^3} \int_{B(x,3R)} |u_h(t,y)|^2 dy \right) dt.$$

Since $R \leq C(\eta_1,\eta_2) 2^{J(\eta_2)}$, $\eta_3 R \ll 1$, (5.6), and (5.7), it follows that

$$\frac{1}{J} \sum_{\substack{1\leq R\leq \\ C(\eta_1,\eta_2)2^{J(\eta_2)}}} (5.71) + (5.72) \lesssim \eta_1^{1/100} W.$$

5.2 Three-Dimensional Energy-Critical Problem

The second component of (5.73), namely

$$\frac{1}{R}\int\int_{|x-y|\leq 2R}|u_h(t,y)|^2|u_h(t,x)|^6 dxdy,$$

is clearly controlled by (5.91). For the first component of (5.73), because $N(t) \geq 1$ for all $t \in I$, (3.60) and (3.61) imply that for all $t \in I$,

$$\int_{|x-x(t)|\geq C(\eta_1)}\int_{|x-y|\leq 2R}|\nabla u_h(t,x)|^2|u_h(t,y)|^2 dxdy$$

$$\lesssim \eta_1\left(\sup_{x\in\mathbf{R}^3}\int_{B(x,2R)}|u_h(t,y)|^2 dy\right).$$

Finally, (3.60), (3.61), and (3.64), imply that for $R \geq 1$,

$$\int_{|x-x(t)|\leq C(\eta_1)}\int_{|x-y|\leq 2R}|\nabla u_h(t,x)|^2|u_h(t,y)|^2 dxdy$$

$$\lesssim \int_{|x-y|\leq C(\eta_1,\eta_2)R}|u_h(t,x)|^6|u_h(t,y)|^2 dxdy.$$

Now

$$\frac{C(\eta_1,\eta_2)}{J}\sum_{\substack{1\leq R\leq \\ C(\eta_1,\eta_2)2^{J(\eta_2)}}}\frac{1}{R}\int\int_I\int_{|x-y|\leq C(\eta_1,\eta_2)R}|u_h(t,y)|^2|u_h(t,x)|^6 dxdy$$

$$\lesssim \frac{C(\eta_1,\eta_2)^2}{J}\sum_{\substack{1\leq R\leq \\ C(\eta_1,\eta_2)^2 2^{J(\eta_2)}}} A_R,$$

where

$$A_R = \frac{1}{R}\int\int_I\int_{|x-y|\leq 2R}|u_h(t,y)|^2|u_h(t,x)|^6 dxdydt.$$

For any R',

$$\sum_{1\leq R\leq R'}A_R \lesssim \int\int_I\int_{|x-y|\leq R'}\frac{|u_h(t,y)|^2|u_h(t,x)|^6}{|x-y|}dxdydt,$$

so, averaging (5.73),

$$\frac{1}{J}\sum_{\substack{1\leq R'\leq \\ C(\eta_1,\eta_2)2^{J(\eta_2)}}}\sum_{1\leq R\leq R'}A_R$$

$$\lesssim \frac{1}{J}\sum_{\substack{1\leq R'\leq \\ C(\eta_1,\eta_2)2^{J(\eta_2)}}}\int\int_I\int_{|x-y|\leq R'}\frac{|u_h(t,y)|^2|u_h(t,x)|^6}{|x-y|}dxdydt \lesssim Y. \quad (5.92)$$

Now choose some $1 \leq J_0(\eta_1,\eta_2) \leq J$. For each $1 \leq R \leq 2^{J-J_0}$, there are $\geq J_0$ values of R' involving that value of R, so (5.92) implies

$$\frac{J_0}{J} \sum_{1 \leq R \leq 2^{J-J_0}} A_R \lesssim Y.$$

Therefore, for $J_0(\eta_1,\eta_2)$ sufficiently large,

$$\frac{C(\eta_1,\eta_2)^2}{J} \sum_{1 \leq R \leq 2^{J-J_0}} A_R \lesssim \eta_1^{1/100} Y.$$

Choosing $J(\eta_1,\eta_2)$ sufficiently large in the final step,

$$\frac{C(\eta_1,\eta_2)^2}{J} \sum_{\substack{2^{J-J_0} \leq R \leq \\ C(\eta_1,\eta_2)2^J}} \frac{1}{R} \int_I \int \int_{|x-y| \leq R} |u_h(t,y)|^2 |u_h(t,x)|^6 \, dx \, dy \, dt$$

$$\lesssim \frac{C(\eta_1,\eta_2,J_0)}{J} \sup_{\substack{1 \leq R \leq \\ C(\eta_1,\eta_2)2^J}} \frac{1}{R} \int_I \int \int_{|x-y| \leq R} |u_h(t,y)|^2 |u_h(t,x)|^6 \, dx \, dy \, dt$$

$$\lesssim \frac{C(\eta_1,\eta_2,J_0)}{J} \sup_{\substack{1 \leq R \leq \\ C(\eta_1,\eta_2)2^J}} \frac{1}{R} \left(\int_I \sup_{x \in \mathbf{R}^3} \left(\int_{B(x,R)} |u_h(t,y)|^2 \, dy \right) dt \right)$$

$$\lesssim \frac{1}{J} C(\eta_1,\eta_2,J_0) W \lesssim \eta_1^{1/100} W. \tag{5.93}$$

Therefore, by Proposition 5.11 and (5.89)–(5.93),

$$\int_I \int |u_h(t,x)|^4 \, dx \, dt + Y \lesssim \eta_1 + \eta_1^{1/100} Y + \eta_1^{1/100} W.$$

The term $\eta_1^{1/100} Y$ may be absorbed into the left-hand side, so to close the bootstrap it only remains to prove $W \lesssim C_0 \eta_1$.

Lemma 5.12 *Suppose that $\psi \in C_0^\infty(\mathbf{R}^3)$. Fix $R > 0$. Suppose u solves*

$$iu_t + \Delta u = F + G, \quad u(0,x) = u_0,$$

with $\|u_0\|_{L^2(\mathbf{R}^3)} + \|F\|_{L_t^2 L_x^{6/5}(J \times \mathbf{R}^3)} \lesssim M_1$, $\|G\|_{L_t^2 L_x^1(J \times \mathbf{R}^3)} \lesssim M_2$. *Then, for any* $\tilde{x}(t) : I \to \mathbf{R}^3$,

$$\frac{1}{R} \int_I \int \psi\left(\frac{x-\tilde{x}(t)}{R}\right) |u(t,x)|^2 \, dx \, dt \lesssim_\psi M_1^2 R + M_2^2.$$

5.2 Three-Dimensional Energy-Critical Problem

Assuming Lemma 5.12 for a moment, recall from the proof of Theorem 5.4 that $P_h F(u) = F + G$, with

$$\|\nabla F\|_{L_t^2 L_x^{6/5}(I \times \mathbf{R}^3)} \lesssim \eta_3^{1/2} \quad \text{and} \quad \|G\|_{L_t^2 L_x^1(I \times \mathbf{R}^3)} \lesssim C_0^{1/2} \eta_1^{1/2}.$$

Therefore,

$$\frac{1}{R}\int_I \psi\left(\frac{x-\tilde{x}(t)}{R}\right)|u_h(t,x)|^2 dx\, dt \lesssim C_0 \eta_1 + R\eta_3$$

and $W \lesssim C_0 \eta_1$. Because of the upper bound on R, $R\eta_3 \ll 1$, which closes the bootstrap and completes the proof of Proposition 5.2 when $d = 3$, assuming that Lemma 5.12 is true. □

Theorem 5.1 is an easy consequence of Proposition 5.2.

Proof of Theorem 5.1 Quickly moving on to step 5, Proposition 5.9 implies $1 \leq N(t) \leq C(\eta_1, \eta_4)$ for all $t \in \mathbf{R}$. Then by Proposition 5.2, and (5.54), if u is nonzero,

$$\int_{\mathbf{R}} N(t)^2 dt \lesssim_{\eta_1, \eta_3, \eta_4} C_0 \eta_1. \tag{5.94}$$

If u is a nonzero, almost-periodic solution, then (5.94) contradicts (4.81). Therefore, the only almost-periodic solution to (5.2) is $u \equiv 0$, which completes the proof of Theorem 5.1. □

Lemma 5.12 was proved in Colliander et al. (2008) and Killip and Visan (2012) using the double Duhamel lemma.

Proof of Lemma 5.12 For any $t \in I$, define the inner product

$$\langle f, g \rangle_\psi = \frac{1}{R}\int \psi\left(\frac{x-\tilde{x}(t)}{R}\right) \overline{f(x)} g(x)\, dx.$$

Remark This inner product does not define a Hilbert space, since $\langle f, f \rangle_\psi$ could be zero for nonzero f.

Now let $I = [t_-, t_+]$ and let $u^\pm$ be the solution to

$$(i\partial_t + \Delta)u^\pm = F, \quad u^\pm(t_\pm) = u(t_\pm).$$

By Strichartz estimates,

$$\|u^\pm\|_{L_t^2 L_x^6(I \times \mathbf{R}^3)} \lesssim M_1. \tag{5.95}$$

By Duhamel's principle, for any $t \in I$,

$$u(t) = u^-(t) - i \int_{t_- < s_- < t} e^{i(t-s_-)\Delta} G(s_-) ds_-$$

and

$$u(t) = u^+(t) + i \int_{t < s_+ < t_+} e^{i(t-s_+)\Delta} G(s_+) ds_+.$$

By (4.105), (5.95), and Hölder's inequality in space,

$$\int_{t_-}^{t_+} \langle u(t), u(t) \rangle_\psi dt$$

$$\lesssim R\|u(t_-)\|_{L_x^2(\mathbf{R}^3)}^2 + R\|u(t_+)\|_{L_x^2(\mathbf{R}^3)}^2 + R\|F\|_{L_t^2 L_x^{6/5}(I \times \mathbf{R}^3)}^2$$

$$+ \int_{t_-}^{t_+} \int_{t_-}^{t} \int_{t}^{t_+} \langle e^{i(t-s_-)\Delta} G(s_-), e^{i(t-s_+)\Delta} G(s_+) \rangle_\psi ds_- ds_+ dt.$$

Define the set

$$\Omega = \{(s_-, s_+) : t_- \leq s_- \leq t - R^2 \text{ or } t + R^2 \leq s_+ \leq t_+\}. \quad (5.96)$$

Using Hölder's inequality in space, the dispersive estimate (1.19), and Young's inequality, we obtain

$$\int_{t_-}^{t_+} \int_\Omega \langle e^{i(t-s_-)\Delta} G(s_-), e^{i(t-s_+)\Delta} G(s_+) \rangle ds_- ds_+ dt$$

$$\lesssim R^2 \int_{t_-}^{t_+} \int_{t_- < s_- \leq t - R^2} \int_{t + R^2 \leq s_+ < t_+} \frac{1}{|t - s_+|^{3/2}} \frac{1}{|t - s_-|^{3/2}}$$

$$\times \|G(s_+)\|_{L^1(\mathbf{R}^3)} \|G(s_-)\|_{L^1(\mathbf{R}^3)} ds_- ds_+ dt$$

$$\lesssim \|G\|_{L_t^2 L_x^1(\mathbf{R}^3)}^2 \lesssim M_2^2.$$

Next, compute the kernel of $e^{i\tau\Delta} \psi\left(\frac{x - \tilde{x}(t)}{R}\right) e^{is\Delta}$ for $0 < s, \tau < R^2$. Without loss of generality suppose $\tilde{x}(t) = 0$. By (1.18),

$$K(\tau, s, x, z) = \frac{c}{R} \frac{1}{\tau^{3/2} s^{3/2}} \int \int e^{-i\frac{|x-y|^2}{4\tau}} \psi\left(\frac{y}{R}\right) e^{-i\frac{|y-z|^2}{4s}} dy.$$

After completing the square in the y variable and then making stationary phase computations,

$$|K(\tau, s, x, z)| \lesssim \frac{1}{R} \frac{1}{(s+\tau)^{3/2}}.$$

Therefore, by Hölder's inequality in space,

$$\int_{t_-}^{t_+}\int_{t-R^2\le s_-\le t}\int_{t\le s_+\le t+R^2}\langle e^{i(t-s_-)\Delta}F(s_-), e^{i(t-s_+)\Delta}F(s_+)\rangle_\psi \, ds_- \, ds_+ \, dt$$

$$\lesssim \frac{1}{R}\int_{t_-<t<t_+}\int_{t-R^2\le s_-\le t}\int_{t\le s_+\le t_+}\frac{1}{(s_+-s_-)^{3/2}}\|F(s_+)\|_{L_x^1(\mathbf{R}^3)}\|F(s_-)\|_{L_x^1(\mathbf{R}^3)},$$

which by Young's inequality,

$$\lesssim \frac{1}{R}\int_{0\le s_-\le R^2}\frac{1}{s_-^{1/2}}\|F\|^2_{L_t^2 L_x^1(J\times\mathbf{R}^3)} \lesssim M_2^2.$$

This completes the proof of Lemma 5.12. □

5.3 The Mass-Critical Problem When $d = 1$

The mass-critical problem in dimensions $d = 1, 2$ should be regarded as extra credit, since the additional difficulties in low dimensions are purely technical in nature. In two dimensions the technical difficulties are due to the lack of an $L_t^2 L_x^\infty$ Strichartz estimate. Indeed, if the endpoint Strichartz estimate in (1.73) also held in two dimensions, the two-dimensional versions of (4.149) and (4.150) would follow exactly as in dimensions $d \ge 3$.

In dimensions $d = 1, 2$, (4.149) will be replaced by a norm based on the U_Δ^2-norm. Such spaces were first introduced for the study of the wave maps problem by Tataru (1998). Koch and Tataru (2007, 2012) applied these spaces to the Schrödinger equation. See Hadac et al. (2009) for a complete description of these spaces and Section 3.2 of Stein (1993) for an example of an atomic space.

Definition 5.13 (U^p space) Suppose $u \in U^p$. We say u is a U^p atom if there exists a sequence $\{t_k\} \nearrow$ satisfying

$$u = \sum_k 1_{[t_k, t_{k+1})} u_k$$

and

$$\sum \|u_k\|^p_{L^2(\mathbf{R}^d)} = 1.$$

Then define the norm

$$\|u(t)\|_{U^p(\mathbf{R}\times\mathbf{R}^d)} = \inf\{\Sigma_\lambda |c_\lambda| : u(t) = c_\lambda u_\lambda(t) \text{ for almost every } t \in \mathbf{R},$$
$$\text{where } u_\lambda(t) \text{ is a } U^p \text{ atom}\}.$$

We then say that U^p is an atomic space.

Theorem 5.14 *The space U^p is a Banach space.*

Proof It is clear from the definition that U^p is a vector space. Moreover, U^p obeys the triangle inequality and is complete. Also, $\|u\|_{U^p} = 0$ if and only if $u = 0$. □

Next, define U_Δ^p spaces.

Definition 5.15 (U_Δ^p spaces) If u is a continuous function in time, we may define the U_Δ^p-norm of u as

$$\|u\|_{U_\Delta^p(\mathbf{R}\times\mathbf{R}^d)} = \|e^{-it\Delta}u\|_{U^p(\mathbf{R}\times\mathbf{R}^d)}.$$

Remark As in the case of $X^{s,b}$ spaces, it is possible to define the space U_S^p for any $S = h(\nabla/i)$, where h is a polynomial. For example, in the KdV case $S = (\partial_x/i)^3$.

Under the conditions specified in (4.143)–(4.147) we now construct a Galilean norm.

Definition 5.16 (Galilean norm) For $0 \le i \le j \le k_0$, if $G_k^j \subset [a,b]$ is part of the partition of $[a,b]$ described in Definition 4.36, write

$$\|P_{\xi(t),i}u\|_{U_\Delta^2(G_k^j\times\mathbf{R}^d)}^2 = \sum_{G_\alpha^i \subset G_k^j} \|P_{\xi(G_\alpha^i),i-2\le\cdot\le i+2}u\|_{U_\Delta^2(G_\alpha^i\times\mathbf{R}^d)}^2.$$

For $i \ge j$, write

$$\|P_{\xi(t),i}u\|_{U_\Delta^2(G_k^j\times\mathbf{R}^d)}^2 = \|P_{\xi(G_k^j),i-2\le\cdot\le i+2}u\|_{U_\Delta^2(G_k^j\times\mathbf{R}^d)}^2. \tag{5.97}$$

It is straightforward to verify that if (p,q) is an admissible pair, then

$$\|u\|_{L_t^p L_x^q(I\times\mathbf{R}^d)} \lesssim_{p,d} \|u\|_{U_\Delta^p(I\times\mathbf{R}^d)}.$$

Indeed, this follows from Strichartz estimates and the definition of the U_Δ^p spaces. Then by (4.155), we have the following lemma.

Lemma 5.17 *For any interval G_k^j, for any integer l, if $(p,q) \in \mathscr{A}_d$ is an admissible pair, then*

$$\|P_{\xi(t),l}u\|_{L_t^p L_x^q(G_k^j\times\mathbf{R}^d)} \lesssim_{p,d} \|P_{\xi(t),l}u\|_{U_\Delta^2(G_k^j\times\mathbf{R}^d)},$$

where the left-hand side is defined as in (4.148).

Now define a variant of (4.149).

5.3 The Mass-Critical Problem When $d = 1$

Definition 5.18 (Long-time Strichartz spaces) For $G_k^j \subset [a,b]$, write

$$\|u\|_{X(G_k^j \times \mathbf{R}^d)}^2 = \sum_{i \leq j} 2^{i-j} \|P_{\xi(t),i}u\|_{U_\Delta^2(G_k^j \times \mathbf{R}^d)}^2 + \sum_{i > j} \|P_{\xi(t),i}u\|_{U_\Delta^2(G_k^j \times \mathbf{R}^d)}^2, \tag{5.98}$$

and for any $0 \leq l \leq k_0$, write

$$\|u\|_{\tilde{X}_l([a,b] \times \mathbf{R}^d)}^2 = \sup_{0 \leq j \leq l} \sup_{G_k^j \subset [a,b]} \|u\|_{X(G_k^j \times \mathbf{R}^d)}^2. \tag{5.99}$$

Define $\|u\|_{\tilde{X}_l(G_k^j \times \mathbf{R}^d)}$ for $l \leq j$ in a similar manner.

Remark The two main differences between Definition 5.18 and (4.149) are that Definition 5.18 uses a U_Δ^2-norm as opposed to the $L_t^2 L_x^{2d/(d-2)}$-norm in (4.149), and that Definition 5.18 has an l^2-norm in frequencies, whereas (4.149) has an l^∞-norm in frequencies.

Theorem 5.19 (Long-time Strichartz estimate) *For any k_0, if u is an almost-periodic solution to (4.142) when $d = 1, 2$, then*

$$\|u\|_{\tilde{X}_{k_0}([a,b] \times \mathbf{R}^d)} \lesssim 1. \tag{5.100}$$

Remark Throughout the proof $A \lesssim B$ means $A \leq C(u) B$. Crucially, the implicit constant in (5.100) does not depend on k_0.

Proof Since $\|u\|_{\tilde{X}_{k_0}([a,b] \times \mathbf{R}^d)}$ is a supremum over many norms, fix G_k^j and compute $\|u\|_{X(G_k^j \times \mathbf{R}^d)}$.

For each $G_\alpha^i \subset G_k^j$, choosing some $a_\alpha^i \in G_\alpha^i$, by Duhamel's principle we have

$$P_{\xi(G_\alpha^i), i-2 \leq \cdot \leq i+2} u(t) = P_{\xi(G_\alpha^i), i-2 \leq \cdot \leq i+2} e^{i(t-a_\alpha^i)\Delta} u(a_\alpha^i)$$
$$- i \int_{a_\alpha^i}^t e^{i(t-\tau)\Delta} P_{\xi(G_\alpha^i), i-2 \leq \cdot \leq i+2} F(u(\tau)) d\tau.$$

Therefore,

$$\left\| P_{\xi(G_\alpha^i), i-2 \leq \cdot \leq i+2} u \right\|_{U_\Delta^2(G_\alpha^i \times \mathbf{R}^d)}$$
$$\lesssim \left\| P_{\xi(G_\alpha^i), i-2 \leq \cdot \leq i+2} u(a_\alpha^i) \right\|_{L^2(\mathbf{R}^d)}$$
$$+ \left\| \int_{a_\alpha^i}^t e^{i(t-\tau)\Delta} P_{\xi(G_\alpha^i), i-2 \leq \cdot \leq i+2} F(u(\tau)) d\tau \right\|_{U_\Delta^2(G_\alpha^i \times \mathbf{R}^d)}.$$

If $a_\alpha^i \in G_\alpha^i$ satisfies

$$\left\| P_{\xi(G_\alpha^i), i-2 \leq \cdot \leq i+2} u(a_\alpha^i) \right\|_{L_x^2(\mathbf{R}^d)} = \inf_{t \in G_\alpha^i} \left\| P_{\xi(G_\alpha^i), i-2 \leq \cdot \leq i+2} u(t) \right\|_{L_x^2(\mathbf{R}^d)}, \tag{5.101}$$

then by (4.152), (4.153), (4.155), and (5.101),

$$\sum_{i \leq j} 2^{i-j} \sum_{G_\alpha^i \subset G_k^j} \left\| P_{\xi(G_\alpha^i), i-2 \leq \cdot \leq i+2} u(a_\alpha^i) \right\|_{L_x^2(\mathbf{R}^d)}^2$$

$$+ \sum_{i > j} \left\| P_{\xi(G_k^j), i-2 \leq \cdot \leq i+2} u(a_k^j) \right\|_{L_x^2(\mathbf{R}^d)}^2 \tag{5.102}$$

$$\lesssim \sum_i \eta_3^{-1} 2^{-j} \int_{G_k^j} \left\| P_{\xi(t), i} u(t) \right\|_{L_x^2(\mathbf{R}^d)}^2 \left(N(t)^3 + \eta_3 \|u(t)\|_{L_x^{2(d+2)/d}}^{2(d+2)/d} \right) dt \lesssim 1. \tag{5.103}$$

It remains to estimate the contribution of the following Duhamel terms:

$$\sum_{0 \leq i \leq j} 2^{i-j} \left\| P_{\xi(G_\alpha^i), i-2 \leq \cdot \leq i+2} \int_{a_\alpha^i}^t e^{i(t-\tau)\Delta} F(u(\tau)) d\tau \right\|_{U_\Delta^2(G_\alpha^i \times \mathbf{R}^d)}$$

$$+ \sum_{i > j} \left\| P_{\xi(G_k^j), i-2 \leq \cdot \leq i+2} \int_{a_k^j}^t e^{i(t-\tau)\Delta} F(u(\tau)) d\tau \right\|_{U_\Delta^2(G_k^j \times \mathbf{R}^d)}.$$

First, suppose G_α^i is an interval satisfying $N(G_\alpha^i) \geq \eta_3^{1/2} 2^{i-5}$. In that case, (4.144) implies that $N(t) \geq 2^{i-6} \eta_3^{1/2}$ for all $t \in G_\alpha^i$. By Strichartz estimates and conservation of mass, if J_l is a small interval and $d = 1, 2$, we have

$$\|F(u)\|_{L_t^1 L_x^2(J_l \times \mathbf{R}^d)} \lesssim 1. \tag{5.104}$$

Then by (4.151)–(4.153),

$$\sum_{i \leq j} 2^{i-j} \sum_{G_\alpha^i \subset G_k^j} \sum_{\substack{J_l \subset G_k^j : N(J_l) \\ \geq 2^{i-6} \eta_3^{1/2}}} \left\| P_{\xi(G_\alpha^i), i-2 \leq \cdot \leq i+2} F(u) \right\|_{L_t^1 L_x^2(J_l \cap G_\alpha^i \times \mathbf{R}^d)}^2$$

$$\lesssim \sum_{i \leq j} 2^{i-j} \sum_{\substack{J_l \subset G_k^j : N(J_l) \\ \geq 2^{i-6} \eta_3^{1/2}}} \left\| P_{\xi(G_\alpha^i), i-2 \leq \cdot \leq i+2} F(u) \right\|_{L_t^1 L_x^2(J_l \times \mathbf{R}^d)}^2$$

$$\lesssim \eta_3^{-1/2} 2^{-j} \sum_{J_l \subset G_k^j} N(J_l) \lesssim 1 \tag{5.105}$$

and

$$\sum_{\substack{i>j \\ \ge 2^{j-6}\eta_3^{1/2}}} \sum_{J_l \subset G_k^j : N(J_l)} \left\| P_{\xi(G_k^j), i-2\le \cdot \le i+2} F(u) \right\|^2_{L_t^1 L_x^2(J_l \cap G_k^j \times \mathbf{R}^d)} \lesssim \sum_{\substack{J_l \subset G_k^j : N(J_l) \\ \ge 2^{j-6}\eta_3^{1/2}}} 1$$

$$\lesssim 1.$$

Next, for any fixed G_k^j, there are at most two small intervals, call them J_1 and J_2, that intersect G_k^j but are not contained in G_k^j:

$$\sum_{i \le j} 2^{i-j} \sum_{G_\alpha^i \subset G_k^j : N(G_\alpha^i) \ge \eta_3^{1/2} 2^{i-5}} \left\| P_{\xi(G_\alpha^i), i-2\le \cdot \le i+2} F(u) \right\|^2_{L_t^1 L_x^2(J_1 \cap G_\alpha^i \times \mathbf{R}^d)}$$

$$+ \sum_{i > j : N(G_k^j) \ge \eta_3^{1/2} 2^{i-5}} \left\| P_{\xi(G_k^j), i-2\le \cdot \le i+2} F(u) \right\|^2_{L_t^1 L_x^2(J_1 \cap G_k^j \times \mathbf{R}^d)} \lesssim 1. \tag{5.106}$$

The estimate for the small interval J_2 is the same.

The contribution of the G_α^i intervals with $N(G_\alpha^i) \le \eta_3^{1/2} 2^{i-5}$ is estimated by induction. Observe that by (4.152), (4.153), and (5.104),

$$\|u\|_{\widetilde{X}_{10}([a,b] \times \mathbf{R}^d)} \lesssim 1,$$

with constant independent of k_0. Then Theorem 5.19 follows from (5.105)–(5.106) and the intermediate theorem.

Theorem 5.20 (Intermediate theorem) *Suppose u is an almost-periodic solution to (4.142) and $d = 1, 2$. Then if $N(G_\alpha^i) \le \eta_3^{1/2} 2^{i-5}$ and $i \ge 10$, there exists $\varepsilon_0 > 0$ small, $c(d), c_1(d) > 0$, such that for $m \ge i-2$, $l \ge i-5$, $l \ge m-5$, the estimate*

$$\left\| \int_{a_\alpha^i}^t e^{i(t-\tau)\Delta} P_m\left((P_l u) u^{4/d}\right)(\tau) d\tau \right\|_{U_\Delta^2(G_\alpha^i \times \mathbf{R}^d)}$$

$$\lesssim \varepsilon_0 2^{c(d)(m-l)} \|P_l u\|_{U_\Delta^2(G_\alpha^i \times \mathbf{R}^d)} \left(\|u\|_{\widetilde{X}_i([a,b] \times \mathbf{R}^d)} + 1\right)^{c_1(d)}, \tag{5.107}$$

holds with implicit constant independent of i.

Then (5.100) is proved by an induction argument on $\widetilde{X}_l$, starting with the base case $l = 0$. Fix $G_k^j \subset [a,b]$. By (5.98), the goal is to estimate

$$\sum_{0 \leq i \leq j} 2^{i-j} \sum_{G_\alpha^i \subset G_k^j} \left\| \int_{a_\alpha^i}^t e^{i(t-\tau)\Delta} P_{\xi(G_\alpha^i), i-2 \leq \cdot \leq i+2} F(u)(\tau) d\tau \right\|^2_{U_\Delta^2(G_\alpha^i \times \mathbf{R}^d)}$$
(5.108)

$$+ \sum_{i > j} \left\| \int_{a_k^j}^t e^{i(t-\tau)\Delta} P_{\xi(G_k^j), i-2 \leq \cdot \leq i+2} F(u)(\tau) d\tau \right\|^2_{U_\Delta^2(G_k^j \times \mathbf{R}^d)}. \quad (5.109)$$

For any $G_\alpha^i \subset G_k^j$, it is possible to make a Galilean transformation translating $\xi(G_\alpha^i)$ to the origin. Having done that, since $4/d$ is even when $d = 1, 2$, for any $m \geq i - 2$, we have

$$P_m(F(P_{\leq i-5} u)) = 0.$$

Then by Young's inequality,

$$\sum_{10 \leq i \leq j} 2^{i-j} \sum_{G_\alpha^i \subset G_k^j} \left(\sum_{l \geq i-5} 2^{c(d)(i-l)} \| P_{\xi(G_\alpha^i), l} u \|_{U_\Delta^2(G_\alpha^i \times \mathbf{R}^d)} \right)^2$$

$$\lesssim \sum_{10 \leq i \leq j} 2^{i-j} \sum_{G_\alpha^i \subset G_k^j} \sum_{i-5 \leq l \leq j} 2^{c(d)(i-l)} \| P_{\xi(G_\alpha^i), l} u \|^2_{U_\Delta^2(G_\alpha^i \times \mathbf{R}^d)}$$

$$+ \sum_{10 \leq i \leq j} 2^{i-j} \sum_{G_\alpha^i \subset G_k^j} \sum_{l \geq j} 2^{c(d)(i-l)} \| P_{\xi(G_\alpha^i), l} u \|^2_{U_\Delta^2(G_\alpha^i \times \mathbf{R}^d)}. \quad (5.110)$$

When $j \geq i$, the interval G_k^j intersects 2^{j-i} intervals G_α^i, so by (4.143)–(4.147) and (5.97),

$$(5.110) \lesssim \sum_{10 \leq i \leq j} \sum_{l \geq j} 2^{c(d)(i-l)} \| P_{\xi(t), l} u \|^2_{U_\Delta^2(G_k^j \times \mathbf{R}^d)} \lesssim \| u \|^2_{X(G_k^j \times \mathbf{R}^d)}.$$

When $i - 5 \leq l \leq j$, an interval at level l overlaps $\sim 2^{l-i} + 1$ intervals at level i. Therefore, by (5.97),

$$\sum_{G_\alpha^i \subset G_k^j} \| P_{\xi(G_\alpha^i), l} u \|^2_{U_\Delta^2(G_\alpha^i \times \mathbf{R}^d)} \lesssim 2^{l-i} \| P_{\xi(t), l} u \|^2_{U_\Delta^2(G_k^j \times \mathbf{R}^d)}.$$

Hence

$$\sum_{10 \leq i \leq j} 2^{i-j} \sum_{G_\alpha^i \subset G_k^j} \sum_{l \geq i-5} 2^{c(d)(i-l)} \| P_{\xi(G_\alpha^i), l} u \|^2_{U_\Delta^2(G_\alpha^i \times \mathbf{R}^d)} \lesssim \| u \|^2_{X(G_k^j \times \mathbf{R}^d)}.$$

Similarly,

$$\sum_{i \geq j} \sum_{l \geq i-5} 2^{c(d)(i-l)} \| P_l u \|^2_{U_\Delta^2(G_k^j \times \mathbf{R}^d)} \lesssim \| u \|^2_{X(G_k^j \times \mathbf{R}^d)}.$$

5.3 The Mass-Critical Problem When $d=1$

Therefore, by (5.105)–(5.106) and (5.107),

$$(5.108) + (5.109) \lesssim 1 + \varepsilon_0 \|u\|_{\tilde{X}_j([a,b] \times \mathbf{R}^d)} \left(\|u\|_{\tilde{X}_j([a,b] \times \mathbf{R}^d)} + 1 \right)^{c_1(d)}.$$

Theorem 5.19 then follows by standard small data-type arguments, induction, and the fact that for any $0 \leq k_* < k_0$,

$$\|u\|_{\tilde{X}_{k_*+1}([0,T] \times \mathbf{R}^d)} \leq 2 \|u\|_{\tilde{X}_{k_*}([0,T] \times \mathbf{R}^d)}. \tag{5.111}$$

Thus for $\varepsilon_0 > 0$ sufficiently small, (5.100) holds for any $k \leq k_0$, with uniform bound. $\square$

When proving Theorem 5.20, the following will be useful.

Theorem 5.21 *For any $p > 2$, if I is an interval with $t_0 \in I$,*

$$\left\| \int_{t_0}^t e^{i(t-\tau)\Delta} F(\tau) d\tau \right\|_{U^2_\Delta(I \times \mathbf{R}^d)} \lesssim_p \sup_{\|G\|_{U^p_\Delta(I \times \mathbf{R}^d)} = 1} \int_I \langle G, F \rangle d\tau. \tag{5.112}$$

Proof To prove this introduce the V^p_Δ space.

Definition 5.22 (V^p_Δ spaces) Suppose $I = [0,T]$ is a compact interval. Define the partition $\mathscr{L} = \{0 = t_0 < t_1 < \cdots < t_n = T\}$. Then for $1 \leq p < \infty$, define the norm

$$\|v\|^p_{V^p(\mathscr{L}; I \times \mathbf{R}^d)} = \sum_{k=1}^n \|v(t_k) - v(t_{k-1})\|^p_{L^2_x(\mathbf{R}^d)}.$$

Then write

$$\|v\|^p_{V^p_\Delta(\mathscr{L}; I \times \mathbf{R}^d)} = \|e^{-it\Delta} v(t)\|^p_{V^p(\mathscr{L}; I \times \mathbf{R}^d)},$$

and define the norm

$$\|v\|_{V^p_\Delta(I \times \mathbf{R}^d)} = \sup_{\mathscr{L}} \|v\|_{V^p_\Delta(\mathscr{L}; I \times \mathbf{R}^d)} + \|v\|_{L^\infty_t L^2_x(I \times \mathbf{R}^d)}.$$

Here $\sup_{\mathscr{L}}$ denotes the supremum over all finite partitions of $I = [0,T]$. One may make the obvious generalization to another compact interval other than $[0,T]$.

Returning to the proof, for any partition $\mathscr{L}$,

$$\left\| \int_0^t e^{i(t-\tau)\Delta} F(\tau) d\tau \right\|^{p'}_{V^{p'}_\Delta(\mathscr{L}; I \times \mathbf{R}^d)} = \sum_k \left\| \int_{t_k}^{t_{k+1}} e^{-i\tau\Delta} F(\tau) d\tau \right\|^{p'}_{L^2_x(\mathbf{R}^d)}. \tag{5.113}$$

Because L^2 is self-dual, for each k, there exists $u_k \in L^2$, $\|u_k\|_{L^2} = 1$, such that

$$(5.113) = \sum_k \left\langle u_k, \int_{t_k}^{t_{k+1}} e^{-i\tau\Delta} F(\tau) d\tau \right\rangle \cdot \left\| \int_{t_k}^{t_{k+1}} e^{-i\tau\Delta} F(\tau) \right\|_{L_x^2}^{p'-1}$$

$$= \sum_k \int_{t_k}^{t_{k+1}} \left\langle e^{i\tau\Delta} u_k, F(\tau) \right\rangle d\tau \cdot \left\| \int_{t_k}^{t_{k+1}} e^{-i\tau\Delta} F(\tau) \right\|_{L_x^2}^{p'-1}. \quad (5.114)$$

Then let

$$u = \sum_k \left\| \int_{t_k}^{t_{k+1}} e^{-i\tau\Delta} F(\tau) d\tau \right\|_{L_x^2(\mathbf{R}^d)}^{p'-1} \cdot \mathbf{1}_{[t_k, t_{k+1})}(t) e^{it\Delta} u_k.$$

By direct computation,

$$\|u\|_{U_\Delta^p(I \times \mathbf{R}^d)} = \left(\sum_k \left\| \int_{t_k}^{t_{k+1}} e^{-i\tau\Delta} F(\tau) d\tau \right\|_{L_x^2(\mathbf{R}^d)}^{p'} \right)^{\frac{p'-1}{p'}}$$

$$= \left\| \int_0^t e^{i(t-\tau)\Delta} F(\tau) d\tau \right\|_{V_\Delta^{p'}(\mathscr{X}, I \times \mathbf{R}^d)}^{p'-1},$$

and therefore by (5.114),

$$\left\| \int_0^t e^{i(t-\tau)\Delta} F(\tau) d\tau \right\|_{V_\Delta^{p'}(\mathscr{X}, I \times \mathbf{R}^d)} \lesssim \sup_{\|G\|_{U_\Delta^p(I \times \mathbf{R}^d)} = 1} \int \langle G, F \rangle d\tau.$$

Similarly, for any $u_0 \in L_x^2(\mathbf{R}^d)$ such that $\|u_0\|_{L_x^2(\mathbf{R}^d)} = 1$, and for any $t \in I$,

$$\left\langle u_0, \int_0^t e^{-i\tau\Delta} F(\tau) d\tau \right\rangle = \int_0^t \left\langle e^{i\tau\Delta} u_0, F(\tau) \right\rangle d\tau \lesssim \sup_{\|G\|_{U_\Delta^p(I \times \mathbf{R}^d)} = 1} \int \langle G, F \rangle d\tau.$$

This implies that for any $k, t \in [t_k, t_{k+1}]$,

$$\left\| \int_{t_k}^t e^{i(t-\tau)\Delta} F(\tau) d\tau \right\|_{L_T^\infty L_x^2(I \times \mathbf{R}^d)} \lesssim_p \sup_{\|G\|_{U_\Delta^p([t_k, t_{k+1}] \times \mathbf{R}^d)} = 1} \int_I \langle G, F \rangle d\tau,$$

and therefore for any $1 < p < \infty$,

$$\left\| \int_0^t e^{i(t-\tau)\Delta} F(\tau) d\tau \right\|_{V_\Delta^{p'}(I \times \mathbf{R}^d)} \lesssim \sup_{\|G\|_{U_\Delta^p(I \times \mathbf{R}^d)} = 1} \int \langle G, F \rangle d\tau.$$

To prove Theorem 5.21, we just need to prove that $V_\Delta^{p'}$ is embedded into U_Δ^r for any $r > p'$, which we do next.

Theorem 5.23 *If $p < q$,*

$$V^p \subset U^q.$$

5.3 The Mass-Critical Problem When $d = 1$

Theorem 5.23 was proved in Hadac et al. (2009).

Proof of Theorem 5.23 First we decompose $v \in V^p$.

Lemma 5.24 *Suppose $v \in V^p$ and $\|v\|_{V^p(I \times \mathbf{R}^d)} = 1$. Then for all integers n, $n \geq 0$,*

1. *there exists a partition $\mathscr{Z}_n$ such that $\mathscr{Z}_0 \subset \mathscr{Z}_1 \subset \cdots$ and $\sharp \mathscr{Z}_n \leq 2^{1+np}$,*
2. *there exists $u_n(t)$ subordinate to $\mathscr{Z}_n$, that is, $u_n(t)$ is constant on each subinterval of $\mathscr{Z}_n$, such that*

$$\|u_n(t)\|_{L^\infty_t L^2_x(I \times \mathbf{R}^d)} \leq 2^{1-n},$$

3. *there exists $v_n(t) \in V^p(I \times \mathbf{R}^d)$ such that*

$$\|v_n(t)\|_{L^\infty_t L^2_x(I \times \mathbf{R}^d)} \leq 2^{-n},$$

4. *and finally we have $v_n = u_{n+1} + v_{n+1}$, $u_0 = 0$, $v_0 = v$.*

Proof The proof is by induction. Without loss of generality suppose $I = [0, T]$. Let $u_0 = 0$, $v_0 = v$, and $\mathscr{Z}_0 = \{0, T\}$. Then $\sharp \mathscr{Z}_0 = 1$.

Next, for any partition

$$\mathscr{Z}_n = \{t_{n,k}\},$$

define the refinement $\mathscr{Z}_{n+1} = \{t_{n+1,k}\}$. Set $t^0_{n+1,k} = t_{n,k}$. For $j \geq 1$, let

$$t^j_{n+1,k} = \inf\left\{t : t^{j-1}_{n+1,k} < t \leq t_{n,k+1} : \|v_n(t) - v_n(t^{j-1}_{n+1,k})\|_{L^2(\mathbf{R}^d)} > 2^{-n-1}\right\}. \quad (5.115)$$

If no such t exists then move on to the next k. Then relabel the $t^j_{n+1,k}$ to obtain the increasing sequence $t_{n+1,k}$. Let $K_n = \sharp \mathscr{Z}_n$,

$$u_{n+1}(t) = \sum_{k=1}^{K_{n+1}} 1_{[t_{n+1,k-1}, t_{n+1,k})} v_n(t_{n+1,k-1}), \quad \text{and} \quad v_{n+1}(t) = v_n(t) - u_{n+1}(t).$$

By induction,

$$\|v_n(t)\|_{L^\infty_t L^2_x(\mathbf{R} \times \mathbf{R}^d)} \leq 2^{-n},$$

so

$$\|u_{n+1}(t)\|_{L^\infty_t L^2_x(\mathbf{R} \times \mathbf{R}^d)} \leq 2^{-n}.$$

Next, by definition of the partition,

$$\|v_{n+1}(t)\|_{L^\infty_t L^2_x(\mathbf{R} \times \mathbf{R}^d)} \leq 2^{-n-1}.$$

Finally, by the definition of the V^p-norm and (5.115),
$$1 = \|v\|_{V^p}^p \geq (K_{n+1} - K_n) 2^{-(n+1)p}.$$

By induction this implies
$$K_{n+1} = \sharp \mathscr{L}_{n+1} \leq 2^{(n+1)p} + K_n \leq 2^{(n+1)p} + 2^{1+np} \leq 2^{1+(n+1)p},$$

which closes the induction, proving Lemma 5.24. $\square$

Then for almost every $t \in \mathbf{R}$, Lemma 5.24 implies
$$v(t) = \sum_{n=0}^{\infty} u_n(t)$$

and
$$\|u_n(t)\|_{U^q(\mathbf{R}\times\mathbf{R}^d)} \leq 2^{\frac{1+np}{q}} 2^{1-n}.$$

Therefore, if $\frac{p}{q} < 1$,
$$\sum_{n=0}^{\infty} \|u_n(t)\|_{U^q(\mathbf{R}\times\mathbf{R}^d)} \leq 4 \sum 2^{n\left(\frac{p}{q}-1\right)} \lesssim_{\frac{p}{q}} 1,$$

which proves Theorem 5.23. $\square$

This completes the proof of Theorem 5.21. $\square$

Now we prove Theorem 5.20 in dimension $d = 1$. First, a lemma.

Lemma 5.25 (Intermediate lemma) *Suppose $l \geq i - 5$. Then when $d = 1$,*
$$\left\| |P_{\xi(G_\alpha^i),l}u| \, |P_{\xi(t),\leq i-10}u|^2 \right\|_{L^2_{t,x}(G_\alpha^i \times \mathbf{R})}$$
$$\lesssim 2^{(i-l)/2} \left\| P_{\xi(G_\alpha^i),l}u \right\|_{U^2_\Delta(G_\alpha^i \times \mathbf{R})} \|u\|_{X(G_\alpha^i \times \mathbf{R})}.$$

Proof Without loss of generality make a Galilean transformation so that $\xi(G_\alpha^i) = 0$. Expanding gives
$$\left\| |P_l u| \, |P_{\xi(t),\leq i-10}u|^2 \right\|^2_{L^2_{t,x}(G_\alpha^i \times \mathbf{R})}$$
$$\lesssim \sum_{m_3 \leq m_2 \leq i-10} \left\| (P_l u)(P_{\xi(t),m_2}u) \right\|_{L^2_{t,x}(G_\alpha^i \times \mathbf{R})}$$
$$\times \left\| (P_l u)(P_{\xi(t),m_3}u) \right\|_{L^2_{t,x}(G_\alpha^i \times \mathbf{R})} \|P_{\xi(t),\leq m_3}u\|^2_{L^\infty_{t,x}(G_\alpha^i \times \mathbf{R})}.$$

5.3 The Mass-Critical Problem When $d = 1$

By the Sobolev embedding theorem, conservation of mass, bilinear Strichartz estimates, and Definition 5.18, we continue:

$$\lesssim \sum_{m_3 \leq m_2 \leq i-10} 2^{m_3-l} \|P_l u\|^2_{U^2_\Delta(G^i_\alpha \times \mathbf{R})} \|P_{\xi(t),m_2} u\|_{U^2_\Delta(G^i_\alpha \times \mathbf{R})} \|P_{\xi(t),m_3} u\|_{U^2_\Delta(G^i_\alpha \times \mathbf{R})}$$

$$\lesssim 2^{i-l} \|P_l u\|^2_{U^2_\Delta(G^i_\alpha \times \mathbf{R})} \|u\|^2_{X(G^i_\alpha \times \mathbf{R})}.$$

$\square$

Remark The same calculations also prove

$$\left\| \left| P_{\xi(G^i_\alpha),l} u \right| \left| P_{C_0 N(t) \leq |\xi - \xi(t)| \leq 2^{i-10}} u \right|^2 \right\|_{L^2_{t,x}(G^i_\alpha \times \mathbf{R})}$$

$$\lesssim 2^{(i-l)/2} \|P_{\xi(G^i_\alpha),l} u\|_{U^2_\Delta(G^i_\alpha \times \mathbf{R})} \|u\|_{X(G^i_\alpha \times \mathbf{R})} \|u_{|\xi-\xi(t)|\geq C_0 N(t)}\|_{L^\infty_t L^2_x(G^i_\alpha \times \mathbf{R})}.$$
(5.116)

Proof of Theorem 5.20 The proof of Theorem 5.20 will be split into two multilinear estimates.

Theorem 5.26 (First multilinear estimate) *Suppose $d = 1$ and fix $G^i_\alpha \subset G^j_k$. For $i \geq 10$, $l \geq i - 5$, $m \geq i - 2$, and $a^i_\alpha \in G^i_\alpha$, if $N(G^i_\alpha) \leq \eta_3^{1/2} 2^{i-5}$,*

$$\left\| \int_{a^i_\alpha}^t e^{i(t-\tau)\Delta} P_m \left((P_l u) \left(P_{\xi(\tau),\geq i-10} u \right) u^3 \right)(\tau) d\tau \right\|_{U^2_\Delta(G^i_\alpha \times \mathbf{R})}$$

$$\lesssim \eta_2^{4/3} 2^{(m-l)/4} \|P_l u\|_{U^2_\Delta(G^i_\alpha \times \mathbf{R})} \|u\|^{8/3}_{X(G^i_\alpha \times \mathbf{R})}$$

$$+ 2^{(i-l)/2} 2^{(i-m)/4} \|P_l u\|_{U^2_\Delta(G^i_\alpha \times \mathbf{R})} \left(\eta_2^{10/3} \|u\|^{2/3}_{X(G^i_\alpha \times \mathbf{R})} + \eta_2^{11/6} \|u\|^{13/6}_{X(G^i_\alpha \times \mathbf{R})} \right).$$
(5.117)

Proof By Theorem 5.21 choose f such that $\|f\|_{U^3_\Delta(G^i_\alpha \times \mathbf{R})} = 1$ and $\hat{f}(t,\xi)$ is supported on $|\xi| \sim 2^m$. By (4.23), since $2^l \gg \eta_3^{-1/4} 2^{-20} N(t)$, for $l \geq 1$,

$$\left\| (P_l u) \left(P_{|\xi-\xi(t)|\leq \eta_3^{-1/4} 2^{-20} N(t)} u \right)^2 \right\|_{L^2_{t,x}(G^i_\alpha \times \mathbf{R})}$$

$$\lesssim \left(\sum_{J_k \cap G^i_\alpha \neq \emptyset} N(J_k) \eta_3^{-1/4} 2^{-20} 2^{-l} \right)^{1/2} \|P_l u\|_{U^2_\Delta(G^i_\alpha \times \mathbf{R})}$$

$$\lesssim \eta_3^{3/8} 2^{(i-l)/2} \|P_l u\|_{U^2_\Delta(G^i_\alpha \times \mathbf{R})}.$$
(5.118)

Then by (4.145), (5.116), and (5.118),

$$\left\| (P_l u) \left(P_{\xi(t),\leq i-10} u \right)^2 \right\|_{L^2_{t,x}(G^i_\alpha \times \mathbf{R})}$$

$$\lesssim 2^{\frac{i-l}{2}} \|P_l u\|_{U^2_\Delta(G^i_\alpha \times \mathbf{R})} \left(\eta_3^{3/8} + \eta_2 \|u\|_{X(G^i_\alpha \times \mathbf{R})} \right).$$
(5.119)

Furthermore, by (5.116) and (5.118), if $\hat{u}_0$ is supported on $|\xi| \sim 2^m$, by Hölder's inequality, then

$$\left\|\left(e^{it\Delta}u_0\right)\left(P_{\xi(t),\leq i-10}u\right)\right\|_{L^3_{t,x}(G^i_\alpha \times \mathbf{R})}$$
$$\lesssim \left\|\left(e^{it\Delta}u_0\right)\left(P_{\xi(t),\leq i-10}u\right)^2\right\|^{1/2}_{L^2_{t,x}(G^i_\alpha \times \mathbf{R})} \left\|e^{it\Delta}u_0\right\|^{1/2}_{L^6_{t,x}(G^i_\alpha \times \mathbf{R})}$$
$$\lesssim 2^{\frac{i-m}{4}} \|u_0\|_{L^2} \left(\eta_3^{3/8} + \eta_2 \|u\|_{X(G^i_\alpha \times \mathbf{R})}\right)^{1/2}, \qquad (5.120)$$

so since $\|f\|_{U^3_\Delta} = 1$, (5.120) implies

$$\left\|f\left(P_{\xi(t),\leq i-10}u\right)\right\|_{L^3_{t,x}(G^i_\alpha \times \mathbf{R})} \lesssim 2^{\frac{i-m}{4}} \left(\eta_3^{3/8} + \eta_2 \|u\|_{X(G^i_\alpha \times \mathbf{R})}\right)^{1/2}. \qquad (5.121)$$

Moreover, by (4.23),

$$\left\|f(P_l u)\right\|_{L^3_{t,x}(G^i_\alpha \times \mathbf{R})} \lesssim 2^{-\frac{1}{4}|m-l|} \|P_l u\|_{U^2_\Delta(G^i_\alpha \times \mathbf{R})}. \qquad (5.122)$$

Also,

$$\left\|P_{\xi(t),\geq i-10}u\right\|^3_{L^6_{t,x}(G^i_\alpha \times \mathbf{R})}$$
$$\lesssim \sum_{i-10\leq m_1\leq m_2\leq m_3} \left\|\left(P_{\xi(t),m_1}u\right)\left(P_{\xi(t),m_3}u\right)\right\|_{L^3_{t,x}(G^i_\alpha \times \mathbf{R})} \left\|P_{\xi(t),m_2}u\right\|_{L^6_{t,x}(G^i_\alpha \times \mathbf{R})}$$
$$\lesssim \eta_2 \|u\|^2_{X(G^i_\alpha \times \mathbf{R})}. \qquad (5.123)$$

Therefore, by (5.119)–(5.123),

$$\int_{G^i_\alpha} \langle f, (P_l u)\left(P_{\xi(t),\geq i-10}u\right) u^3\rangle dt$$
$$\lesssim \left\|P_{\xi(t),\geq i-10}u\right\|^4_{L^6_{t,x}} \left\|(P_l u)f\right\|_{L^3_{t,x}}$$
$$+ \left\|(P_l u)\left(P_{\xi(t),\leq i-10}u\right)^2\right\|_{L^2_{t,x}} \left\|f\left(P_{\xi(t),\leq i-10}u\right)\right\|_{L^3_{t,x}} \left\|P_{\xi(t),\geq i-10}u\right\|_{L^6_{t,x}}$$
$$\lesssim 2^{-\frac{|m-l|}{4}} \eta_2^{4/3} \|P_l u\|_{U^2_\Delta(G^i_\alpha \times \mathbf{R})} \|u\|^{8/3}_{X(G^i_\alpha \times \mathbf{R})}$$
$$+ 2^{\frac{(i-l)}{2} + \frac{i-m}{4}} \|P_l u\|_{U^2_\Delta(G^i_\alpha \times \mathbf{R})} \left(\eta_2^{10/3} \|u\|^{2/3}_{X(G^i_\alpha \times \mathbf{R})} + \eta_2^{11/6} \|u\|^{13/6}_{X(G^i_\alpha \times \mathbf{R})}\right).$$

This proves Theorem 5.26. $\square$

The estimate (5.117) is in the form of (5.107), so Theorem 5.20 follows directly from (5.117) and the following second multilinear estimate.

Theorem 5.27 (Second multilinear estimate) *For $d=1$, $|l-m|\leq 5$, $l\geq i-5$,*

5.3 The Mass-Critical Problem When $d = 1$

and $N(G_\alpha^i) \leq \eta_3^{1/2} 2^{i-5}$,

$$\left\| \int_{a_\alpha^i}^t e^{i(t-\tau)\Delta} P_{\xi(G_\alpha^i),m}\left(P_{\xi(G_\alpha^i),l}u\right)\left(P_{\xi(\tau),\leq i-10}u\right)^4 d\tau \right\|_{U_\Delta^2(G_\alpha^i \times \mathbf{R})}$$

$$\lesssim 2^{(i-l)/6} \eta_2 \|P_l u\|_{U_\Delta^2(G_\alpha^i \times \mathbf{R})} \|u\|_{\tilde{X}_{k_0}([0,T] \times \mathbf{R})}^3.$$

Indeed, decompose

$$P_m\left((P_l u) u^4\right) = P_m\left((P_l u)\left(P_{\xi(t),\leq i-10}u\right)^4\right) + P_m O\left((P_l u)\left(P_{\xi(t),\geq i-10}u\right) u^3\right).$$

Thus, with Theorem 5.27, the proof of Theorem 5.20 will be complete.

Proof of Theorem 5.27 Because Theorem 5.27 concerns only the interval G_α^i and not G_k^j, to economize notation it is convenient to relabel and consider the interval G_k^j and prove

$$\left\| \int_{a_k^j}^t e^{i(t-\tau)\Delta} P_m\left(P_{\xi(G_k^j),l}u\right)\left(P_{\xi(\tau),\leq i-10}u\right)^4 d\tau \right\|_{U_\Delta^2(G_k^j \times \mathbf{R})}$$

$$\lesssim \eta_2 2^{(j-l)/6} \left\|P_{\xi(G_k^j),l}u\right\|_{U_\Delta^2(G_k^j \times \mathbf{R})} \|u\|_{\tilde{X}_{k_0}([0,T] \times \mathbf{R})}^3$$

when $N(G_k^j) \leq \eta_3^{1/2} 2^{j-5}$. In this case, $\xi(G_k^j) = 0$, but $\xi(G_\alpha^{m_3})$ need not be zero for $G_\alpha^{m_3} \subset G_k^j$.

Lemma 5.28 *For any $0 \leq m_3 \leq j - 10$, let $\mathscr{Z}_{m_3}$ be a partition of G_k^j that is the union of the endpoints of the subintervals $G_\alpha^{m_3} \subset G_k^j$ and a_k^j. Then if u is replaced by $P_{|\xi - \xi(t)| \geq 2^{-20} \eta_3^{-1/4} N(t)} u$,*

$$\sum_{\substack{0 \leq m_3 \leq m_2 \\ \leq j - 10}} \left\| \int_{a_k^j}^t e^{i(t-\tau)\Delta} (P_l u)\left(P_{\xi(\tau),m_2}u\right) \right.$$

$$\left. \times \left(P_{\xi(\tau),m_3}u\right)\left(P_{\xi(\tau),\leq m_3}u\right)^2 d\tau \right\|_{V_\Delta^1(\mathscr{Z}_{m_3}; G_k^j \times \mathbf{R})}$$

$$\lesssim 2^{j-l} \eta_2^2 \|P_l u\|_{U_\Delta^2(G_k^j \times \mathbf{R})} \|u\|_{\tilde{X}_j([0,T] \times \mathbf{R})}^2.$$

Also,

$$\sum_{J_l \cap G_k^j \neq \emptyset} \left\| \int_{J_l} e^{-i\tau\Delta}(P_l u)\left(P_{|\xi - \xi(\tau)| \leq \eta_3^{-1/4} 2^{-20} N(\tau)}u\right)^4 d\tau \right\|_{L_x^2(\mathbf{R})}$$

$$\lesssim 2^{j-l} \eta_3^{3/8} \|P_l u\|_{U_\Delta^2(G_k^j \times \mathbf{R})}.$$

Proof First replace u with $P_{|\xi-\xi(t)|\geq 2^{-20}\eta_3^{-1/4}N(t)}u$. If $\|u_0\|_{L^2}=1$ and u_0 has Fourier transform supported on $|\xi|\sim 2^l$, then by (4.144) and (4.145),

$$\left\langle u_0, \int_{G_\alpha^{m_3}} e^{-i\tau\Delta}(P_l u)(P_{\xi(\tau),m_2}u)(P_{\xi(\tau),m_3}u)(P_{\xi(\tau),\leq m_3}u)^2 d\tau\right\rangle$$

$$\lesssim \left\|(e^{it\Delta}u_0)\left(P_{\xi(G_\alpha^{m_3}),m_3-2\leq\cdot\leq m_3+2}u\right)\left(P_{\xi(G_\alpha^{m_3}),\leq m_3+2}u\right)\right\|_{L^2_{t,x}(G_\alpha^{m_3}\times\mathbf{R})}$$

$$\times \left\|(P_l u)(P_{\xi(t),m_2}u)\left(P_{\xi(G_\alpha^{m_3}),\leq m_3+2}u\right)\right\|_{L^2_{t,x}(G_\alpha^{m_3}\times\mathbf{R})}$$

$$\lesssim \eta_2 2^{(m_3-l)/2}\left\|P_{\xi(G_\alpha^{m_3}),m_3-2\leq\cdot\leq m_3+2}u\right\|_{U^2_\Delta(G_\alpha^{m_3}\times\mathbf{R})}$$

$$\times \left\|(P_l u)(P_{\xi(t),m_2}u)\left(P_{\xi(G_\alpha^{m_3}),\leq m_3+2}u\right)\right\|_{L^2_{t,x}(G_\alpha^{m_3}\times\mathbf{R})}.$$

The last line used the Sobolev embedding theorem and (4.23). Summing up, using the Cauchy–Schwarz inequality and the definition of the $X(G_k^j\times\mathbf{R})$-norms, combined with an argument similar to the proof of Lemma 5.25 (remember that u was replaced with $P_{|\xi-\xi(t)|\geq 2^{-20}\eta_3^{-1/4}N(t)}u$),

$$\sum_{\substack{m_3\leq m_2\\\leq j-10}}\sum_{G_\alpha^{m_3}\subset G_k^j}\eta_2 2^{(m_3-l)/2}\left\|P_{\xi(G_\alpha^{m_3}),m_3-2\leq\cdot\leq m_3+2}u\right\|_{U^2_\Delta(G_\alpha^{m_3}\times\mathbf{R})}$$

$$\times \left\|(P_l u)(P_{\xi(t),m_2}u)\left(P_{\xi(G_\alpha^{m_3}),\leq m_3+2}u\right)\right\|_{L^2_{t,x}(G_\alpha^{m_3}\times\mathbf{R})}$$

$$\lesssim \eta_2 \sum_{\substack{m_3\leq m_2\\\leq j-10}} 2^{(m_3-l)/2}\left\|P_{\xi(t),m_3}u\right\|_{U^2_\Delta(G_k^j\times\mathbf{R})}$$

$$\times \left\|(P_l u)(P_{\xi(t),m_2}u)(P_{\xi(t),\leq m_3}u)\right\|_{L^2_{t,x}(G_k^j\times\mathbf{R})}$$

$$\lesssim \eta_2\|u\|_{X(G_k^j\times\mathbf{R})}\left(\sum_{m_3\leq j-10}\left(\sum_{m_3\leq m_2\leq j-10}\|(P_l u)(P_{\xi(t),m_2}u)\right.\right.$$

$$\left.\left.\times(P_{\xi(t),\leq m_3}u)\|_{L^2_{t,x}(G_k^j\times\mathbf{R})}\right)^2\right)^{1/2}$$

$$\lesssim \eta_2^2\|u\|_{X(G_k^j\times\mathbf{R})}\left(\sum_{m_3\leq j-10}\left(\sum_{m_3\leq m_2\leq j-10}2^{(m_3-l)/2}\|P_l u\|_{U^2_\Delta(G_k^j\times\mathbf{R})}\right.\right.$$

$$\left.\left.\times\|P_{\xi(t),m_2}u\|_{U^2_\Delta(G_k^j\times\mathbf{R})}\right)^2\right)^{1/2}$$

$$\lesssim \eta_2^2\|P_l u\|_{U^2_\Delta(G_k^j\times\mathbf{R})}\|u\|^2_{X(G_k^j\times\mathbf{R})}.$$

Next, combining a bilinear estimate of $\|(P_l u)(P_{|\xi-\xi(t)|\leq\eta_3^{-1/4}2^{-20}N(t)}u)^2\|_{L^2_{t,x}}$

with the Sobolev embedding theorem gives

$$\sum_{J_l \cap G_k^j \neq \emptyset} \left\| \int_{J_l} e^{-i\tau\Delta} (P_l u) \left(P_{\xi(\tau), \leq \eta_3^{-1/4} 2^{-20} N(\tau)} u \right)^4 d\tau \right\|_{L_x^2(\mathbf{R})} \quad (5.124)$$

$$\lesssim \left(N\left(G_k^j\right) \eta_3^{-1/4} + \sum_{J_l \subset G_k^j} N(J_l) \eta_3^{-1/4} \right) \|P_l u\|_{U_\Delta^2(G_k^j \times \mathbf{R})} \quad (5.125)$$

$$\lesssim \eta_3^{3/4} 2^{j-l} \|P_l u\|_{U_\Delta^2(G_k^j \times \mathbf{R})}. \quad (5.126)$$

Inequality (5.126) follows from (4.151), which implies

$$\sum_{J_l \subset G_k^j} N(J_l) \lesssim \eta_3 2^j, \quad (5.127)$$

as well as the fact that there may be two intervals J_l that overlap G_k^j but are not contained in G_k^j. However, because $N(t) \leq N(G_k^j) \leq \eta_3^{1/2} 2^{j-5}$ for any $t \in G_k^j$, their contribution is controlled by $N(J_l) \lesssim \eta_3^{1/2} 2^{j-5}$. □

Because $V_\Delta^1 \subset U_\Delta^2$, to complete the proof of Theorem 5.27, it suffices to prove the following.

Lemma 5.29 *If u is replaced by $P_{|\xi - \xi(t)| \geq 2^{-20} \eta_3^{-1/4} N(t)} u$, then*

$$\sum_{m_3 \leq m_2 \leq j-10} \left(\sum_{G_\alpha^{m_3} \subset G_k^j} \left\| \int_{a_\alpha^{m_3}}^t e^{i(t-\tau)\Delta} (P_l u) \left(P_{\xi(\tau), m_2} u \right) \left(P_{\xi(\tau), m_3} u \right) \right. \right.$$

$$\left. \left. \times \left(P_{\xi(\tau), \leq m_3} u \right)^2 d\tau \right\|_{U_\Delta^2(G_\alpha^{m_3} \times \mathbf{R})}^2 \right)^{1/2} \quad (5.128)$$

$$\lesssim \eta_2 2^{\frac{3(j-l)}{4}} \|P_l u\|_{U_\Delta^2(G_k^j \times \mathbf{R})} \|u\|_{X(G_k^j \times \mathbf{R})}^3 \quad (5.129)$$

and

$$\left(\sum_{J_l \cap G_k^j \neq \emptyset} \left\| \int e^{i(t-\tau)\Delta} (P_l u) \left(P_{|\xi - \xi(\tau)| \leq \eta_3^{-1/4} 2^{-20} N(\tau)} u \right)^4 d\tau \right\|_{U_\Delta^2(J_l \times \mathbf{R})}^2 \right)^{1/2}$$

$$\lesssim \eta_3^{3/8} 2^{\frac{j-l}{2}} \|P_l u\|_{U_\Delta^2(G_k^j \times \mathbf{R})}. \quad (5.130)$$

Proof First suppose u is replaced with $P_{|\xi - \xi(t)| \geq 2^{-20} \eta_3^{-1/4} N(t)} u$. By Theorem 5.21 and the Cauchy–Schwarz inequality, there exists some f such that $\hat{f}$ is supported on $|\xi| \sim 2^l$ and $\|f\|_{U^{12/5}_\Delta(G^{m_3}_\alpha \times \mathbf{R})} = 1$ for some $G^{m_3}_\alpha \subset G^j_k$, such that

$$\sum_{G^{m_3}_\alpha} \left\| \int_{G^{m_3}_\alpha} e^{i(t-\tau)\Delta} (P_l u) (P_{\xi(\tau),m_2} u) (P_{\xi(\tau),m_3} u) (P_{\xi(\tau),\leq m_3} u)^2 d\tau \right\|^2_{U^2_\Delta(G^{m_3}_\alpha \times \mathbf{R})} \tag{5.131}$$

$$\lesssim \left(\sup_{G^{m_3}_\alpha} \left\| f \left(P_{\xi(G^{m_3}_\alpha), m_3-2\leq \cdot \leq m_3+2} u \right) \left(P_{\xi(G^{m_3}_\alpha), \leq m_3+2} u \right) \right\|^2_{L^2_{t,x}(G^{m_3}_\alpha \times \mathbf{R})} \right)$$

$$\times \left\| (P_l u) (P_{\xi(t),m_2} u) (P_{\xi(t),\leq m_3} u) \right\|^2_{L^2_{t,x}(G^j_k \times \mathbf{R})}. \tag{5.132}$$

By an argument similar to the proof of Lemma 5.25,

$$\left\| (P_l u) (P_{\xi(t),m_2} u) (P_{\xi(t),\leq m_3} u) \right\|^2_{L^2_{t,x}(G^j_k \times \mathbf{R})}$$
$$\lesssim \eta_2^2 2^{j-l} 2^{m_3-m_2} \|P_l u\|^2_{U^2_\Delta(G^j_k \times \mathbf{R})} \|u\|^2_{X(G^j_k \times \mathbf{R})}.$$

To estimate

$$\left\| f \left(P_{\xi(t),m_3} u \right) \left(P_{\xi(t),\leq m_3} u \right) \right\|^2_{L^2_{t,x}(G^{m_3}_\alpha \times \mathbf{R})},$$

use

$$\sum_{i \leq m_3} \|P_{\xi(t),i} u\|_{L^{12}_{t,x}(G^{m_3}_\alpha \times \mathbf{R})} \lesssim \sum_{i \leq m_3} 2^{i/4} \|P_{\xi(t),i} u\|_{U^2_\Delta(G^{m_3}_\alpha \times \mathbf{R})}$$
$$\lesssim 2^{m_3/4} \|u\|_{X(G^{m_3}_\alpha \times \mathbf{R})}. \tag{5.133}$$

Therefore, by (5.133) and bilinear Strichartz estimates,

$$\left\| f \left(P_{\xi(G^{m_3}_\alpha), m_3-2\leq \cdot \leq m_3+2} u \right) (P_{\xi(t),\leq m_3} u) \right\|^2_{L^2_{t,x}(G^{m_3}_\alpha \times \mathbf{R})}$$
$$\lesssim \left\| f \left(P_{\xi(G^{m_3}_\alpha), m_3-2\leq \cdot \leq m_3+2} u \right) \right\|^2_{L^{12/5}_{t,x}(G^{m_3}_\alpha \times \mathbf{R})} \|P_{\xi(t),\leq m_3} u\|^2_{L^{12}_{t,x}(G^{m_3}_\alpha \times \mathbf{R})}$$
$$\lesssim 2^{\frac{m_3-l}{2}} \left\| P_{\xi(G^{m_3}_\alpha), m_3-2\leq \cdot \leq m_3+2} u \right\|^2_{U^2_\Delta(G^{m_3}_\alpha \times \mathbf{R})} \|u\|^2_{X(G^{m_3}_\alpha \times \mathbf{R})}$$
$$\lesssim 2^{\frac{m_3-l}{2}} \|u\|^4_{X(G^{m_3}_\alpha \times \mathbf{R})}.$$

Therefore,

$$(5.131)^{1/2} \lesssim \eta_2 2^{(m_3-l)/4} 2^{(m_3-m_2)/2} 2^{(j-l)/2} \|P_l u\|_{U^2_\Delta(G^j_k \times \mathbf{R})} \|u\|^3_{\tilde{X}_j([0,T] \times \mathbf{R})}.$$

Summing over $0 \leq m_3 \leq m_2 \leq j - 10$ yields (5.129).

Similarly, to derive (5.130), recall that the proof of (5.126) implies

$$\left\| (P_l u) \left(P_{\xi(t), \leq 2^{-20} \eta_3^{-1/4} N(t)} u \right)^2 \right\|_{L^2_{t,x}(G^j_k \times \mathbf{R})}^2$$
$$\lesssim \eta_3^{3/4} 2^{j-l} \|P_j u\|_{U^2_\Delta(G^j_k \times \mathbf{R})}^2 \|u\|_{X(G^j_k \times \mathbf{R})}^2.$$

Because J_l is a small interval,

$$\left\| f\left(P_{\xi(t), \leq 2^{-20} \eta_3^{-1/4} N(t)} u \right)^2 \right\|_{L^2_{t,x}(J_l \times \mathbf{R})} \lesssim 1.$$

This proves (5.130) and Lemma 5.29. □

This completes the proof of Theorem 5.27. □

The proof of Theorem 5.20 is therefore now complete. □

As in dimensions $d \geq 3$, the long-time Strichartz estimates of Theorem 5.19 yield frequency-localized interaction Morawetz estimates.

Theorem 5.30 *If u is an almost-periodic solution to* (4.142) *in dimension $d = 1$ on $\mathbf{R}$ with $N(t) \leq 1$ for all $t \in \mathbf{R}$, and $[0, T]$ is an interval satisfying*

$$\int_0^T N(t)^3 \, dt = K, \tag{5.134}$$

then

$$\int_0^T \int \left| P_{\leq \eta_3^{-1} K} u(t,x) \right|^8 dx \, dt + \int_0^T \int \left(\partial_x \left| P_{\leq \eta_3^{-1} K} u(t,x) \right|^2 \right)^2 dx \, dt \lesssim o(K), \tag{5.135}$$

where $o(K)$ is a quantity satisfying $\frac{o(K)}{K} \to 0$ as $K \nearrow \infty$.

Proof Suppose $[0, T]$ is an interval such that $\|u\|_{L^6_{t,x}([0,T] \times \mathbf{R})}^6 = 2^{k_0}$. After rescaling,

$$u(t,x) \mapsto \lambda^{1/2} u(\lambda^2 t, \lambda x), \quad \text{with } \lambda = \frac{\eta_3 2^{k_0}}{K}. \tag{5.136}$$

Inequality (5.135) is equivalent to proving

$$\int_0^T \int |P_{\leq k_0} u(t,x)|^8 dx \, dt + \int_0^T \int \left(\partial_x |P_{\leq k_0} u(t,x)|^2 \right)^2 dx \, dt \lesssim o(K) \frac{\eta_3 2^{k_0}}{K}.$$

Now by (4.10) and (4.18) with $\mu = 1$ and $u = v$, if u solves (4.142), then

$$\int_0^T \int |u(t,x)|^8 dx \, dt + \int_0^T \int \left(\partial_x |u(t,x)|^2 \right)^2 dx \, dt$$
$$\lesssim \left\| e^{-ix \cdot \xi(t)} u \right\|_{L^\infty_t \dot{H}^1([0,T] \times \mathbf{R})} \|u\|_{L^\infty_t L^2_x([0,T] \times \mathbf{R})}^3.$$

Therefore, letting $I = P_{\leq k_0}$,

$$\int_0^T \int |Iu(t,x)|^8 dx dt + \int_0^T \left(\partial_x |Iu(t,x)|^2\right)^2 dx dt$$

$$\lesssim \left\|e^{-ix\cdot\xi(t)}Iu\right\|_{L_T^\infty \dot{H}^1([0,T]\times\mathbf{R})} \|Iu\|^3_{L_T^\infty L_x^2([0,T]\times\mathbf{R})} + \int_0^T \mathcal{E}(t) dt,$$

where, as in (4.169),

$\mathcal{E}(t)$

$$= \int\int \frac{(x-y)}{|x-y|} |Iu(t,y)|^2 \operatorname{Re}\left[\left(IF(\bar{u}) - F(\overline{Iu})\right)(\partial_x - i\xi(t))Iu\right](t,x) dx dy$$

(5.137)

$$- \int\int \frac{(x-y)}{|x-y|} |Iu(t,y)|^2 \operatorname{Re}\left[\overline{Iu}(\partial_x - i\xi(t))(F(Iu) - IF(u))\right](t,x) dx dy$$

$$+ 2\int\int \frac{(x-y)}{|x-y|} \operatorname{Im}\left[\overline{Iu}IF(u)\right](t,y) \operatorname{Im}\left[\overline{Iu}(\partial_x - i\xi(t))Iu\right](t,x) dx dy.$$

Also, as in Lemma 4.38,

$$\left\|e^{-ix\cdot\xi(t)}Iu\right\|_{L_T^\infty \dot{H}^1([0,T]\times\mathbf{R})} \|Iu\|^3_{L_T^\infty L_x^2([0,T]\times\mathbf{R})} \lesssim \frac{\eta_3 2^{k_0}}{K} o(K).$$

Let $\tilde{\partial}_x = \partial_x - i\xi(t)$.

Lemma 5.31 *If u is an almost-periodic solution to (4.142) that satisfies the above conditions, then*

$$\int_0^T \int\int \frac{(x-y)}{|x-y|} |Iu(t,y)|^2 \operatorname{Re}\left[\left(IF(\bar{u}) - F(\overline{Iu})\right)(\partial_x - i\xi(t))Iu\right](t,x) dx dy dt$$

$$\lesssim o(K) \frac{\eta_3 2^{k_0}}{K}.$$

(5.138)

Proof Using the Galilean Littlewood–Paley projection, let $u_l = u_{\xi(t),\leq k_0-5}$ and $u_h + u_l = u$. By Fourier support arguments,

$$IF(u_l) - F(Iu_l) = 0. \tag{5.139}$$

Next, since I is a Fourier multiplier whose symbol satisfies $|\partial_\xi m(\xi)| \lesssim \frac{1}{|\xi|}$, for $|\eta| \leq |\xi|$ we have

$$|m(\xi+\eta) - m(\xi)| \lesssim \frac{|\eta|}{|\xi|}, \tag{5.140}$$

5.3 The Mass-Critical Problem When $d = 1$

so

$$\left\| \left[I\left(u_h u_l^4\right) - (Iu_h) u_l^4 \right] \left(\widetilde{\partial}_x Iu\right) \right\|_{L^1_{t,x}([0,T] \times \mathbf{R})}$$
$$\lesssim 2^{-k_0} \left\| (u_h) u_l^2 \right\|_{L^2_{t,x}([0,T] \times \mathbf{R})} \left\| \left(\widetilde{\partial}_x u_l\right) u_l^2 \right\|_{L^2_{t,x}([0,T] \times \mathbf{R})}^{1/2}$$
$$\times \left\| \widetilde{\partial}_x u_l \right\|_{L^\infty_t L^2_x([0,T] \times \mathbf{R})}^{1/2} \left\| \widetilde{\partial}_x Iu \right\|_{L^4_t L^\infty_x([0,T] \times \mathbf{R})}. \tag{5.141}$$

By Theorem 5.19, Lemma 5.25, (5.134), and (5.136),

$$\left\| u_h u_l^2 \right\|_{L^2_{t,x}([0,T] \times \mathbf{R})} \lesssim 1. \tag{5.142}$$

Similarly, partitioning $[0, T]$ into G^j_α intervals and applying Lemma 5.25 and the definition of the $X\left(G^j_\alpha \times \mathbf{R}\right)$-norm on each subinterval gives

$$\left\| \left(P_{\xi(t), j} u\right) u_l^2 \right\|_{L^2_{t,x}([0,T] \times \mathbf{R})} \lesssim 2^{k_0/2} 2^{-j/2}. \tag{5.143}$$

Therefore, again by Theorem 5.19,

$$\left\| \left(\widetilde{\partial}_x Iu\right) u_l^2 \right\|_{L^2_{t,x}([0,T] \times \mathbf{R})} \lesssim \sum_{j \leq k_0} 2^{\frac{k_0-j}{2}} 2^j \lesssim 2^{k_0}. \tag{5.144}$$

Also, by definition of X-norms and the fact that $N(t) \leq \frac{\eta_3 2^{k_0}}{K}$ for all $t \in [0, T]$,

$$\left\| \widetilde{\partial}_x Iu \right\|_{L^4_t L^\infty_x([0,T] \times \mathbf{R})} \lesssim 2^{k_0} \quad \text{and} \quad \left\| \widetilde{\partial}_x u_l \right\|_{L^\infty_t L^2_x([0,T] \times \mathbf{R})} \lesssim o(K) \frac{\eta_3 2^{k_0}}{K}. \tag{5.145}$$

Combining (5.141)–(5.145) we obtain

$$(5.141) \lesssim o(K) \frac{\eta_3 2^{k_0}}{K}.$$

Next,

$$\left\| u_h^2 u_l^3 \left(\widetilde{\partial}_x Iu\right) \right\|_{L^1_{t,x}([0,T] \times \mathbf{R})}$$
$$\lesssim \left\| u_h u_l^2 \right\|_{L^2_{t,x}([0,T] \times \mathbf{R})}^{3/2} \left\| u_h \right\|_{L^6_{t,x}([0,T] \times \mathbf{R})}^{1/2} \left\| \widetilde{\partial}_x Iu \right\|_{L^4_t L^\infty_x([0,T] \times \mathbf{R})}^{2/3} \left\| \widetilde{\partial}_x Iu \right\|_{L^\infty_t L^2_x([0,T] \times \mathbf{R})}^{1/3}$$
$$\lesssim o(K) \frac{\eta_3 2^{k_0}}{K},$$

$$\left\| u_h^3 u_l^2 \left(\widetilde{\partial}_x Iu\right) \right\|_{L^1_{t,x}([0,T] \times \mathbf{R})}$$
$$\lesssim \left\| u_h u_l^2 \right\|_{L^2_{t,x}([0,T] \times \mathbf{R})} \left\| u_h \right\|_{L^6_{t,x}([0,T] \times \mathbf{R})}^{2} \left\| \widetilde{\partial}_x Iu \right\|_{L^4_t L^\infty_x([0,T] \times \mathbf{R})}^{2/3} \left\| \widetilde{\partial}_x Iu \right\|_{L^\infty_t L^2_x([0,T] \times \mathbf{R})}^{1/3}$$
$$\lesssim o(K) \frac{\eta_3 2^{k_0}}{K}.$$

Finally,

$$\|u_h^4 u(\widetilde{\partial}_x Iu)\|_{L^1_{t,x}([0,T]\times\mathbf{R})}$$
$$\lesssim \|u_h\|^4_{L^5_t L^{10}_x([0,T]\times\mathbf{R})} \|u\|_{L^\infty_t L^2_x([0,T]\times\mathbf{R})} \|\widetilde{\partial}_x Iu\|^{4/5}_{L^4_t L^\infty_x([0,T]\times\mathbf{R})} \|\widetilde{\partial}_x Iu\|^{1/5}_{L^\infty_t L^2_x([0,T]\times\mathbf{R})}$$
$$\lesssim o(K)\frac{\eta_3 2^{k_0}}{K}.$$

This completes the proof of (5.138). $\square$

Turning to the second term in (5.137), we have the following lemma.

Lemma 5.32

$$\int_0^T \int\int \frac{(x-y)}{|x-y|} |Iu(t,y)|^2 \operatorname{Re}\left[\overline{Iu}(\partial_x - i\xi(t))(F(Iu) - IF(u))\right](t,x)\,dx\,dy\,dt$$
$$\lesssim o(K)\frac{\eta_3 2^{k_0}}{K} + \left(o(K)\frac{\eta_3 2^{k_0}}{K}\right)^{1/4} \|Iu\|^6_{L^8_{t,x}([0,T]\times\mathbf{R})}. \tag{5.146}$$

Proof Integrating by parts,

$$(5.146) = -\int_0^T \int\int \frac{(x-y)}{|x-y|} |Iu(t,y)|^2 \operatorname{Re}\left[(F(Iu) - IF(u))\right.$$
$$\left. \times (\partial_x + i\xi(t))\overline{Iu}\right](t,x)\,dx\,dy\,dt \tag{5.147}$$

$$-\int_0^T \int |Iu(t,x)|^2 \operatorname{Re}\left[(F(Iu) - IF(u))\overline{Iu}\right](t,x)\,dx\,dt. \tag{5.148}$$

Expression (5.147) is equal to (5.138), so by Lemma 5.31, we find $(5.147) \lesssim o(K)\frac{\eta_3 2^{k_0}}{K}$. Since $IF(u_l) - F(Iu_l) = 0$,

$$|IF(u) - F(Iu)| \lesssim |u_l|^5 |u_h| + |u_h|^6,$$

so

$$(5.148) \lesssim \|Iu\|^2_{L^\infty_{t,x}([0,T]\times\mathbf{R})} \|u_h\|^6_{L^6_{t,x}([0,T]\times\mathbf{R})}$$
$$+ \|Iu\|^6_{L^8_{t,x}([0,T]\times\mathbf{R})} \|u_h\|_{L^{\frac{16}{3}}_t L^8_x([0,T]\times\mathbf{R})} \|Iu\|_{L^{16}_t L^8_x([0,T]\times\mathbf{R})}. \tag{5.149}$$

By the Sobolev embedding theorem,

$$\|Iu\|^2_{L^\infty_{t,x}([0,T]\times\mathbf{R})} \lesssim \|\widetilde{\partial}_x Iu\|_{L^\infty_t L^2_x([0,T]\times\mathbf{R})} \|Iu\|_{L^\infty_t L^2_x([0,T]\times\mathbf{R})}$$
$$\lesssim o(K)\frac{\eta_3 2^{k_0}}{K}, \tag{5.150}$$

5.3 The Mass-Critical Problem When $d = 1$

and therefore

$$\|Iu\|_{L^\infty_{t,x}([0,T]\times\mathbf{R})}^2 \|u_h\|_{L^6_{t,x}([0,T]\times\mathbf{R})}^6 \lesssim o(K)\frac{\eta_3 2^{k_0}}{K}.$$

Again using the Sobolev embedding theorem and also interpolation, we get

$$\|Iu\|_{L^{16}_t L^8_x([0,T]\times\mathbf{R})} \lesssim \|\widetilde{\partial_x}Iu\|_{L^4_t L^\infty_x([0,T]\times\mathbf{R})}^{1/4} \|Iu\|_{L^\infty_t L^2_x([0,T]\times\mathbf{R})}^{3/4}$$

$$\lesssim \left(o(K)\frac{\eta_3 2^{k_0}}{K}\right)^{1/4}. \tag{5.151}$$

This completes the derivation of (5.146). □

Finally, take the third term in (5.137).

Lemma 5.33

$$\int_0^T \int\int \frac{(x-y)}{|x-y|} \operatorname{Im}\left[\overline{Iu}IF(u)\right](t,y) \operatorname{Im}\left[\overline{Iu}(\partial_x - i\xi(t))Iu\right](t,x) \, dx\, dy\, dt$$

$$\lesssim o(K)\frac{\eta_3 2^{k_0}}{K} + \left(o(K)\frac{\eta_3 2^{k_0}}{K}\right)^{1/4} \|Iu\|_{L^8_{t,x}([0,T]\times\mathbf{R})}^6.$$

Proof Because $\overline{Iu}F(Iu) = |Iu|^6$, it follows that

$$\operatorname{Im}\left[\overline{Iu}IF(u)\right] = \operatorname{Im}\left[\overline{Iu}[IF(u) - F(Iu)]\right].$$

By a careful analysis of Fourier supports, we obtain

$$I\left(u_h u_l^4\right) - \left(u_l^4 Iu_h\right) = I\left(\left(P_{>k_0-2}u\right)u_l^4\right) - u_l^4\left(P_{>k_0-2}Iu\right).$$

Therefore,

$$P_{\leq k_0-10}\left[\overline{u_l}\left(I\left(u_h u_l^4\right) - \left(u_l^4 Iu_h\right)\right)\right] = 0,$$

and integrating by parts yields

$$\int_0^T \int\int \frac{(x-y)}{|x-y|} \operatorname{Im}\left[\overline{Iu}\cdot\left(I\left(u_l^4 u_h\right) - (Iu_h)u_l^4\right)\right](t,y)$$

$$\times \operatorname{Im}\left[\overline{Iu}(\partial_x - i\xi(t))Iu\right](t,x) \, dx\, dy\, dt$$

$$= -\int_0^T \int \frac{\partial_y}{\Delta_y} P_{>k_0-10} \operatorname{Im}\left[\overline{Iu}[I\left(u_l^4 u_h\right) - (Iu_h)u_l^4]\right](t,y)$$

$$\times \operatorname{Im}\left[\overline{Iu}(\partial_x - i\xi(t))Iu\right](t,y) \, dy\, dt. \tag{5.152}$$

Using analysis similar to that for (5.149)–(5.151),

$$\text{RHS of (5.152)} \lesssim \left(o(K)\frac{\eta_3 2^{k_0}}{K}\right)^{1/4} \|Iu\|_{L^8_{t,x}([0,T]\times\mathbf{R})}^6.$$

For terms with at least two u_h, by Lemma 5.25 we have

$$\|u_h^2 u^4\|_{L_{t,x}^1([0,T]\times\mathbf{R})} \lesssim \|u_h u_l^2\|_{L_{t,x}^2([0,T]\times\mathbf{R})}^2 + \|u_h\|_{L_{t,x}^6([0,T]\times\mathbf{R})}^6 \lesssim 1.$$

Using the second inequality in (5.150) completes the proof of Lemma 5.33. □

Combining Lemmas 5.31–5.33 proves Theorem 5.30. □

Corollary 5.34 *There exists no nonzero almost-periodic solution to* (4.142) *with* $N(t) \leq 1$ *on* $\mathbf{R}$ *and*

$$\int_{-\infty}^{\infty} N(t)^3 \, dt = \infty.$$

Proof If u is an almost-periodic solution to (4.142) with nonzero mass, then by (4.167), and (5.135), if u is nonzero,

$$K = \int_0^T N(t)^3 \, dt \lesssim \int_0^T \int \left(\partial_x |P_{\leq \eta_3^{-1}K} u(t,x)|^2\right)^2 dx \, dt \lesssim o(K),$$

with bound independent of T. This implies an upper bound on K as $T \nearrow \infty$. The negative time direction follows from time-reversal symmetry. □

Theorem 5.35 *Suppose u is an almost-periodic solution to* (4.142) *when $d = 1$, on $\mathbf{R}$ with $N(t) \leq 1$ for all $t \in \mathbf{R}$. Suppose also that*

$$\int_{-\infty}^{\infty} N(t)^3 \, dt < \infty. \tag{5.153}$$

Then $u \equiv 0$.

Proof By (4.144), (5.153), and the fundamental theorem of calculus, there exist $\xi_+, \xi_- \in \mathbf{R}$ such that $\xi(t) \to \xi_\pm$ as $t \to \pm\infty$ respectively.

Suppose $\int_T^\infty N(t)^3 \, dt = K$. Again by (4.144), $|\xi(t) - \xi_+| \lesssim \eta_1^{-1} K$ for all $t \in [T, \infty)$. Without loss of generality make a Galilean transformation mapping $\xi_+ \mapsto 0$.

We now need the following intermediate theorem.

Theorem 5.36 *If $\int_T^\infty N(t)^3 \, dt = K$ and u is an almost-periodic solution to* (4.142) *when $d = 1$, and $\xi_+ = 0$, then*

$$\|u\|_{L_t^\infty \dot H_x^{1/2}([T,\infty)\times\mathbf{R}^d)} \lesssim K^{1/2}.$$

5.3 The Mass-Critical Problem When $d = 1$

Proof For any $T_1 < \infty$, $\int_T^{T_1} N(t)^3 dt < K$. Choose T_1 such that for some integer $k_0 > 0$,

$$\int_T^{T_1} \int |u(t,x)|^6 dx\, dt = 2^{k_0}.$$

Taking $\lambda = \eta_3 2^{k_0} / \int_T^{T_1} N(t)^3 dt$ and rescaling $u \mapsto u_\lambda$ gives

$$\|u_\lambda\|_{\widetilde{X}_{k_0}\left(\left[\frac{T}{\lambda^2}, \frac{T_1}{\lambda^2}\right] \times \mathbf{R}\right)} \lesssim 1.$$

Furthermore, Theorem 5.20 implies that for $k \geq k_0$, we have

$$\|P_{>k} u_\lambda\|_{U^2_\Delta\left(\left[\frac{T}{\lambda^2}, \frac{T_1}{\lambda^2}\right] \times \mathbf{R}\right)}$$
$$\lesssim \left\|P_{>k} u_\lambda\left(\frac{T_1}{\lambda^2}\right)\right\|_{L^2_x(\mathbf{R})} + \varepsilon \|P_{>k-5} u_\lambda\|_{U^2_\Delta\left(\left[\frac{T}{\lambda^2}, \frac{T_1}{\lambda^2}\right] \times \mathbf{R}\right)}. \quad (5.154)$$

Rescaling back, since U^2_Δ and L^2 are scale-invariant norms, if $2^k \geq \eta_3^{-1} K$,

$$\|P_{>k} u\|_{U^2_\Delta([T,T_1] \times \mathbf{R})} \lesssim \|P_{>k} u_\lambda(T_1)\|_{L^2_x(\mathbf{R})} + \varepsilon \|P_{>k-5} u_\lambda\|_{U^2_\Delta([T,T_1] \times \mathbf{R})}. \quad (5.155)$$

Now, $N(t) \to 0$ and $\xi(t) \to 0$ as $t \to \infty$, so by (4.80),

$$\lim_{T_1 \to +\infty} \|P_{>k} u(T_1)\|_{L^2_x(\mathbf{R})} = 0. \quad (5.156)$$

Since the $\varepsilon > 0$ in (5.154) may be made arbitrarily small, in particular so that $C\varepsilon \leq 2^{-5}$, where C is the implicit constant in (5.155), for any $t \in [T, \infty)$, it follows that (5.155) and (5.156) imply that for $2^k \geq \eta_3^{-1} K$,

$$\|P_{>k} u(t)\|_{L^2_x(\mathbf{R}^d)} \lesssim 2^{-k} K.$$

This combined with conservation of mass proves Theorem 5.36. Note that $|\xi(t) - \xi_+| \leq \eta_1^{-1} K$ also implies

$$\|e^{-ix\cdot \xi(t)} u(t)\|_{\dot{H}^{1/2}(\mathbf{R})} \lesssim K^{1/2}, \quad (5.157)$$

for $\xi(t) : \mathbf{R} \to \mathbf{R}$ and $\xi(0) = 0$. $\square$

By the dominated convergence theorem,

$$\lim_{T \to \infty} \int_T^\infty N(t)^3 dt = 0,$$

so for a general $\xi_+ \in \mathbf{R}$, (5.157) implies

$$\lim_{t \to +\infty} \|e^{-ix\cdot \xi(t)} u(t)\|_{\dot{H}^{1/2}(\mathbf{R}^d)} = 0.$$

The same analysis can be done as $t \to -\infty$. Therefore, by (4.18),

$$\left\| \partial_x \left(|u|^2 \right) \right\|_{L^2_{t,x}(\mathbf{R}\times\mathbf{R})} = 0,$$

and therefore $u \equiv 0$, completing the proof of Theorem 5.35. □

5.4 The Two-Dimensional Mass-Critical Problem

Proof of scattering for the two-dimensional, mass-critical problem will also use the function space from Definition 5.18. In addition, the proof will utilize a seminorm that measures the size of an almost-periodic solution at frequency scales well separated from the almost-periodic scale.

Definition 5.37 If $G_k^j \subset [a,b]$ define

$$\|u\|_{Y(G_k^j \times \mathbf{R}^2)} = \sum_{5 < i \leq j} 2^{i-j} \sum_{\substack{G_\alpha^i \subset G_k^j: \\ N(G_\alpha^i) \leq 2^{i-5} \eta_3^{1/2}}} \left\| P_{\xi(G_\alpha^i), i-2 \leq \cdot \leq i+2} u \right\|^2_{U_\Delta^2(G_\alpha^i \times \mathbf{R}^2)}$$

$$+ \sum_{\substack{i > \max\{j,5\}: \\ N(G_k^j) \leq 2^{i-5} \eta_3^{1/2}}} \left\| P_{\xi(G_k^j), i-2 \leq \cdot \leq i+2} u \right\|^2_{U_\Delta^2(G_k^j \times \mathbf{R}^2)}.$$

Define $\|u\|_{\widetilde{Y}_I([a,b] \times \mathbf{R}^2)}$ in the manner analogous to (5.99).

By a calculation similar to (5.102)–(5.103), since $2^i \geq 32\eta_3^{-1/2} N(t)$ for all $t \in G_\alpha^i$, if a_α^i is as in (5.101), then (4.145) implies

$$\sum_{5<i\leq j} 2^{i-j} \sum_{\substack{G_\alpha^i \subset G_k^j: \\ N(G_\alpha^i) \leq 2^{i-5} \eta_3^{1/2}}} \left\| P_{\xi(G_\alpha^i), i-2 \leq \cdot \leq i+2} u\left(a_\alpha^i\right) \right\|^2_{L^2_x(\mathbf{R}^d)}$$

$$+ \sum_{\substack{i > \max\{j,5\}: \\ N(G_k^j) \leq 2^{i-5} \eta_3^{1/2}}} \left\| P_{\xi(G_k^j), i-2 \leq \cdot \leq i+2} u\left(a_k^j\right) \right\|^2_{L^2_x(\mathbf{R}^d)} \lesssim \eta_2^2. \qquad (5.157)$$

Remark Because of the projective properties of P_0, compared with those of P_i for $i > 0$ (see Definition 4.32), it is important that $i = 0$ not be included in the Y-seminorm. If $i = 0$ were included, the bound for (5.157) would be 1.

Let us resume the two-dimensional analysis that began in the previous section and paused after the proof of Theorem 5.23.

5.4 The Two-Dimensional Mass-Critical Problem

Theorem 5.38 *When $d = 2$ and u is an almost-periodic solution to (4.142),*

$$\|u\|^2_{X(G^j_k \times \mathbf{R}^d)}$$
$$\lesssim 1 + \sum_{\substack{i > j : N(G^j_k) \\ \leq 2^{i-5} \eta_3^{1/2}}} \left\| \int_{a^j_k}^t e^{i(t-\tau)\Delta} P_{\xi(G^j_k), i-2 \leq \cdot \leq i+2} F(u(\tau)) d\tau \right\|^2_{U^2_\Delta(G^j_k \times \mathbf{R}^d)}$$
$$+ \sum_{0 \leq i \leq j} 2^{i-j} \sum_{\substack{G^i_\alpha \subset G^j_k : \\ N(G^i_\alpha) \leq 2^{i-5} \eta_3^{1/2}}} \left\| \int_{a^i_\alpha}^t e^{i(t-\tau)\Delta} P_{\xi(G^i_\alpha), i-2 \leq \cdot \leq i+2} F(u(\tau)) d\tau \right\|^2_{U^2_\Delta(G^i_\alpha \times \mathbf{R}^d)}$$

and

$$\|u\|^2_{Y(G^j_k \times \mathbf{R}^d)}$$
$$\lesssim \eta_2^2 + \sum_{\substack{i > \max\{j, 5\}: \\ N(G^j_k) \leq 2^{i-5} \eta_3^{1/2}}} \left\| \int_{a^j_k}^t e^{i(t-\tau)\Delta} P_{\xi(G^j_k), i-2 \leq \cdot \leq i+2} F(u(\tau)) d\tau \right\|^2_{U^2_\Delta(G^j_k \times \mathbf{R}^d)} \quad (5.158)$$

$$+ \sum_{5 < i \leq j} 2^{i-j} \sum_{\substack{G^i_\alpha \subset G^j_k : \\ N(G^i_\alpha) \leq 2^{i-5} \eta_3^{1/2}}} \left\| \int_{a^i_\alpha}^t e^{i(t-\tau)\Delta} P_{\xi(G^i_\alpha), i-2 \leq \cdot \leq i+2} F(u(\tau)) d\tau \right\|^2_{U^2_\Delta(G^i_\alpha \times \mathbf{R}^d)}. \quad (5.159)$$

Proof The terms in (5.105)–(5.106) will appear only in the X-norm and not in the Y-norm. □

Thus, in two dimensions the proof of Theorem 5.19 again reduces to proving the intermediate theorem, Theorem 5.20. In this case $\varepsilon_0 > 0$ small in (5.107) will require $\|u\|_{\widetilde{Y}_l}$ to be small, which will follow along from Theorem 5.20, (5.158), (5.159), and the fact that analogously to (5.111),

$$\|u\|_{\widetilde{Y}_{l+1}([0,T] \times \mathbf{R}^2)} \leq 2 \|u\|_{\widetilde{Y}_l([0,T] \times \mathbf{R}^2)}. \quad (5.160)$$

First compute

$$\left\| \int_{a^i_\alpha}^t e^{i(t-\tau)\Delta} P_m \left((P_l u) \left(P_{\xi(\tau), \geq i-10} u \right) u \right) d\tau \right\|_{U^2_\Delta(G^i_\alpha \times \mathbf{R}^2)}. \quad (5.161)$$

By (5.112), to estimate (5.161) it suffices to estimate the inner product of the Duhamel term with a generic f such that $\hat{f}(t, \xi)$ is supported on $|\xi| \sim 2^m$ and $\|f\|_{U^{5/2}_\Delta(G^i_\alpha \times \mathbf{R}^2)} = 1$.

Make a bilinear Strichartz estimate and use (4.145) and $N(G_\alpha^i) \leq 2^{i-5}\eta_3^{1/2}$ to give, if $i \geq 11$,

$$\|f(P_l u)(P_{\xi(t), \geq i-10}u)u\|_{L^1_{t,x}(G_\alpha^i \times \mathbf{R}^2)}$$
$$\lesssim \|f(P_l u)\|^{1/2}_{L^{5/2}_t L^{5/3}_x(G_\alpha^i \times \mathbf{R}^2)} \|f\|^{1/2}_{L^{5/2}_t L^{10}_x(G_\alpha^i \times \mathbf{R}^2)}$$
$$\times \|P_l u\|^{1/2}_{L^{5/2}_t L^{10}_x(G_\alpha^i \times \mathbf{R}^2)} \|P_{\xi(t), \geq i-10}u\|^{1/2}_{L^{5/2}_t L^{10}_x(G_\alpha^i \times \mathbf{R}^2)} \|u\|_{L^\infty_t L^2_x(G_\alpha^i \times \mathbf{R}^2)}$$
$$\lesssim 2^{-\frac{|m-l|}{5}} \|P_l u\|_{U^2_\Delta(G_\alpha^i \times \mathbf{R}^2)} \|P_{\xi(t), \geq i-10}u\|^{5/6}_{L^{25/12}_t L^{50}_x(G_\alpha^i \times \mathbf{R}^2)}$$
$$\times \|P_{\xi(t), \geq i-10}u\|^{1/6}_{L^\infty_t L^2_x(G_\alpha^i \times \mathbf{R}^2)}$$
$$\lesssim 2^{-\frac{|m-l|}{5}} \eta_2^{1/6} \|P_l u\|_{U^2_\Delta(G_\alpha^i \times \mathbf{R}^2)} \|u\|^{5/6}_{X(G_\alpha^i \times \mathbf{R}^2)}. \tag{5.162}$$

For $0 \leq i \leq 10$,

$$(P_l u)(P_{\xi(t), \geq i-10}u)u = (P_l u)u^2.$$

Since $m \geq i - 2$ and $\|u\|_{L^4_{t,x}(G_\alpha^i \times \mathbf{R}^2)} \lesssim 1$, by Strichartz estimates and (4.145),

$$\|P_{\xi(G_\alpha^i), m}((P_l u)u^2)\|_{L^1_t L^2_x(G_\alpha^i \times \mathbf{R}^2)} \lesssim \|P_{|\xi-\xi(t)|>2^{m-5}}u\|^{1/2}_{L^\infty_t L^2_x} \|u\|^{5/2}_{L^{5/2}_t L^{10}_x}$$
$$\lesssim \eta_2^{1/2}.$$

Meanwhile, by the analysis in (5.162),

$$\|f(P_l u)(P_{\xi(t), \geq i-10}u)u\|_{L^1_{t,x}(G_\alpha^i \times \mathbf{R}^2)} \lesssim 2^{-\frac{|m-l|}{5}} \|P_l u\|_{U^2_\Delta(G_\alpha^i \times \mathbf{R}^2)}.$$

Then by interpolation,

$$\|f(P_l u)(P_{\xi(t), \geq i-10}u)u\|_{L^1_{t,x}(G_\alpha^i \times \mathbf{R}^2)} \lesssim 2^{-\frac{|m-l|}{10}} \eta_2^{1/4}\left(1 + \|P_l u\|_{U^2_\Delta(G_\alpha^i \times \mathbf{R}^2)}\right).$$

Therefore, after making the same relabeling and Galilean transform as in the proof of Theorem 5.27, it only remains to estimate, for $j \geq 11$,

$$\left\|\int_{a_k^j}^t e^{i(t-\tau)\Delta}(P_l u)(P_{\xi(\tau), <j-10}u)^2 d\tau\right\|_{U^2_\Delta(G_k^j \times \mathbf{R}^2)}. \tag{5.163}$$

The estimate of (5.163) will utilize the bilinear estimate

$$\||P_l u||P_{\xi(t), <j-10}u|\|_{L^2_{t,x}(G_k^j \times \mathbf{R}^2)}$$
$$\lesssim 2^{\frac{j-l}{2}} \|P_l u\|_{U^2_\Delta(G_k^j \times \mathbf{R}^2)}\left(1 + \|u\|_{\widetilde{X}_j([0,T] \times \mathbf{R}^2)}\right)^2,$$

5.4 The Two-Dimensional Mass-Critical Problem

which, by the definition of U_Δ^2 atoms, follows directly from

$$\left\|v\left(\overline{P_{\xi(t),<j-10}u}\right)\right\|_{L^2_{t,x}\left(G_k^j \times \mathbf{R}^2\right)} \lesssim 2^{\frac{j-l}{2}} \|v_0\|_{L^2(\mathbf{R}^2)} \left(1 + \|u\|_{\widetilde{X}_j([0,T]\times\mathbf{R}^2)}\right)^2,$$

when v solves

$$i\partial_t v - \Delta v = 0, \quad v(0,x) = v_0, \tag{5.164}$$

and $\hat{v}_0(\xi)$ is supported on $|\xi| \sim 2^l$, $l \geq j-5$.

Theorem 5.39 (Two-dimensional bilinear Strichartz estimate) *Suppose $v_0 \in L^2(\mathbf{R}^2)$ has Fourier transform $\hat{v}_0(\xi)$ supported on $|\xi| \sim 2^l$, $l \geq j-5$, and $j \geq 11$. Also suppose that $\|u\|_{\widetilde{Y}_j([0,T]\times\mathbf{R}^2)}$ is small. Then for any $0 \leq l_2 \leq j-10$, $G_\beta^{l_2} \subset G_k^j$,*

$$\left\|v\left(\overline{P_{\xi(t),\leq l_2}u}\right)\right\|^2_{L^2_{t,x}\left(G_\beta^{l_2} \times \mathbf{R}^2\right)} \lesssim 2^{l_2-l} \|v_0\|^2_{L^2(\mathbf{R}^2)} \left(1 + \|u\|_{\widetilde{X}_{l_2}\left(G_k^j \times \mathbf{R}^2\right)}\right)^4. \tag{5.165}$$

Proof First observe that by (4.144) and the fact that $\xi\left(G_\alpha^{l_2}\right) = 0$,

$$v(t,x)\left(\overline{P_{\xi(t),\leq l_2}u}\right) \tag{5.166}$$

is supported on $|\xi| \gtrsim 2^l$. Therefore, following the analysis in (4.33)–(4.35),

$$\left\|v\left(\overline{P_{\xi(t),\leq l_2}u}\right)\right\|^2_{L^2_{t,x}\left(G_\beta^{l_2}\times\mathbf{R}^2\right)}$$

$$\lesssim 2^{l_2-2l} \int_{G_\beta^{l_2}} \int_{S^{d-1}} \int \int_{x_\omega = y_\omega} \left|\partial_\omega\left(v(t,x)\left(\overline{P_{\xi(t),\leq l_2}u(t,y)}\right)\right)\right|^2 dx\,dy\,d\omega\,dt.$$

To simplify notation let $w(t,x) = P_{\xi(t),\leq l_2}u$. Also by (4.25),

$$\int f(y) \frac{(x-y)_\omega}{|(x-y)_\omega|} [g(x)\partial_\omega h(x)] d\omega = C \int f(y) \frac{(x-y)}{|x-y|} \cdot [g(x)\nabla h(x)].$$

Therefore, by $\mu = 1$, (4.26)–(4.30), (5.164), the direct calculation

$$\partial_t w = \partial_t\left(P_{\xi(t),\leq l_2}u\right) = P_{\xi(t),\leq l_2}\left(i\Delta u - iF(u)\right) + \frac{d}{dt}\left(P_{\xi(t),\leq l_2}\right)u$$

$$= i\Delta w - iF(w) + iF\left(P_{\xi(t),\leq l_2}u\right) - iP_{\xi(t),\leq l_2}F(u) + \frac{d}{dt}\left(P_{\xi(t),\leq l_2}\right)u,$$

and the fundamental theorem of calculus,

$$2^{l_2-2l}\int_{G_\beta^{l_2}}\int\int\int_{x_\omega=y_\omega}|\partial_\omega(v(t,y)\overline{w(t,x)})|^2\,dxdyd\omega\,dt$$

$$\lesssim 2^{l_2-2l}\sup_{t\in G_\beta^{l_2}}\left|\frac{d}{dt}M(t)\right| \tag{5.167}$$

$$+2^{l_2-2l}\int_{G_\beta^{l_2}}\int\int|v(t,y)|^2\frac{(x-y)}{|x-y|}\cdot\operatorname{Re}\left[\overline{\mathscr{N}}(\nabla-i\xi(t))w\right](t,x)\,dxdydt \tag{5.168}$$

$$+2^{l_2-2l}\int_{G_\beta^{l_2}}\int\int|v(t,y)|^2\frac{(x-y)}{|x-y|}\cdot\operatorname{Re}\left[\bar{w}(\nabla-i\xi(t))\mathscr{N}\right](t,x)\,dxdydt \tag{5.169}$$

$$+2^{l_2-2l}\int_{G_\beta^{l_2}}\int\int\operatorname{Im}\left[\bar{w}\mathscr{N}\right](t,y)\frac{(x-y)}{|x-y|}\cdot\operatorname{Im}\left[\bar{v}(\nabla-i\xi(t))v\right](t,x)\,dxdydt, \tag{5.170}$$

where

$$\mathscr{N} = iF\left(P_{\xi(t),\leq l_2}u\right) - iP_{\xi(t),\leq l_2}F(u) + \frac{d}{dt}\left(P_{\xi(t),\leq l_2}\right)u$$

$$= \mathscr{N}_1 + \frac{d}{dt}\left(P_{\xi(t),\leq l_2}\right)u = \mathscr{N}_1 + \mathscr{N}_2$$

and

$$\partial_t M(t) = \int\int|w(t,y)|^2\frac{(x-y)}{|x-y|}\cdot\operatorname{Im}\left[\bar{v}(\nabla-i\xi(t))v\right](t,x)\,dxdy$$

$$+\int\int|v(t,y)|^2\frac{(x-y)}{|x-y|}\cdot\operatorname{Im}\left[\bar{w}(\nabla-i\xi(t))w\right](t,x)\,dxdy$$

$$\lesssim 2^l\|w\|^2_{L_t^\infty L_x^2}\|v_0\|^2_{L_x^2} \lesssim 2^l\|v_0\|^2_{L^2}. \tag{5.171}$$

Remark Here, there is again an abuse of the notation in (4.26)–(4.30) since (5.171) need not be the time derivative of any quantity that is in the form of the $M(t)$ in (4.26)–(4.30).

It is easy to see that $\sup_{t\in G_\alpha^i}2^{l_2-2l}|\partial_t M(t)|$ is bounded by the right-hand side of (5.165).

Now for (5.168)–(5.170), to simplify notation, every $L_t^p L_x^q$-norm will be taken over $G_\beta^{l_2}\times\mathbf{R}^2$ unless otherwise stated. Since $P_{\xi(t),\leq l_2}$ is a Fourier multiplier whose symbol is $\phi\left(\frac{\xi-\xi(t)}{2^{l_2}}\right)$, we have

$$\frac{d}{dt}\phi\left(\frac{\xi-\xi(t)}{2^{l_2}}\right) = \nabla\phi\left(\frac{\xi-\xi(t)}{2^{l_2}}\right)\cdot\frac{\xi'(t)}{2^{l_2}}, \tag{5.172}$$

5.4 The Two-Dimensional Mass-Critical Problem

so

(5.168)

$$\lesssim 2^{-2l} \|v\|_{L_t^\infty L_x^2}^2 \left(\int_{G_\beta^{l_2}} |\xi'(t)| \|P_{\xi(t), l_2-5 \leq \cdot \leq l_2+5} u(t)\|_{L^2} \|(\nabla - i\xi(t))w\|_{L^2} dt \right) \tag{5.173}$$

$$+ 2^{l_2-2l} \|v\|_{L_t^\infty L_x^2}^2 \|\mathcal{N}_1\|_{L_t^{3/2} L_x^{6/5}} \|(\nabla - i\xi(t))w\|_{L_t^3 L_x^6}. \tag{5.174}$$

By (4.144), (4.147), and conservation of mass,

$$(5.173) \lesssim 2^{2l_2 - 2l} \|v_0\|_{L_x^2}^2.$$

Next, by definition of w, again adopting the convention $\widetilde{\nabla} = \nabla - i\xi(t)$,

$$\|\widetilde{\nabla} w\|_{L_t^3 L_x^6 (G_\beta^{l_2} \times \mathbf{R}^2)} \lesssim \left(\sum_{0 \leq k \leq l_2} 2^k \|P_{\xi(t), k} u\|_{L_t^{5/2} L_x^{10} (G_\beta^{l_2} \times \mathbf{R}^2)} \right)^{5/6} \|\widetilde{\nabla} w\|_{L_t^\infty L_x^2}^{1/6}$$

$$\lesssim 2^{l_2} \|u\|_{\widetilde{X}_{l_2}}. \tag{5.175}$$

Therefore,

$$(5.168) \lesssim 2^{2l_2 - 2l} \|v_0\|_{L^2}^2 \|u\|_{\widetilde{X}_{l_2}(G_\beta^{l_2} \times \mathbf{R}^2)}$$
$$\times \|P_{\xi(t), \leq l_2} F(u) - F(P_{\xi(t), \leq l_2} u)\|_{L_t^{3/2} L_x^{6/5}}.$$

Split u as $u_h + u_l$, where $u_l = P_{\xi(t), \leq l_2 - 5} u$. Following the same analysis as in (5.139),

$$P_{\xi(t), \leq l_2} F(u_l) - F(P_{\xi(t), \leq l_2} u_l) = 0. \tag{5.176}$$

Next, as in (5.140)–(5.141),

$$\|P_{\xi(t), \leq l_2} (u_l^2 u_h) - (P_{\xi(t), \leq l_2} u_l)^2 (P_{\xi(t), \leq l_2} u_h)\|_{L_t^{3/2} L_x^{6/5}}$$
$$\lesssim 2^{-l_2} \|u_h\|_{L_t^3 L_x^6} \|\widetilde{\nabla} w\|_{L_t^3 L_x^6} \|u\|_{L_t^\infty L_x^2} \lesssim \|u\|_{\widetilde{X}_{l_2}}^2. \tag{5.177}$$

Finally,

$$\|u u_h^2\|_{L_t^{3/2} L_x^{6/5}} \lesssim \|u_h\|_{L_t^3 L_x^6}^2 \|u\|_{L_t^\infty L_x^2} \lesssim \|u\|_{\widetilde{X}_{l_2}}^2. \tag{5.178}$$

Therefore,

$$(5.174) \lesssim 2^{2l_2 - 2l} \|v_0\|_{L^2(\mathbf{R}^2)}^2 \left(1 + \|u\|_{\widetilde{X}_{l_2}(G_\beta^{l_2} \times \mathbf{R}^2)}\right)^3$$

and

$$(5.168) \lesssim 2^{2l_2 - 2l} \|v_0\|_{L^2}^2 \left(1 + \|u\|_{\widetilde{X}_j(G_k^j \times \mathbf{R}^2)}\right)^3. \tag{5.179}$$

The right-hand side of this term is clearly bounded by the right-hand side of (5.165).

Turning now to (5.169), integrating by parts in space gives

$$(5.169) = (5.168) - 2^{l_2 - 2l} \int_{G_\beta^{l_2}} \int \int |v(t,y)|^2 \frac{1}{|x-y|} \operatorname{Re}[\bar{w}\mathcal{N}](t,x)\,dx\,dy\,dt.$$

Using the Hardy–Littlewood–Sobolev theorem, Hardy's inequality, the Sobolev embedding theorem, (4.144), (4.147), (5.172), (5.177), and (5.178), we get

$$2^{l_2 - 2j} \int_{G_\beta^{l_2}} \int \int \frac{1}{|x-y|} |v(t,y)|^2 \operatorname{Re}[\bar{w}\mathcal{N}](t,x)\,dx\,dy\,dt$$

$$\lesssim 2^{-2j} \int_{G_\beta^{l_2}} |\xi'(t)| \|v(t)\|_{L^2}^2 \|\tilde{\nabla} w(t)\|_{L^2} \|P_{\xi(t),l_2-5 \leq \cdot \leq l_2+5} w(t)\|_{L^2}\,dt$$

$$+ 2^{l_2 - 2j} \|v\|_{L_T^\infty L_x^2} \|v\|_{L_T^3 L_x^6} \|\mathcal{N}_1\|_{L_t^{3/2} L_x^2} \|w\|_{L_t^\infty L_x^3}$$

$$\lesssim 2^{2l_2 - 2j} \|v_0\|_{L^2}^2 + \|v_0\|_{L^2}^2 \|\tilde{\nabla}\mathcal{N}_1\|_{L_t^{3/2} L_x^{6/5}}^{2/3} \|\mathcal{N}_1\|_{L_t^{3/2} L_x^{6/5}}^{1/3} \|\tilde{\nabla} w\|_{L_t^\infty L_x^2}^{1/3} \|w\|_{L_t^\infty L_x^3}^{2/3}$$

$$\lesssim 2^{2l_2 - 2j} \|v_0\|_{L^2}^2 \left(1 + \|u\|_{\tilde{X}_{l_2}}\right)^3. \tag{5.180}$$

This term is also clearly bounded by the right-hand side of (5.165).

Now consider (5.170). Again by conservation of mass, (4.144), and (4.147),

$$2^{l_2 - 2j} \int_{G_\beta^{l_2}} \int \int \operatorname{Im}[\mathcal{N}_2 \bar{w}](t,y) \frac{(x-y)}{|x-y|} \cdot \operatorname{Im}[\bar{v}(\nabla - i\xi(t))v](t,x)\,dx\,dy\,dt$$

$$\lesssim 2^{-j} \|v_0\|_{L^2}^2 \left(\int_{G_\beta^{l_2}} |\xi'(t)| \|P_{\xi(t),l_2-5 \leq \cdot \leq l_2+5} u(t)\|_{L^2} \|w(t)\|_{L^2}\,dt\right)$$

$$\lesssim 2^{l_2 - j} \|v_0\|_{L^2}^2. \tag{5.181}$$

To analyze the contribution of $\mathcal{N}_1$, first observe that

$$\operatorname{Im}\left[P_{\xi(t),\leq l_2} F(u) \overline{P_{\xi(t),\leq l_2} w}\right]$$
$$= \operatorname{Im}\left[\left(P_{\xi(t),\leq l_2} F(u) - F\left(P_{\xi(t),\leq l_2} u\right)\right) \left(\overline{P_{\xi(t),\leq l_2} u}\right)\right].$$

Recall from (5.176) that

$$P_{\xi(t),\leq l_2} F(u_l) - F\left(P_{\xi(t),\leq l_2} u_l\right) = 0. \tag{5.182}$$

Also, as in the one-dimensional case,

$$\left[P_{\xi(t),\leq l_2}\left(u_l^2 u_h\right) - \left(P_{\xi(t),\leq l_2} u_l\right)^2 \left(P_{\xi(t),\leq l_2} u_h\right)\right] \left(\overline{P_{\xi(t),\leq l_2} u_l}\right)$$

5.4 The Two-Dimensional Mass-Critical Problem   221

is supported on $|\xi| \sim 2^{l_2}$, so as in (5.152), using the Hardy–Littlewood–Sobolev inequality and the Sobolev embedding theorem, yields

$$2^{l_2-2l} \int_{G_\beta^{l_2}} \int \int \mathrm{Im}\left[[P_{\xi(t),\leq l_2}\left(u_l^2 u_h\right)\right.$$

$$\left. - \left(P_{\xi(t),\leq l_2}u_l\right)^2 \left(P_{\xi(t),\leq l_2}u_h\right)]\left(\overline{P_{\xi(t),\leq l_2}u_l}\right)\right](t,y)$$

$$\times \frac{(x-y)}{|x-y|} \cdot \mathrm{Im}\left[\bar{v}(\nabla - i\xi(t))v\right](t,x)\,dx\,dy\,dt$$

$$\lesssim 2^{-2l} \int_{G_\beta^{l_2}} \int \int |u_h(t,y)||u_l(t,y)|^3 \frac{1}{|x-y|}|v(t,x)||\widetilde{\nabla}v(t,x)|\,dx\,dy\,dt$$

$$\lesssim 2^{-2l} \|u_h\|_{L_t^3 L_x^6} \|u_l\|_{L_{t,x}^6}^3 \|v\|_{L_t^\infty L_x^2} \|\widetilde{\nabla}v\|_{L_t^3 L_x^6}^{1/2} \|\widetilde{\nabla}v\|_{L_t^\infty L_x^2}^{1/2}$$

$$\lesssim 2^{-2l} \|u_h\|_{L_t^3 L_x^6} \||\widetilde{\nabla}|^{1/2} u_l\|_{L_{t,x}^4}^2 \|u_l\|_{L_t^\infty L_x^2} \|v\|_{L_t^\infty L_x^2} \|\widetilde{\nabla}v\|_{L_t^3 L_x^6}^{1/2} \|\widetilde{\nabla}v\|_{L_t^\infty L_x^2}^{1/2}$$

$$\lesssim 2^{l_2-l} \|v_0\|_{L^2}^2 \|u\|_{\widetilde{X}_{l_2}}^3. \tag{5.183}$$

This term is bounded by the right-hand side of (5.165). Then

$$2^{l_2-2l} \int_{G_\beta^{l_2}} \int \int \mathrm{Im}\left[u_h^2 u^2\right](t,y)\frac{(x-y)}{|x-y|} \cdot \mathrm{Im}\left[\bar{v}(\nabla - i\xi(t))v\right](t,x)\,dx\,dy\,dt$$

$$\lesssim 2^{l_2-2l}\left(\|u_h u_l\|_{L_{t,x}^2(G_\beta^{l_2}\times \mathbf{R}^2)}^2 + \|u_h\|_{L_{t,x}^4(G_\beta^{l_2}\times \mathbf{R}^2)}^4\right)$$

$$\times \|v\|_{L_t^\infty L_x^2(G_\beta^{l_2}\times \mathbf{R}^2)}\|\widetilde{\nabla}v\|_{L_t^\infty L_x^2(G_\beta^{l_2}\times \mathbf{R}^2)}$$

$$\lesssim 2^{l_2-l}\|v_0\|_{L^2}^2\left(\|u_h u_l\|_{L_{t,x}^2(G_\beta^{l_2}\times \mathbf{R}^2)}^2 + \|u_h\|_{L_{t,x}^4(G_\beta^{l_2}\times \mathbf{R}^2)}^4\right). \tag{5.184}$$

By direct computation,

$$2^{l_2-l}\|v_0\|_{L^2}^2 \|u_h\|_{L_{t,x}^4(G_\beta^{l_2}\times \mathbf{R}^2)}^4 \lesssim 2^{l_2-l}\|v_0\|_{L^2}^2\|u\|_{\widetilde{X}_{l_2}}^4 \tag{5.185}$$

and

$$2^{l_2-l}\|v_0\|_{L^2}^2 \|u_h^2\left(P_{\xi(t),\geq j-15}u\right)^2\|_{L_{t,x}^1(G_\beta^{l_2}\times \mathbf{R}^2)}$$

$$\lesssim 2^{l_2-l}\|v_0\|_{L^2}^2 \|u_h\|_{L_{t,x}^4}^2 \|P_{\xi(t),\geq j-15}u\|_{L_{t,x}^4}^2 \lesssim 2^{l_2-l}\|v_0\|_{L^2}^2 \|u\|_{\widetilde{X}_{l_2}}^4. \tag{5.186}$$

Therefore, fixing $l_2 = j - 10$ and $G_\beta^{l_2} = G_k^j$ (which is an acceptable abuse of notation since G_k^j is the union of a bounded number, 2^{10}, of G_β^{j-10} intervals) and collecting (5.179), (5.180), (5.181), (5.184), (5.185), and (5.186) implies

$$2^{l-j}\left\|(P_l u)\left(P_{\xi(t),\leq j-10}u\right)\right\|^2_{L^2_{t,x}(G^j_k\times\mathbf{R}^2)}$$

$$\lesssim \|P_l u\|^2_{U^2_\Delta(G^j_k\times\mathbf{R}^2)}\left(1+\|u\|_{\widetilde{X}_j(G^j_k\times\mathbf{R}^2)}\right)^4$$

$$+\|P_l u\|^2_{U^2_\Delta(G^j_k\times\mathbf{R}^2)}\left\|\left(P_{\xi(t),\geq j-5}u\right)\left(P_{\xi(t),\leq j-15}u\right)\right\|^2_{L^2_{t,x}(G^j_k\times\mathbf{R}^2)}$$

$$\lesssim \|P_l u\|^2_{U^2_\Delta(G^j_k\times\mathbf{R}^2)}\left(1+\|u\|_{\widetilde{X}_j(G^j_k\times\mathbf{R}^2)}\right)^4$$

$$+\|P_l u\|^2_{U^2_\Delta(G^j_k\times\mathbf{R}^2)}\left(\sup_{\substack{0\leq l_2\\ \leq j-5}}\sup_{G^{l_2}_\alpha\subset G^j_k}\left\|\left(P_{\xi(t),\geq l_2}u\right)\left(P_{\xi(t),\leq l_2-10}u\right)\right\|^2_{L^2_{t,x}(G^{l_2}_\alpha\times\mathbf{R}^2)}\right).$$

Therefore,

$$\sup_{0\leq j\leq k_0}\sup_{\substack{G^j_k\subset[0,T]:\\ N(G^j_k)\leq 2^{j-10}\eta_3^{1/2}}}\sup_{l\geq j-5} 2^{l-j}\left\|\left(P_{\xi(G^j_k),l}u\right)\left(P_{\xi(t),\leq j-10}u\right)\right\|^2_{L^2_{t,x}(G^j_k\times\mathbf{R}^2)}$$

$$\lesssim \|u\|^2_{\widetilde{Y}_{k_0}([0,T]\times\mathbf{R}^2)}\left(1+\|u\|_{\widetilde{X}_{k_0}([0,T]\times\mathbf{R}^2)}\right)^4+\|u\|^2_{\widetilde{Y}_{k_0}([0,T]\times\mathbf{R}^2)}$$

$$\times\left(\sup_{\substack{0\leq l_2\leq\\ k_0-5}}\sup_{G^{l_2}_\alpha\subset[0,T]}\sup_{k\geq l_2-5} 2^{k-l_2}\left\|\left(P_{\xi(G^{l_2}_\alpha),k}u\right)\left(P_{\xi(t),\leq l_2-10}u\right)\right\|^2_{L^2_{t,x}(G^{l_2}_\alpha\times\mathbf{R}^2)}\right).$$

(5.187)

Also, trivially, by (4.144), (4.145), and (4.151), if $N(G^{l_2}_\alpha)\geq 2^{l_2-10}\eta_3^{1/2}$, then

$$\|u_h u_l\|^2_{L^2_{t,x}(G^{l_2}_\alpha\times\mathbf{R}^2)}\lesssim \|u\|^4_{L^4_{t,x}(G^{l_2}_\alpha\times\mathbf{R}^2)}\lesssim 1. \tag{5.188}$$

Plugging (5.188) into the right-hand side of (5.187) if $N(G^{l_2}_\alpha)\geq 2^{l_2-10}\eta_3^{1/2}$, then a bootstrap argument combined with (5.186)–(5.188) implies that when $\|u\|_{\widetilde{Y}_j([0,T]\times\mathbf{R}^2)}$ is small we have

$$\sup_{0\leq j\leq k_0}\sup_{G^j_k\subset[0,T]}\sup_{l\geq j-5} 2^{l-j}\left\|\left(P_{\xi(G^j_k),l}u\right)\left(P_{\xi(t),\leq j-10}u\right)\right\|^2_{L^2_{t,x}(G^j_k\times\mathbf{R}^2)}$$

$$\lesssim \left(1+\|u\|_{\widetilde{X}_{k_0}([0,T]\times\mathbf{R}^2)}\right)^4. \tag{5.189}$$

Combining (5.189) and (5.184) completes the proof of Theorem 5.39. □

Theorem 5.39 may be easily upgraded to an l^2 summation theorem.

5.4 The Two-Dimensional Mass-Critical Problem

Theorem 5.40 (Second bilinear Strichartz estimate) *For any $0 \leq j \leq k_0$,*

$$\sum_{0 \leq l_2 \leq j-10} \left\| \left(e^{it\Delta} v_0\right) \left(P_{\xi(t), \leq l_2} u\right) \right\|^2_{L^2_{t,x}(G^j_k \times \mathbf{R}^2)}$$
$$\lesssim 2^{j-l} \|v_0\|^2_{L^2} \left(1 + \|u\|_{\widetilde{X}_j(G^j_k \times \mathbf{R}^2)}\right)^6.$$

Proof Again use the bilinear interaction Morawetz estimate. Recall (5.167)–(5.170). It is straightforward to verify that as in (5.171),

$$\sum_{0 \leq l_2 \leq j-10} 2^{l_2 - 2l} \sup_{t \in G^j_k} |\partial_t M(t)| \lesssim \sum_{0 \leq l_2 \leq j-10} 2^{l_2 - l} \|v_0\|^2_{L^2} \lesssim 2^{j-l} \|v_0\|^2_{L^2}.$$

Next, since there are 2^{j-l_2} intervals $G^{l_2}_\alpha$ contained in G^j_k,

$$\sum_{0 \leq l_2 \leq j-10} \sum_{G^{l_2}_\alpha \subset G^j_k} (5.179) \lesssim 2^{j-l} \|v_0\|^2_{L^2} \left(1 + \|u\|_{\widetilde{X}_j([0,T] \times \mathbf{R}^2)}\right)^3$$

and

$$\sum_{0 \leq l_2 \leq j-10} \sum_{G^{l_2}_\alpha \subset G^j_k} (5.180) \lesssim \|v_0\|^2_{L^2} \left(1 + \|u\|_{\widetilde{X}_j([0,T] \times \mathbf{R}^2)}\right)^3.$$

This takes care of the terms generated by (5.168) and (5.169).

Now consider the terms generated by (5.170). First take (5.184). By Definition 5.18, since $l \geq j - 5$,

$$\sum_{0 \leq l_2 \leq j-10} 2^{l_2 - l} \|v_0\|^2_{L^2} \|P_{\xi(t), \geq l_2 - 5} u\|^4_{L^4_{t,x}(G^j_k \times \mathbf{R}^2)} \lesssim \|v_0\|^2_{L^2} \|u\|^4_{\widetilde{X}_j([0,T] \times \mathbf{R}^2)}.$$

Also, by (5.165), the fact that there are 2^{m-l_2} intervals $G^{l_2}_\alpha \subset G^m_\beta$ when $m \leq j$, Definition 5.18, and Young's inequality,

$$\sum_{0 \leq l_2 \leq j-10} 2^{l_2 - l} \|v_0\|^2_{L^2} \left\| \left(P_{\xi(t), \geq l_2 - 5} u\right) \left(P_{\xi(t), \leq l_2 - 5} u\right) \right\|^2_{L^2_{t,x}(G^j_k \times \mathbf{R}^2)}$$
$$\lesssim \sum_{0 \leq l_2 \leq j-10} \|v_0\|^2_{L^2} \sum_{m \geq l_2 - 5} 2^{l_2 - l} \|P_{\xi(t), m} u\|^2_{U^2_\Delta(G^j_k \times \mathbf{R}^2)} \left(1 + \|u\|_{\widetilde{X}_j([0,T] \times \mathbf{R}^2)}\right)^4$$
$$\lesssim 2^{j-l} \|v_0\|^2_{L^2} \left(1 + \|u\|_{\widetilde{X}_j([0,T] \times \mathbf{R}^2)}\right)^6.$$

This estimates (5.184). Next, summing (5.183) gives

$$2^{-2l} \sum_{0 \leq l_2 \leq j-10} \|P_{\xi(t), \geq l_2-5} u\|_{L_t^3 L_x^6} \|P_{\xi(t), \leq l_2-5} u\|_{L_{t,x}^6}^3 \|\widetilde{\nabla} v\|_{L_t^6 L_x^3} \|v\|_{L_t^\infty L_x^2}$$

$$\lesssim 2^{-l} \sum_{0 \leq l_2 \leq j-10} \|v_0\|_{L^2}^2 \|P_{\xi(t), \geq l_2-5} u\|_{L_t^3 L_x^6(G_k^j \times \mathbf{R}^2)} \|P_{\xi(t), \leq l_2-5} u\|_{L_{t,x}^6(G_k^j \times \mathbf{R}^2)}^3$$

$$\lesssim 2^{-l} \sum_{0 \leq l_2 \leq j-10} \|v_0\|_{L^2}^2 \sum_{l_2-5 \leq m \leq j} \sum_{G_\alpha^m \subset G_k^j} \|P_{\xi(t),m} u\|_{L_t^3 L_x^6(G_\alpha^m \times \mathbf{R}^2)}$$

$$\times \|P_{\xi(t), \leq l_2-5} u\|_{L_{t,x}^6(G_\alpha^m \times \mathbf{R}^2)}^3$$

$$+ 2^{-l} \sum_{0 \leq l_2 \leq j-10} \|v_0\|_{L^2}^2 \|P_{\xi(t), \geq j} u\|_{L_t^3 L_x^6(G_k^j \times \mathbf{R}^2)} \|P_{\xi(t), \leq l_2-5} u\|_{L_{t,x}^6(G_\alpha^m \times \mathbf{R}^2)}^3$$

$$\lesssim 2^{-l} \sum_{0 \leq l_2 \leq j-10} \|v_0\|_{L^2}^2 \sum_{l_2-5 \leq m \leq j} \|P_{\xi(t),m} u\|_{U_\Delta^2(G_k^j \times \mathbf{R}^2)} \|P_{\xi(t), \leq l_2-5} u\|_{L_{t,x}^6(G_k^j \times \mathbf{R}^2)}^3$$

$$+ 2^{-l} \sum_{0 \leq l_2 \leq j-10} \|v_0\|_{L^2}^2 \|P_{\xi(t), \geq j} u\|_{U_\Delta^2(G_k^j \times \mathbf{R}^2)} \|P_{\xi(t), \leq l_2-5} u\|_{L_{t,x}^6(G_k^j \times \mathbf{R}^2)}^3.$$

(5.190)

Now by Young's inequality and Definition 5.18,

$$2^{-l} \sum_{0 \leq l_2 \leq j-10} 2^{l_2} \|v_0\|_{L^2}^2 \left(\sum_{l_2-5 \leq m \leq j} \|P_{\xi(t),m} u\|_{U_\Delta^2(G_k^j \times \mathbf{R}^2)} \right)^2$$

$$\lesssim 2^{j-l} \|v_0\|_{L^2}^2 \|u\|_{\widetilde{X}_j([0,T] \times \mathbf{R}^2)}^2$$

and

$$2^{-l} \sum_{0 \leq l_2 \leq j-10} 2^{-l_2} \|v_0\|_{L^2}^2 \left(\sum_{0 \leq m \leq l_2-5} 2^{m/3} \|P_{\xi(t),m} u\|_{U_\Delta^2(G_k^j \times \mathbf{R}^2)}^{1/3} \right.$$

$$\left. \times \sup_{G_\alpha^m \subset G_k^j} \|P_{\xi(t),m} u\|_{U_\Delta^2(G_\alpha^m \times \mathbf{R}^2)}^{1/3} \|P_{\xi(t),m} u\|_{L_t^\infty L_x^2} \right)^6$$

$$\lesssim 2^{j-l} \|v_0\|_{L^2}^2 \|u\|_{\widetilde{X}_j([0,T] \times \mathbf{R}^2)}^6.$$

Therefore, by the Cauchy–Schwarz inequality and the Sobolev embedding theorem,

$$(5.190) \lesssim 2^{j-l} \|v_0\|_{L^2}^2 \|u\|_{\widetilde{X}_j([0,T] \times \mathbf{R}^2)}^6.$$

It only remains to calculate

$$\sum_{0 \leq l_2 \leq j-10} 2^{l_2-2l} \int_{G_k^j} \int \int \mathrm{Im} \left[\mathcal{N}_2 \left(\overline{P_{\xi(t), \leq l_2} u} \right) \right](t,y) \frac{(x-y)}{|x-y|}$$

$$\times \mathrm{Im} \left[\bar{v}(\nabla - i\xi(t)) v \right](t,x) \, dx \, dy \, dt.$$

5.4 The Two-Dimensional Mass-Critical Problem

Now by (5.172),
$$\text{Im}\left[\mathcal{N}_2\left(\overline{P_{\xi(t),\leq l_2-10}u}\right)\right]$$
is supported on $|\xi| \sim 2^{l_2}$. Therefore, again as in (5.183),

$$\sum_{0\leq l_2\leq j-10} 2^{l_2-2l} \int_{G_k^j} \int \int \text{Im}\left[\mathcal{N}_2\left(\overline{P_{\xi(t),\leq l_2-10}u}\right)\right](t,y) \frac{(x-y)}{|x-y|}$$
$$\times \text{Im}\left[\bar{v}(\nabla - i\xi(t))v\right](t,x) \, dx \, dy \, dt$$
$$\lesssim \sum_{0\leq l_2\leq j-10} 2^{-2l} \int_{G_k^j} \int \int \frac{1}{|x-y|} |\mathcal{N}_2(t,y)| \, |P_{\xi(t),\leq l_2-10}u(t,y)|$$
$$\times |v(t,x)| \, |(\nabla - i\xi(t))v(t,x)| \, dx \, dy \, dt,$$

which, by Hardy's inequality and (5.172),

$$\lesssim 2^{-l} \sum_{\substack{0\leq l_2\leq \\ j-10}} \|v_0\|_{L^2}^2 \int_{G_k^j} \|\widetilde{\nabla}P_{\xi(t),\leq l_2-10}u(t)\|_{L^2} \|P_{\xi(t),l_2-5\leq\cdot\leq l_2+5}u(t)\|_{L^2} \, dt$$
$$\lesssim 2^{j-l} \|v_0\|_{L^2}^2.$$

Also by Bernstein's inequality,

$$\sum_{0\leq l_2\leq j-10} 2^{l_2-2l} \int_{G_k^j} \int \int \text{Im}\left[\mathcal{N}_2\left(\overline{P_{\xi(t),l_2-10\leq\cdot\leq}u}\right)\right](t,y) \frac{(x-y)}{|x-y|}$$
$$\times \text{Im}[\bar{v}(\nabla - i\xi(t))v](t,x) \, dx \, dy \, dt$$
$$\lesssim 2^{-l} \sum_{0\leq l_2\leq j-10} \|v_0\|_{L^2}^2 \int_{G_k^j} \|\widetilde{\nabla}P_{\xi(t),\leq l_2-10}u(t)\|_{L^2}$$
$$\times \|P_{\xi(t),l_2-5\leq\cdot\leq l_2+5}u(t)\|_{L^2} \, dt$$
$$\lesssim 2^{j-l} \|v_0\|_{L^2}^2.$$

Therefore, by the Cauchy–Schwarz inequality,

$$\sum_{0\leq l_2\leq j-10} (5.181) \lesssim 2^{j-l} \|v_0\|_{L^2}^2.$$

This completes the proof of Theorem 5.40. $\square$

Turning now to estimating (5.163), choose a constant C_0 sufficiently large and $n(t) \in \mathbb{Z}_{\geq 0}$ so that if $i \leq n(t) + C_0$, then $2^i \leq 64N(t)\eta_3^{-1/2}$, $2^{n(t)} \sim N(t)$, and $C_0 \sim \eta_3^{-1/2}$. Then decompose:

$$\left(P_{\xi(t),\leq j-10}u\right)^2 = \sum_{0\leq l_2\leq j-10} \left(P_{\xi(t),l_2}u\right) \cdot \left(P_{\xi(t),\leq l_2}u\right)$$
$$= \sum_{n(t)+C_0 < l_2 \leq j-10} \left(P_{\xi(t),l_2}u\right) \cdot \left(P_{\xi(t),\leq l_2}u\right) + \left(P_{\xi(t),\leq n(t)+C_0}u\right)^2.$$

For each t, let
$$v(t) = \sum_{\substack{\max\{n(t)+C_0,5\} \\ <l_2 \leq j-10}} \left(P_{l_2} u(t)\right).$$

Then compute the size of the Duhamel term.

Lemma 5.41 *As in Lemma 5.28 let $\mathscr{X}_{l_2}$ be the partition of G_k^j that includes a_k^j and the endpoints of $G_{\bar{\alpha}}^{l_2} \subset G_k^j$. Then*

$$\sum_{0 \leq l_2 \leq j-10} \left\| \int_{a_k^j}^t e^{i(t-\tau)\Delta} \left(P_l u\right) \left(P_{\xi(t),l_2} v\right) \left(P_{\xi(t),\leq l_2} u\right) d\tau \right\|_{V_\Delta^1(\mathscr{X}_{l_2}; G_k^j \times \mathbf{R}^2)}$$
$$\lesssim 2^{j-l} \|P_l u\|_{U_\Delta^2(G_k^j \times \mathbf{R}^2)} \|u\|_{\widetilde{Y}_j([0,T] \times \mathbf{R}^2)} \left(1 + \|u\|_{\widetilde{X}_j([0,T] \times \mathbf{R}^2)}\right)^3. \quad (5.191)$$

Proof To compute (5.191), it suffices to compute the inner product of the Duhamel term with a generic u_0 satisfying $\|u_0\|_{L^2} = 1$ and $\hat{u}_0(\xi)$ supported on $|\xi| \sim 2^l$. Making a bilinear estimate,

$$\int_{G_{\bar{\alpha}}^{l_2}} \langle e^{it\Delta} u_0, (P_l u) \left(P_{\xi(t),l_2} v\right) \left(P_{\xi(t),\leq l_2} u\right) \rangle dt$$
$$\lesssim 2^{\frac{l_2-l}{2}} \|u_0\|_{L^2(\mathbf{R}^2)} \|P_{\xi(t),l_2} v\|_{U_\Delta^2(G_{\bar{\alpha}}^{l_2} \times \mathbf{R}^2)} \|(P_l u)\left(P_{\xi(t),\leq l_2} u\right)\|_{L^2_{t,x}(G_{\bar{\alpha}}^{l_2} \times \mathbf{R}^2)}.$$

Now recall that if $N(G_\alpha^i) \geq 2^{i-5} \eta_3^{1/2}$ then $N(t) \geq 2^{i-6} \eta_3^{1/2}$ for all $t \in G_\alpha^i$. Therefore, on the support in time and Fourier space of v, by Theorem 5.40 and the Cauchy–Schwarz inequality,

$$(5.191) \lesssim \sum_{0 \leq l_2 \leq j-10} 2^{\frac{l_2-l}{2}} \|P_{\xi(t),l_2} u\|_{U_\Delta^2(G_{\bar{\alpha}}^{l_2} \times \mathbf{R}^2)} \|(P_l u)\left(P_{\xi(t),\leq l_2} u\right)\|_{L^2_{t,x}(G_{\bar{\alpha}}^{l_2} \times \mathbf{R}^2)}$$
$$\lesssim 2^{j-l} \|P_l u\|_{U_\Delta^2(G_k^j \times \mathbf{R}^2)} \|u\|_{\widetilde{Y}_j([0,T] \times \mathbf{R}^2)} \left(1 + \|u\|_{\widetilde{X}_j([0,T] \times \mathbf{R}^2)}\right)^3.$$

This proves Lemma 5.41. □

Lemma 5.42 *Let $\mathscr{X}$ be a partition of G_k^j that includes a_k^j and the endpoints of the small intervals J_m such that $J_m \cap G_k^j \neq \emptyset$ and $N(G_k^j) \leq \eta_3^{1/2} 2^{j-5}$. Then*

$$\left\| \int_{a_k^j}^t e^{i(t-\tau)\Delta} \left(P_l u\right) \left(P_{\xi(t), \leq \max\{n(t)+C_0,5\}} u\right)^2 d\tau \right\|_{V_\Delta^1(\mathscr{X}; G_k^j \times \mathbf{R}^2)}$$
$$\lesssim \eta_2^2 2^{j-l} \|P_l u\|_{U_\Delta^2(G_k^j \times \mathbf{R}^2)}. \quad (5.192)$$

5.4 The Two-Dimensional Mass-Critical Problem

Remark Previously, the small intervals were usually denoted J_l. To avoid confusion with P_l, in the proof of Lemma 5.41 they will be written as J_m.

Proof By a computation similar to (5.124)–(5.127) using bilinear Strichartz estimates, if $5 \leq n(t) + C_0$, we have

$$\int_{J_m} \left\langle u_0, e^{i(t-\tau)\Delta} (P_l u) \left(P_{\xi(t), \leq n(t)+C_0} u \right)^2 d\tau \right\rangle$$

$$\lesssim \eta_3^{-1/2} 2^{-l} N(J_m) \|u_0\|_{L^2} \|P_l u\|_{U_\Delta^2(G_k^j \times \mathbf{R}^2)}.$$

By (4.151) and (4.152), summing over $J_m \subset G_k^j$ is bounded by the right-hand side of (5.192).

If $n(t) + C_0 \leq 5$ then splitting, we find

$$\int_{J_m} \left\langle u_0, e^{i(t-\tau)\Delta} (P_l u) \left(P_{\xi(t), \leq 5} u \right)^2 d\tau \right\rangle$$

$$\lesssim \left\| (e^{it\Delta} u_0) \left(P_{|\xi - \xi(t)| \leq \eta_3^{-1/2} N(t)} u \right) \right\|_{L^2_{t,x}(J_m \times \mathbf{R}^2)}$$

$$\times \left\| (P_l u) \left(P_{|\xi - \xi(t)| \leq \eta_3^{-1/2} N(t)} u \right) \right\|_{L^2_{t,x}(J_m \times \mathbf{R}^2)}$$

$$+ \left\| (e^{it\Delta} u_0) \left(P_{\eta_3^{-1/2} N(t) \leq |\xi - \xi(t)| \leq 32} u \right) \right\|_{L^2_{t,x}(J_m \times \mathbf{R}^2)}$$

$$\times \left\| (P_l u) \left(P_{\eta_3^{-1/2} N(t) \leq |\xi - \xi(t)| \leq 32} u \right) \right\|_{L^2_{t,x}(J_m \times \mathbf{R}^2)}$$

$$\lesssim \eta_3^{-1/2} 2^{-l} N(J_m) \|u_0\|_{L^2} \|P_l u\|_{U_\Delta^2(G_k^j \times \mathbf{R}^2)}$$

$$+ \left\| (e^{it\Delta} u_0) \left(P_{\eta_3^{-1/2} N(t) \leq |\xi - \xi(t)| \leq 32} u \right) \right\|_{L^2_{t,x}(J_m \times \mathbf{R}^2)}$$

$$\times \left\| (P_l u) \left(P_{\eta_3^{-1/2} N(t) \leq |\xi - \xi(t)| \leq 32} u \right) \right\|_{L^2_{t,x}(J_m \times \mathbf{R}^2)}. \tag{5.193}$$

By Duhamel's formula, (4.144), and (4.145), when $n(t) + C_0 \leq 5$,

$$\left\| P_{\eta_3^{-1/2} N(t) \leq |\xi - \xi(t)| \leq 32} u \right\|_{U_\Delta^2(J_m \times \mathbf{R}^2)}$$

$$\lesssim \eta_2 + \|u\|^2_{L_t^{8/3} L_x^8(J_m \times \mathbf{R}^2)} \left\| P_{|\xi - \xi(t)| \geq \eta_3^{-1/2} 2^{-20} N(t)} u \right\|_{L_t^\infty L_x^2(J_m \times \mathbf{R}^2)}$$

$$\lesssim \eta_2. \tag{5.194}$$

Plugging (5.194) into (5.193), by the bilinear Strichartz estimate, the final term in (5.193) is bounded by

$$\eta_2^2 2^{-l} \|u_0\|_{L^2} \|P_l u\|_{U_\Delta^2(G_k^j \times \mathbf{R}^2)}. \tag{5.195}$$

Again, (4.151) and (4.152) imply that summing over $J_m \subset G_k^j$ is bounded by the right-hand side of (5.192).

Now consider an interval $J_m \cap G_k^j \neq \emptyset$, but where J_m is not contained in G_k^j. There are at most two such intervals, so if $n(t) + C_0 \leq 5$ then use (5.193)–(5.195) to estimate the two overlapping intervals. If $n(t) + C_0 \geq 5$, then since $N(G_k^j) \leq \eta_3^{1/2} 2^{j-5}$, by (4.144), on each such interval J_m, $2^{n(t) + \frac{C_0}{2}} \sim \eta_3^{-3/4} 2^{j-5}$, so

$$\int_{J_m} \left\langle u_0, e^{i(t-\tau)\Delta}(P_l u) \left(P_{\xi(t), \leq n(t) + \frac{C_0}{2}} u\right)^2 d\tau \right\rangle$$

$$\lesssim 2^{j-l} \eta_3^{1/4} \|u_0\|_{L^2} \|P_l u\|_{U_\Delta^2(G_k^j \times \mathbf{R}^2)}.$$

Finally, as in (5.194),

$$\int_{J_m} \left\langle u_0, e^{i(t-\tau)\Delta}(P_l u) \left(P_{\xi(t), n(t) + \frac{C_0}{2} \leq \cdot \leq n(t) + C_0} u\right)^2 d\tau \right\rangle$$

$$\lesssim 2^{j-l} \eta_2^2 \|u_0\|_{L^2} \|P_l u\|_{U_\Delta^2(G_k^j \times \mathbf{R}^2)}.$$

$\square$

Next, by Theorem 5.21,

$$\sum_{\substack{0 \leq l_2 \\ \leq j-10}} \left(\sum_{\substack{G_\alpha^{l_2} \subset G_k^j : \\ N(G_\alpha^{l_2}) \leq \eta_3^{1/2} 2^{l_2-5}}} \left\| \int_{a_\alpha^{l_2}}^t e^{i(t-\tau)\Delta}(P_l u)(P_{\xi(t), l_2} u) \right.\right.$$

$$\left.\left. \times (P_{\xi(t), \leq l_2} u) \, d\tau \right\|_{U_\Delta^2(G_\alpha^{l_2} \times \mathbf{R}^2)}^2 \right)^{1/2}$$

$$\lesssim \sum_{\substack{0 \leq l_2 \\ \leq j-10}} \|(P_l u)(P_{\xi(t), \leq l_2} u)\|_{L_{t,x}^2(G_k^j \times \mathbf{R}^2)}$$

$$\times \left(\sup_{\substack{G_\alpha^{l_2} \subset G_k^j : \\ N(G_\alpha^{l_2}) \leq \eta_3^{1/2} 2^{l_2-5}}} \|f(P_{\xi(t), l_2} u)\|_{L_{t,x}^2(G_\alpha^{l_2} \times \mathbf{R}^2)} \right), \tag{5.196}$$

for some $\hat{f}(t, \xi)$ that is supported on $|\xi| \sim 2^l$ and $\|f\|_{U_\Delta^{8/3}(G_\alpha^{l_2} \times \mathbf{R}^2)} = 1$. Summing up, by Theorem 5.40,

(5.196)
$$\lesssim \sum_{0\leq l_2\leq j-10} \left\|(P_l u)\left(P_{\xi(t),\leq l_2}u\right)\right\|_{L^2_{t,x}\left(G^j_k\times\mathbf{R}^2\right)}$$

$$\times \left(\sup_{\substack{G^{l_2}_\alpha\subset G^j_k:N\left(G^{l_2}_\alpha\right)\\ \leq \eta_3^{1/2}2^{l_2-5}}} 2^{\frac{l_2-l}{4}}\|f\|_{U^{8/3}_\Delta\left(G^{l_2}_\alpha\times\mathbf{R}^2\right)}\left\|P_{\xi(t),l_2}u\right\|_{U^{8/3}_\Delta\left(G^{l_2}_\alpha\times\mathbf{R}^2\right)}\right)$$

$$\lesssim \sum_{0\leq l_2\leq j-10} 2^{\frac{l_2-l}{4}}\|P_l u\|_{U^2_\Delta\left(G^j_k\times\mathbf{R}^2\right)}\|u\|_{\widetilde{Y}_j([0,T]\times\mathbf{R}^2)}\left(1+\|u\|_{\widetilde{X}_j([0,T]\times\mathbf{R}^2)}\right)^3$$

$$\lesssim 2^{\frac{j-l}{4}}\|P_l u\|_{U^2_\Delta\left(G^j_k\times\mathbf{R}^2\right)}\|u\|_{\widetilde{Y}_j([0,T]\times\mathbf{R}^2)}\left(1+\|u\|_{\widetilde{X}_j([0,T]\times\mathbf{R}^2)}\right)^3.$$

Finally, following (5.193),

$$\sum_{J_m\cap G^j_k\neq\emptyset}\left(\int_{J_m}\langle f,(P_l u)\left(P_{\xi(t),\leq C_0+n(t)}u\right)^2 d\tau\rangle\right)^2$$

$$\lesssim \left\|(P_l u)\left(P_{\xi(t),\leq C_0+n(t)}u\right)\right\|^2_{L^2_{t,x}\left(G^j_k\times\mathbf{R}^2\right)}$$

$$\times \left(\sup_{J_m\cap G^j_k\neq\emptyset}\left\|f\left(P_{\xi(t),\leq C_0+n(t)}u\right)\right\|_{L^2_{t,x}(J_m\times\mathbf{R}^2)}\right)^2$$

$$\lesssim 2^{j-l}\eta_2\|P_l u\|^2_{U^2_\Delta\left(G^j_k\times\mathbf{R}^2\right)}.$$

Therefore, we have proved the following result.

Theorem 5.43 *For $j\geq 11$, $l\geq j-5$,*

$$\left\|\int_{a^j_k}^t e^{i(t-\tau)\Delta}(P_l u)\left(P_{\xi(\tau),\leq j-10}u\right)^2 d\tau\right\|_{U^2_\Delta\left(G^j_k\times\mathbf{R}^2\right)}$$

$$\lesssim 2^{\frac{j-l}{2}}\eta_2\|P_l u\|_{U^2_\Delta\left(G^j_k\times\mathbf{R}^2\right)}$$

$$+ 2^{\frac{j-l}{5}}\|P_l u\|_{U^2_\Delta\left(G^j_k\times\mathbf{R}^2\right)}\|u\|_{\widetilde{Y}_j([0,T]\times\mathbf{R}^2)}\left(1+\|u\|_{\widetilde{X}_j([0,T]\times\mathbf{R}^2)}\right)^3. \quad (5.197)$$

The right-hand side of (5.197) satisfies Theorem 5.20 provided that $\|u\|_{\widetilde{Y}_j([0,T]\times\mathbf{R}^2)}$ is small.

Proof of smallness To prove that $\|u\|_{\widetilde{Y}_j([0,T]\times\mathbf{R}^2)}$ is small, observe that by

(5.158)–(5.159), (5.162), and (5.197),

$$\sum_{5<i\leq j} 2^{i-j} \sum_{\substack{G_\alpha^i \subset G_k^j: \\ N(G_\alpha^i) \leq 2^{i-10}\eta_3^{1/2}}} \left\| \int_{a_\alpha^i}^t e^{i(t-\tau)\Delta} P_{\xi(G_\alpha^i), i-2\leq \cdot \leq i+2} F(u(\tau)) d\tau \right\|_{U_\Delta^2(G_\alpha^i \times \mathbf{R}^d)}^2$$

$$+ \sum_{\substack{i>\max\{j,5\}: \\ N(G_k^j) \leq 2^{i-10}\eta_3^{1/2}}} \left\| \int_{a_k^j}^t e^{i(t-\tau)\Delta} P_{\xi(G_k^j), i-2\leq \cdot \leq i+2} F(u(\tau)) d\tau \right\|_{U_\Delta^2(G_k^j \times \mathbf{R}^d)}^2$$

$$\lesssim \eta_2^{1/3} \|u\|_{\widetilde{X}_j([0,T]\times\mathbf{R}^2)}^4 + \|u\|_{\widetilde{Y}_j([0,T]\times\mathbf{R}^2)}^4 \left(1 + \|u\|_{\widetilde{X}_j([0,T]\times\mathbf{R}^2)}\right)^6. \tag{5.198}$$

Next, by Strichartz estimates,

$$\sum_{5<i\leq j} 2^{i-j} \sum_{\substack{G_\alpha^i \subset G_k^j: \\ 2^{i-10}\eta_3^{1/2} \leq \\ N(G_\alpha^i) \leq 2^{i-5}\eta_3^{1/2}}} \left\| \int_{a_\alpha^i}^t e^{i(t-\tau)\Delta} P_{\xi(G_\alpha^i), i-2\leq \cdot \leq i+2} F(u(\tau)) d\tau \right\|_{U_\Delta^2(G_\alpha^i \times \mathbf{R}^d)}^2$$

$$+ \sum_{\substack{i>\max\{j,5\}: \\ \eta_3^{1/2} 2^{i-10} \leq \\ N(G_k^j) \leq 2^{i-5}\eta_3^{1/2}}} \left\| \int_{a_k^j}^t e^{i(t-\tau)\Delta} P_{\xi(G_k^j), i-2\leq \cdot \leq i+2} F(u(\tau)) d\tau \right\|_{U_\Delta^2(G_k^j \times \mathbf{R}^d)}^2$$

$$\lesssim \sum_{5<i\leq j} 2^{i-j} \sum_{\substack{G_\alpha^i \subset G_k^j: \\ 2^{i-10}\eta_3^{1/2} \leq \\ N(G_\alpha^i) \leq 2^{i-5}\eta_3^{1/2}}} \|u\|_{L_t^{5/2}L_x^{10}(G_\alpha^i \times \mathbf{R}^2)}^5 \left\| P_{\xi(G_\alpha^i), \geq i-5} u \right\|_{L_t^\infty L_x^2(G_\alpha^i \times \mathbf{R}^2)}$$

$$+ \sum_{\substack{i>\max\{j,5\}: \\ 2^{i-10}\eta_3^{1/2} \leq \\ N(G_k^j) \leq 2^{i-5}\eta_3^{1/2}}} \left\| P_{\xi(G_k^j), i-2\leq \cdot \leq i+2} F(u) \right\|_{L_t^1 L_x^2(G_k^j \times \mathbf{R}^2)}^2.$$

Now if $2^{i-10}\eta_3^{1/2} \leq N(G_k^j) \leq 2^{i-5}\eta_3^{1/2}$, then by (4.81), (4.145), and (4.151),

$$\sum_{\substack{i>\max\{j,5\}: 2^{i-10}\eta_3^{1/2} \\ \leq N(G_k^j) \leq 2^{i-5}\eta_3^{1/2}}} \left\| P_{\xi(G_k^j), i-2\leq \cdot \leq i+2} F(u) \right\|_{L_t^1 L_x^2(G_k^j \times \mathbf{R}^2)}^2$$

$$\lesssim \left\| P_{\xi(G_k^j), \geq j-2} u \right\|_{L_t^\infty L_x^2(G_k^j \times \mathbf{R}^2)} \|u\|_{L_t^{5/2} L_x^{10}(G_k^j \times \mathbf{R}^2)}^5 \lesssim \eta_2. \tag{5.199}$$

5.4 The Two-Dimensional Mass-Critical Problem

Also,

$$\sum_{5<i\leq j} 2^{i-j} \sum_{\substack{G_\alpha^i \subset G_k^j : 2^{i-10}\eta_3^{1/2} \\ \leq N(G_\alpha^i) \leq 2^{i-5}\eta_3^{1/2}}} \|u\|_{L_t^{5/2}L_x^{10}(G_\alpha^i \times \mathbf{R}^2)}^5 \|P_{\xi(G_\alpha^i),\geq i-5}u\|_{L_t^\infty L_x^2(G_\alpha^i \times \mathbf{R}^2)}$$

$$\lesssim \sum_{i\leq j} 2^{i-j} \left(\#\{J_l \subset G_k^j : N(J_l) \sim 2^{i-5}\eta_3^{1/2}\} + 2 \right) \cdot \eta_2 \lesssim \eta_2. \quad (5.200)$$

Therefore, by (5.158), (5.159), (5.198), (5.199), and (5.200),

$$\|u\|_{\widetilde{Y}_j([0,T]\times\mathbf{R}^2)} \lesssim \eta_2^{1/2} + \eta_2^{1/6}\|u\|_{\widetilde{X}_j([0,T]\times\mathbf{R}^2)}^2$$

$$+ \|u\|_{\widetilde{Y}_j([0,T]\times\mathbf{R}^2)}^2 \left(1 + \|u\|_{\widetilde{X}_j([0,T]\times\mathbf{R}^2)}\right)^3.$$

Thus $\widetilde{Y}_j$ retains the needed smallness, which completes the proof of Theorem 5.19 in dimension $d = 2$. □

It only remains to use the long-time Strichartz estimates to rule out the existence of nonzero almost-periodic solutions to (4.142) when $d = 2$. The good news is that much of the work has already been done, either in the analysis of the one-dimensional case or in the proof of Theorem 5.39.

Theorem 5.44 *If u is an almost-periodic solution to (4.142) in dimension $d = 2$ on $\mathbf{R}$ with $N(t) \leq 1$ for all $t \in \mathbf{R}$, and $[0,T]$ is an interval satisfying*

$$\int_0^T N(t)^3 \, dt = K,$$

then

$$\left\| ||\nabla|^{1/2} P_{\leq \eta_3^{-1}K} u(t,x)|^2 \right\|_{L_{t,x}^2([0,T]\times\mathbf{R}^2)}^2 \lesssim o(K), \quad (5.201)$$

where $o(K)$ is a quantity satisfying $\frac{o(K)}{K} \to 0$ as $K \nearrow \infty$.

Proof Again, if $[0,T]$ is an interval such that $\|u\|_{L_{t,x}^4([0,T]\times\mathbf{R})}^4 = 2^{k_0}$, then after rescaling,

$$u(t,x) \mapsto \lambda^{1/2} u(\lambda^2 t, \lambda x), \quad \text{with } \lambda = \frac{\eta_3 2^{k_0}}{K},$$

proving (5.201) is equivalent to proving

$$\left\| ||\nabla|^{1/2} P_{\leq k_0} u|^2 \right\|_{L_{t,x}^2([0,T]\times\mathbf{R}^2)}^2 \lesssim o(K) \frac{\eta_3 2^{k_0}}{K}. \quad (5.202)$$

Now by (4.18), (4.30), (4.31), and (4.32), if u solves (4.142),

$$\left\| ||\nabla|^{1/2} |u|^2 \right\|_{L_{t,x}^2([0,T]\times\mathbf{R}^2)}^2 \lesssim \|e^{-ix\cdot\xi(t)}u\|_{L_t^\infty \dot{H}^1([0,T]\times\mathbf{R}^2)} \|u\|_{L_t^\infty L_x^2([0,T]\times\mathbf{R}^2)}^3.$$

Therefore, if $P_{\leq k_0} = I$,

$$\left\||\nabla|^{1/2}|Iu|^2\right\|^2_{L^2_{t,x}([0,T]\times \mathbf{R}^2)}$$
$$\lesssim \left\|e^{-ix\cdot\xi(t)}Iu\right\|_{L^\infty_T \dot H^1([0,T]\times \mathbf{R}^2)}\|Iu\|^3_{L^\infty_T L^2_x([0,T]\times \mathbf{R}^2)} + \int_0^T \mathscr{E}(t)\,dt,$$

where

$$\mathscr{E}(t)$$
$$= \iint \frac{(x-y)}{|x-y|}|Iu(t,y)|^2 \cdot \mathrm{Re}\left[\left(IF(\bar u) - F(\overline{Iu})\right)(\nabla - i\xi(t))Iu\right](t,x)\,dx\,dy \tag{5.203}$$

$$- \iint \frac{(x-y)}{|x-y|}|Iu(t,y)|^2 \cdot \mathrm{Re}\left[\overline{Iu}(\nabla - i\xi(t))(F(Iu) - IF(u))\right](t,x)\,dx\,dy \tag{5.204}$$

$$+ 2\iint \frac{(x-y)}{|x-y|} \cdot \mathrm{Im}\left[\overline{Iu}IF(u)\right](t,y)\,\mathrm{Im}\left[\overline{Iu}(\nabla - i\xi(t))Iu\right](t,x)\,dx\,dy. \tag{5.205}$$

Again, as in Lemma 4.38,

$$\left\|e^{-ix\cdot\xi(t)}Iu\right\|_{L^\infty_T \dot H^1([0,T]\times \mathbf{R}^2)}\|Iu\|^3_{L^\infty_T L^2_x([0,T]\times \mathbf{R}^2)} \lesssim \frac{\eta_3 2^{k_0}}{K}o(K).$$

Next, modifying the analysis of (5.168), then (5.175), (5.178), and (5.179) imply

$$\int_0^T \iint \frac{(x-y)}{|x-y|}|Iu(t,y)|^2 \cdot \mathrm{Re}\left[\left(IF(\bar u) - F(\overline{Iu})\right)(\nabla - i\xi(t))Iu\right](t,x)\,dx\,dy\,dt$$
$$\lesssim \|u\|^2_{\tilde X_{k_0}([0,T]\times \mathbf{R}^2)}\|\tilde\nabla Iu\|^{5/6}_{L^{5/2}_T L^{10}_x([0,T]\times \mathbf{R}^2)}\|\tilde\nabla Iu\|^{1/6}_{L^\infty_T L^2_x([0,T]\times \mathbf{R}^2)}$$
$$\lesssim \frac{\eta_3 2^{k_0}}{K}o(K).$$

Next, integrating by parts,

(5.204)
$$= (5.203) - \int_0^T \iint \frac{1}{|x-y|}|Iu(t,y)|^2 \cdot \mathrm{Re}\left[\overline{Iu}(F(Iu) - IF(u))\right](t,x)\,dx\,dy\,dt.$$

By the Hardy–Littlewood–Sobolev inequality, the Sobolev embedding

5.4 The Two-Dimensional Mass-Critical Problem

theorem, and interpolation,

$$\int_0^T \int \int \frac{1}{|x-y|} |Iu(t,y)|^2 \cdot \text{Re}\left[\overline{Iu}(F(Iu) - IF(u))\right](t,x) \, dx\, dy\, dt$$

$$\lesssim \|Iu\|_{L_t^\infty L_x^2} \|Iu\|_{L_t^6 L_x^{12}}^2 \|F(Iu) - IF(u)\|_{L_t^{3/2} L_x^{6/5}}$$

$$\lesssim \|Iu\|_{L_t^\infty L_x^2}^2 \|\widetilde{\nabla} Iu\|_{L_t^3 L_x^6} \|F(Iu) - IF(u)\|_{L_t^{3/2} L_x^{6/5}}$$

$$\lesssim \frac{\eta_3 2^{k_0}}{K} o(K).$$

Finally, collecting (5.184), Theorem 5.19, and Theorem 5.39,

$$(5.205) \quad \lesssim \frac{\eta_3 2^{k_0}}{K} o(K) + \|\widetilde{\nabla} Iu\|_{L_t^\infty L_x^2}^2 \|Iu\|_{L_t^\infty L_x^2} \|u_h u\|_{L_{t,x}^2}^2 \lesssim \frac{\eta_3 2^{k_0}}{K} o(K).$$

This gives (5.202), which proves Theorem 5.44. □

Corollary 5.45 *There does not exist an almost-periodic solution to* (4.142) *when* $d = 2$ *with* $N(t) \leq 1$ *on* **R** *and*

$$\int_{-\infty}^{\infty} N(t)^3 \, dt = \infty.$$

Proof By (4.167), if u is an almost-periodic solution to (4.142) with nonzero mass, then

$$\int_0^T N(t)^3 \, dt = K \lesssim \left\| |\nabla|^{1/2} |u|^2 \right\|_{L_{t,x}^2([0,T] \times \mathbf{R}^2)}^2 \lesssim o(K),$$

with bound independent of T. This implies an upper bound on K as $T \nearrow \infty$. The bound for the negative time direction follows from time reversal symmetry. □

The proof that rapid cascade solutions must be $u \equiv 0$ in dimension $d = 2$ is nearly identical to the proof of the corresponding result in dimension $d = 1$ (Theorem 5.35).

Theorem 5.46 *Suppose u is an almost-periodic solution to* (4.142) *when* $d = 2$, *on* **R** *with* $N(t) \leq 1$ *for all* $t \in \mathbf{R}$. *Suppose also that*

$$\int_{-\infty}^{\infty} N(t)^3 \, dt < \infty. \tag{5.206}$$

Then $u \equiv 0$.

Proof The proof is identical to the proof of Theorem 5.35. Indeed, by the fundamental theorem of calculus, (4.144), and (5.206), there exist $\xi_+, \xi_- \in \mathbf{R}^2$ such that $\xi(t) \to \xi_\pm$ as $t \to \pm\infty$ respectively. Now suppose $\int_T^\infty N(t)^3 \, dt = K$.

Again by (4.144) and (5.206),

$$|\xi(t) - \xi_+| \lesssim \eta_1^{-1} K, \qquad (5.207)$$

for all $t \in [T, \infty)$. Make a Galilean transformation so that $\xi_+ \mapsto 0$.

Theorem 5.47 *If $\int_T^\infty N(t)^3 \, dt = K$, u is an almost-periodic solution to (4.142) when $d = 2$, and $\xi_+ = 0$, then*

$$\|u\|_{L_t^\infty \dot H_x^{1/2}([T,\infty) \times \mathbf{R}^2)} \lesssim K^{1/2}.$$

Proof Since $\xi_+ = 0$, (5.207) implies $|\xi(t)| \leq \eta_1^{-1} K$ when $t > T$. Choose T_1 such that

$$\int_T^{T_1} \int |u(t,x)|^4 \, dx \, dt = 2^{k_0}.$$

Taking $\lambda = \eta_3 2^{k_0} / \left(\int_T^{T_1} N(t)^3 \, dt \right)$ and rescaling $u \mapsto u_\lambda$,

$$\|u_\lambda\|_{\tilde X_{k_0}\left(\left[\frac{T}{\lambda^2}, \frac{T_1}{\lambda^2}\right] \times \mathbf{R}^2\right)} \lesssim 1. \qquad (5.208)$$

Furthermore, from the proof of Theorem 5.20 when $d = 2$, for $k \geq k_0$,

$$\|P_{>k} u_\lambda\|_{U_\Delta^2\left(\left[\frac{T}{\lambda^2}, \frac{T_1}{\lambda^2}\right] \times \mathbf{R}^2\right)}$$
$$\lesssim \left\|P_{>k} u_\lambda\left(\frac{T_1}{\lambda^2}\right)\right\|_{L_x^2(\mathbf{R}^2)} + \varepsilon \|P_{>k-5} u_\lambda\|_{U_\Delta^2\left(\left[\frac{T}{\lambda^2}, \frac{T_1}{\lambda^2}\right] \times \mathbf{R}^2\right)}. \qquad (5.209)$$

Rescaling back, since U_Δ^2 and L^2 are scale-invariant norms and $\int_T^{T_1} N(t)^3 \, dt < K$, if $2^k \geq \eta_3^{-1} K$,

$$\|P_{>k} u\|_{U_\Delta^2([T, T_1] \times \mathbf{R}^2)} \lesssim \|P_{>k} u_\lambda(T_1)\|_{L_x^2(\mathbf{R}^2)} + \varepsilon \|P_{>k-5} u_\lambda\|_{U_\Delta^2([T, T_1] \times \mathbf{R}^2)}. \qquad (5.210)$$

Since $N(t) \to 0$ and $\xi(t) \to 0$ as $t \to \infty$ then by (4.80),

$$\lim_{T_1 \to +\infty} \|P_{>k} u(T_1)\|_{L_x^2(\mathbf{R})} = 0. \qquad (5.211)$$

Since the $\varepsilon > 0$ in (5.209) may be made to be arbitrarily small, in particular so that $C\varepsilon \leq 2^{-5}$, where C is the implicit constant in (5.210), then for any $t \in [T, \infty)$, (5.210) and (5.211) imply that for $2^k \geq \eta_3^{-1} K$,

$$\|P_{>k} u(t)\|_{L_x^2(\mathbf{R}^d)} \lesssim 2^{-k} K.$$

This combined with conservation of mass proves Theorem 5.47. □

For general $\xi_+ \in \mathbf{R}^2$, Theorem 5.47 and (4.144) imply

$$\|e^{-ix \cdot \xi(t)} u\|_{L_t^\infty \dot H^{1/2}([T,\infty) \times \mathbf{R}^2)} \lesssim K^{1/2}.$$

5.4 The Two-Dimensional Mass-Critical Problem

Now by the dominated convergence theorem,

$$\lim_{T \to \infty} \int_T^\infty N(t)^3 \, dt = 0,$$

so

$$\lim_{t \to +\infty} \left\| e^{-ix \cdot \xi(t)} u(t) \right\|_{\dot{H}^{1/2}(\mathbf{R}^2)} = 0.$$

The same analysis can be done as $t \to -\infty$. Therefore, by (4.18), (4.31), (4.32), and the Sobolev embedding theorem,

$$\left\| u \right\|_{L_t^4 L_x^8 (\mathbf{R} \times \mathbf{R}^2)}^4 \lesssim \left\| |\nabla|^{1/2} |u|^2 \right\|_{L_{t,x}^2 (\mathbf{R} \times \mathbf{R}^2)}^2 = 0.$$

Therefore, $u \equiv 0$. $\square$

References

Berestycki, H. and P. L. Lions (1979). Existence d'ondes solitaires dans des problèmes nonlinéaires du type Klein–Gordon. *C. R. Acad. Sci. Paris Sér. A–B*, **288**(7), A395–A398.

Bergh, J. and J. Löfstrom (1976). *Interpolation Spaces. An Introduction*, Grundlehren der Mathematischen Wissenschaften, vol. 223, Berlin-New York: Springer.

Bourgain, J. (1998). Refinements of Strichartz' inequality and applications to 2D-NLS with critical nonlinearity. *Internat. Math. Res. Notices*, **1998**(5), 253–283.

Bourgain, J. (1999). Global well-posedness of defocusing critical nonlinear Schrödinger equation in the radial case. *J. Amer. Math. Soc.*, **12**(1), 145–171.

Cazenave, T. and F. B. Weissler (1988). The Cauchy problem for the nonlinear Schrödinger equation in H^1. *F.B. Manuscripta Math.*, **61**, 477–494.

Christ, M., J. Colliander, and T. Tao (2003). Asymptotics, frequency modulation, and low regularity ill-posedness for canonical defocusing equations. *Amer. J. Math.*, **125**(6), 1235–1293.

Christ, M., J. Colliander, and T. Tao (2008). A priori bounds and weak solutions for the nonlinear Schrödinger equation in Sobolev spaces of negative order. *J. Funct. Anal.*, **254**(2), 368–395.

Christ, M. and A. Kiselev (2001). Maximal functions associated to filtrations. *J. Funct. Anal.*, **179**(2), 409–425.

Colliander, J., M. Grillakis, and N. Tzirakis (2007). Improved interaction Morawetz inequalities for the cubic nonlinear Schrödinger equation on $\mathbf{R}^2$. *Internat. Math. Res. Notices*, **23**(1), 90–119.

Colliander, J., M. Grillakis, and N. Tzirakis (2009). Tensor products and correlation estimates with applications to nonlinear Schrödinger equations. *Comm. Pure Appl. Math.*, **62**(7), 920–968.

Colliander, J., M. Keel, G. Staffilani, H. Takaoka, and T. Tao (2004). Global existence and scattering for rough solutions of a nonlinear Schrödinger equation on $\mathbf{R}^3$. *Comm. Pure Appl. Math.*, **21**, 987–1014.

Colliander, J., M. Keel, G. Staffilani, H. Takaoka, and T. Tao (2008). Global well-posedness and scattering for the energy-critical nonlinear Schrödinger equation on $\mathbf{R}^3$. *Ann. of Math. (2)*, **167**, 767–865.

Colliander, J. and T. Roy (2011). Bootstrapped Morawetz estimates and resonant decomposition for low regularity global solutions of cubic NLS on $\mathbf{R}^2$. *Commun. Pure Appl. Anal.*, **10**(2), 397–414.

Conway, J. B. (1990). *A Course in Functional Analysis*, 2nd edn, Graduate Texts in Mathematics, vol. 96, New York: Springer.

Dodson, B. (2011). Improved almost Morawetz estimates for the cubic nonlinear Schrödinger equation. *Commun. Pure Appl. Anal.*, **10**(1), 127–140.

Dodson, B. (2012). Global well-posedness and scattering for the defocusing L^2-critical nonlinear Schrödinger equation when $d \geq 3$. *J. Amer. Math. Soc.*, **25**(2), 429–463.

Dodson, B. (2015). Global well-posedness and scattering for the mass critical nonlinear Schrödinger equation with mass below the mass of the ground state. *Adv. Math.*, **285**, 1589–1618.

Dodson, B. (2016a). Global well-posedness and scattering for the defocusing L^2-critical nonlinear Schrödinger equation when $d = 1$. *Amer. J. Math.*, **138**(2), 531–569.

Dodson, B. (2016b). Global well-posedness and scattering for the defocusing L^2-critical nonlinear Schrödinger equation when $d = 2$. *Duke Math. J.*, **165**(18), 3435–3516.

Duyckaerts, T., J. Holmer, and S. Roudenko (2008). Scattering for the non-radial 3D cubic nonlinear Schrödinger equation. *Math. Res. Lett.*, **15**(5–6), 1233–1250.

Ginibre, J. and G. Velo (1992). Smoothing properties and retarded estimates for some dispersive evolution equations. *Comm. Math. Phys.*, **144**(1), 163–188.

Grafakos, L. (2004). *Classical Fourier Analysis*, 3rd edn, Graduate Texts in Mathematics, vol. 249, New York: Springer.

Grillakis, M. (2000). On nonlinear Schrödinger equations. *Comm. Partial Differential Equations* **25**(9–10), 1827–1844.

Guevara, C. (2014). Global behavior of finite energy solutions to the focusing nonlinear Schrödinger equation in d dimensions. *Appl. Math. Res. Express*, **2014**, 177–243.

Hadac, M., S. Herr, and H. Koch (2009). Well-posedness and scattering for the KP-II equation in a critical space. *Ann. Inst. H. Poincaré Anal. Non Linéaire*, **26**(3), 917–941.

Holmer, J., R. Platte, and S. Roudenko (2010). Blow-up criteria for the 3D cubic nonlinear Schrödinger equation. *Nonlinearity*, **23**(4), 977–1030.

Holmer, J. and S. Roudenko (2008). A sharp condition for scattering of the radial 3D cubic nonlinear Schrödinger equation. *Comm. Math. Phys.*, **282**(2), 435–467.

Keel, M. and T. Tao (1998). Endpoint Strichartz estimates. *Amer. J. Math.*, **120**(4–6), 945–957.

Kenig, C. and F. Merle (2006). Global well-posedness, scattering, and blow-up for the energy-critical, focusing nonlinear Schrödinger equation in the radial case. *Invent. Math.*, **166**(3), 645–675.

Kenig, C. and F. Merle (2010). Scattering for $\dot{H}^{1/2}$ bounded solutions to the cubic, defocusing NLS in 3 dimensions. *Trans. Amer. Math. Soc.*, **362**(4), 1937–1962.

Keraani, S. (2001). On the defect of compactness for the Strichartz estimates of the Schrödinger equations. *J. Differential Equations*, **175**(2), 353–392.

Killip, R., T. Tao, and M. Visan (2009). The cubic nonlinear Schrödinger equation in two dimensions with radial data. *J. Eur. Math. Soc. (JEMS)*, **11**(6), 1203–1258.

Killip, R. and M. Visan (2010). The focusing energy-critical nonlinear Schrödinger equation in dimensions five and higher. *Amer. J. Math.*, **132**(2), 361–424.

Killip, R. and M. Visan (2012). Global well-posedness and scattering for the defocusing quintic NLS in three dimensions. *Anal. PDE*, **5**(4), 855–885.

Killip, R. and M. Visan (2013). Nonlinear Schrödinger equations at critical regularity. In *Clay Math. Proceedings*, vol. 17, Providence, RI: American Mathematical Society, pp. 325–437.

Killip, R., M. Visan, and X. Zhang (2008). The mass-critical nonlinear Schrödinger equation with radial data in dimensions three and higher. *Anal. PDE*, **1**(2), 229–266.

Koch, H. and D. Tataru (2007). A priori bounds for the 1D cubic NLS in negative Sobolev spaces. *Internat. Math. Res. Notices*, **16**, Art. ID rnm053, 36 pp.

Koch, H. and D. Tataru (2012). Energy and local energy bounds for the 1-D cubic NLS equation in $H^{-1/4}$. *Ann. Inst. H. Poincaré Anal. Non Linéaire*, **29**(6), 955–988.

Lieb, E. and M. Loss (2001). *Analysis*, 2nd edn, Graduate Studies in Mathematics, vol. 14, Providence, RI: American Mathematical Society.

Lin, J. and W. Strauss (1978). Decay and scattering of solutions of a nonlinear Schrödinger equation. *J. Funct. Anal.*, **30**(2), 245–263.

Montgomery-Smith, J. S., (1998). Time decay for the bounded mean oscillation of solutions of the Schrödinger and wave equations. *Duke Math. J.*, **91**(2), 393–408.

Murphy, J. (2014). Inter-critical NLS: critical $\dot{H}^s$ bounds imply scattering. *SIAM J. Math. Anal.*, **46**(1), 939–997.

Murphy, J. (2015). The radial, defocusing, nonlinear Schrödinger equation in three dimensions. *Comm. Partial Differential Equations*, **40**(2), 265–308.

Muscalu, C. and W. Schlag (2013). *Classical and Multilinear Harmonic Analysis*, Cambridge Studies in Advanced Mathematics, vol. 137, Cambridge: Cambridge University Press.

Planchon, F. and L. Vega (2009). Bilinear virial identities and applications. *Ann. Sci. École Norm. Sup. (4)*, **42**(2), 261–290.

Ryckman, E. and M. Visan (2007). Global well-posedness and scattering for the defocusing energy-critical nonlinear Schrödinger equation in $\mathbb{R}^{1+4}$. *Amer. J. Math.*, **129**(1), 1–60.

Shao, S. (2009). A note on the restriction conjecture in the cylindrically symmetric case. *Proc. Amer. Math. Soc.*, **137**(1), 135–143.

Smith, H. F. and C. D. Sogge (2000). Global Strichartz estimates for nontrapping perturbations of the Laplacian. *Comm. Partial Differential Equations*, **25**(11–12), 2171–2183.

Sogge, C. D. (1993). *Fourier Integrals in Classical Analysis*, Cambridge Tracts in Mathematics, vol. 105, Cambridge: Cambridge University Press.

Stein, E. M. (1970). *Singular Integrals and Differentiability Properties of Functions*, Princeton, NJ: Princeton University Press.

Stein, E. M. (1993). *Harmonic Analysis: Real-Variable Methods, Orthogonality, and Oscillatory Integrals*, Princeton, NJ: Princeton University Press.

Strichartz, R. S. (1977). Restrictions of Fourier transforms to quadratic surfaces and decay of solutions of wave equations. *Duke Math. J.*, **44**(3), 705–714.

Sulem, C. and P. L. Sulem (1999). *The Nonlinear Schrödinger Equation*, Applied Mathematical Sciences, vol. 139, New York: Springer.

Tao, T. (2000). Spherically averaged endpoint Strichartz estimates for the two-dimensional Schrödinger equation. *Comm. Partial Differential Equations*, **25**(7–8), 1471–1485.

Tao, T. (2003). A sharp bilinear restrictions estimate for paraboloids. *Geom. Funct. Anal.*, **13**(6), 1359–1384.

Tao, T. (2005). Global well-posedness and scattering for the higher-dimensional energy-critical nonlinear Schrödinger equation for radial data. *New York J. Math.*, **11**, 57–80.

Tao, T. (2006). *Nonlinear Dispersive Equations. Local and Global Analysis*, CBMS Regional Conference Series in Mathematics, vol. 104, Washington, DC: Conference Board of the Mathematical Sciences.

Tao, T. and M. Visan (2005). Stability of energy-critical nonlinear Schrödinger equations in high dimensions. *Electron. J. Differential Equations*, **118**, 28 pp.

Tao, T., M. Visan, and X. Zhang (2007a). The nonlinear Schrödinger equation with combined power-type nonlinearities. *Comm. Partial Differential Equations*, **32**(7–9), 1281–1343.

Tao, T., M. Visan, and X. Zhang (2007b). Global well-posedness and scattering for the defocusing mass-critical nonlinear Schrödinger equation for radial data in high dimensions. *Duke Math. J.*, **140**(1), 165–202.

Tao, T., M. Visan, and X. Zhang (2008). Minimal-mass blowup solutions of the mass-critical NLS. *Forum Math.*, **20**(5), 881–919.

Tataru, D. (1998). Local and global results for wave maps I. *Comm. Partial Differential Equations*, **23**(9–10), 1781–1793.

Taylor, M. E. (2000). *Tools for PDE. Pseudodifferential Operators, Paradifferential Operators, and Layer Potentials*, Mathematical Surveys and Monographs, vol. 81, Providence, RI: American Mathematical Society.

Taylor, M. E. (2006). *Measure Theory and Integration*, Graduate Studies in Mathematics, vol. 6, Providence, RI: American Mathematical Society.

Taylor, M. E. (2011). *Partial Differential Equations I–III*, 2nd edn, Applied Mathematical Sciences, vol. 115, New York: Springer.

Visan, M. (2006). The defocusing energy-critical nonlinear Schrödinger equation in dimensions five and higher. Ph.D. thesis, UCLA.

Visan, M. (2007). The defocusing energy-critical nonlinear Schrödinger equation in higher dimensions. *Duke Math. J.*, **138**, 281–374.

Visan, M. (2012). Global well-posedness and scattering for the defocusing cubic nonlinear Schrödinger equation in four dimensions. *Internat. Math. Res. Notices* 2, **5**, 1037–1067.

Weinstein, M. (1982). Nonlinear Schrödinger equations and sharp interpolation estimates. *Comm. Math. Phys.*, **87**(4), 567–576.

Weinstein, M. (1989). The nonlinear Schrödinger equation–singularity formation, stability and dispersion. In *The Connection between Infinite-Dimensional and Finite-Dimensional Dynamical Systems (Boulder, CO, 1987)*, Contemporary Mathematics, vol. 99, Providence, RI: American Mathematical Society, pp. 213–232.

Yajima, K. (1987). Existence of solutions for Schrödinger evolution equations. *Comm. Math. Phys.*, **110**(3), 415–426.

Yosida, K. (1980). *Functional Analysis*, 6th edn, Grundlehren der Mathematischen Wissenschaften [Fundamental Principles of Mathematical Sciences], vol. 123, Berlin: Springer.

Index

N space, 23
S space, 22
U^p spaces, 191
U^p_Δ spaces, 192
V^p_Δ spaces, 197
admissible pair, 12
almost-periodic solution
   energy-critical, 85
   mass-critical, 122
asymptotic orthogonality, 77

bilinear Strichartz estimate, 56
   first proof, 106
   second proof, 107

conservation of energy, 36
conservation of mass, 34
conservation of momentum, 35
core, 77
critical Sobolev space, 41

dispersive estimate, 5
double Duhamel lemma, 93
dyadic cubes, 111
dyadic interval, 18

first bilinear estimate in two dimensions, 217
Fourier inversion for Schwartz functions, 3
Fourier transform, 1
fractional product rule, 42
frequency-localized interaction Morawetz estimate, 166

Galilean invariant bilinear estimate, 104
Galilean Littlewood–Paley projection, 147
Galilean norm, 192
Galilean transformation, 9

interaction Morawetz estimate, 96
inverse Fourier transform, 2

linear Schrödinger equation, 1
Littlewood–Paley decomposition, 6
Littlewood–Paley kernel, 8
Littlewood–Paley theorem, 7
local conservation, 34
local well-posedness, 70
long-time perturbations, 71
long-time Strichartz estimate, 148
long-time Strichartz seminorm, 148
long-time Strichartz spaces, 193

mass bracket, 34
mass density, 34
momentum bracket, 34
momentum density, 34
Morawetz estimate, 49

Parseval identity, 4
perturbation lemma, 43, 67
perturbation lemma for the mass-critical problem, 26
Plancherel identity, 4
profile decomposition, energy-critical, 77
pseudoconformal conservation law, 37
pseudoconformal transformation, 39

Radon transform, 107
rapid cascade solution, 127

scale, 77
scaling, 10, 41
scattering, 25
scattering size function, energy-critical, 73
scenarios, energy-critical, 86
scenarios, mass-critical, 126
Schwartz space, 2
second bilinear estimate in two dimensions, 223
self-similar solution, 126

small data scattering for the energy-critical
   problem, 64
small intervals, 148
soliton-like solution, 127

Strichartz estimates, 14
Strichartz space, 22

well-posedness, 24
Whitney decomposition, 18